Modern Birkhäuser Classics

Many of the original research and survey monographs, as well as textbooks, in pure and applied mathematics published by Birkhäuser in recent decades have been groundbreaking and have come to be regarded as foundational to the subject. Through the MBC Series, a select number of these modern classics, entirely uncorrected, are being re-released in paperback (and as eBooks) to ensure that these treasures remain accessible to new generations of students, scholars, and researchers.

Topics in the Mathematical Modelling of Composite Materials

Andrej V. Cherkaev
Robert Kohn
Editors

Reprint of the 1997 Edition

 Birkhäuser

Editors
Andrej V. Cherkaev
Department of Mathematics
University of Utah
Salt Lake City, Utah, USA

Robert Kohn
Courant Mathematical Institute
New York University
New York, USA

ISSN 2197-1803   ISSN 2197-1811   (electronic)
Modern Birkhäuser Classics
ISBN 978-3-319-97183-4   ISBN 978-3-319-97184-1   (eBook)
https://doi.org/10.1007/978-3-319-97184-1

Library of Congress Control Number: 2018950852

This book is published under the imprint Birkhäuser, www.birkhauser-science.com by the registered company Springer Nature Switzerland AG.
The registered company address is: Gewerbestrasse 11, 6330 Cham, Switzerland

Topics in the Mathematical Modelling of Composite Materials

Andrej Cherkaev
Robert Kohn
Editors

Springer Science+Business Media, LLC

Andrej Cherkaev
Department of Mathematics
University of Utah
Salt Lake City, UT 84112

Robert Kohn
Courant Mathematical Institute
New York University
New York, NY 10012

Library of Congress Cataloging-in-Publication Data

Topics in the mathematical modelling of composite materials / Andrej
  Cherkaev, Robert Kohn, editors.
     p.   cm. -- (Progress in nonlinear differential equations and
  their applications ; v. 31)
   Includes bibliographical references.
   ISBN 978-1-4612-7390-5   ISBN 978-1-4612-2032-9 (eBook)
   DOI 10.1007/978-1-4612-2032-9
   1. Composite materials--Mathematical models. I. Cherkaev,
  Andrej. 1950- . II. Kohn, Robert V.   III. Series.
  TA418.9.C6T676 1997
  620.1'18'015118--dc21   97-180
                                                           CIP

Printed on acid-free paper

© Springer Science+Business Media New York 1997
Originally published by Birkhäuser Boston in 1997
Softcover reprint of the hardcover 1st edition 1997

ISBN 978-1-4612-7390-5
Reformatted and typeset from disks by Texniques, Inc., Boston, MA.

9 8 7 6 5 4 3 2 1

Contents

Introduction

Andrej V. Cherkaev and Robert V. Kohn

In the past twenty years we have witnessed a renaissance of theoretical work on the macroscopic behavior of microscopically heterogeneous materials. This activity brings together a number of related themes, including: (1) the use of weak convergence as a rigorous yet general language for the discussion of macroscopic behavior; (2) interest in new types of questions, particularly the "G-closure problem," motivated in large part by applications of optimal control theory to structural optimization; (3) the introduction of new methods for bounding effective moduli, including one based on "compensated compactness"; and (4) the identification of deep links between the analysis of microstructures and the multidimensional calculus of variations. This work has implications for many physical problems involving optimal design, composite materials, and coherent phase transitions. As a result it has received attention and support from numerous scientific communities, including engineering, materials science, and physics as well as mathematics.

There is by now an extensive literature in this area. But for various reasons certain fundamental papers were never properly published, circulating instead as mimeographed notes or preprints. Other work appeared in poorly distributed conference proceedings volumes. Still other work was published in standard books or journals, but written in Russian or French. The net effect is a sort of "gap" in the literature, which has made the subject unnecessarily difficult for newcomers to penetrate.

The present book aims to help fill this gap by assembling a coherent selection of this work in a single, readily accessible volume, in English translation. We do not claim that these articles represent the last word — or the first word — on their respective topics. But we do believe they represent fundamental work, well worth reading and studying today. They form the foundation upon which subsequent progress has been built.

The decision of what to include in a volume such as this is difficult and necessarily somewhat arbitrary. We have restricted ourselves to work originally written in Russian or French, by a handful of authors with different but related viewpoints. It would have been easy to add other fundamental work. We believe, however, that our choice has a certain coherence. This book will interest scientists working in the area, and those who wish to enter it. The book contains papers we want our Ph.D. students to study, to which they have not until now had ready access.

We now list the chapters in this book, and comment briefly on each one. They are presented, here and in the book, in chronological order.

1. *On the control of coefficients in partial differential equations* by F. Murat and L. Tartar. This is a translation of a paper published in *Lecture Notes in Economics and Mathematical Systems 107 (Control Theory, Numerical Methods, and Computer System Modelling)*, Springer-Verlag, 1975, 420–426. The translation was arranged by Doina Cioranescu with cooperation from the authors. The article is reproduced with permission from Springer-Verlag. It represents some of the earliest work recognizing the ill-posedness of optimal control problems when the "control" is the coefficient of a PDE. Other early work of a similar type is described in the review article by Lurie and Cherkaev (see the paper by Lurie and Cherkaev, this volume).

2. *Estimation of homogenized coefficients* by L. Tartar. This is a translation of a paper published in *Lecture Notes in Mathematics 704 (Computing Methods in Applied Sciences and Engineering)*, Springer-Verlag, 1977, 364–373. The translation was arranged by Doina Cioranescu with cooperation from the authors. It is reproduced in this form with the permission of Springer-Verlag. This is one of the earliest applications of weak convergence as a tool for bounding the effective moduli of composite materials.

3. *H-Convergence* by F. Murat and L. Tartar. This is a translation of notes that have circulated in mimeographed form since 1978, based on lectures given by F. Murat at the University of Algiers. The translation was done by Gilles Francfort with cooperation from the authors. The theory of H-convergence provides a mathematical framework for analysis of composites in complete generality, without any need for geometrical hypotheses such as periodicity or randomness. When specialized to the self-adjoint case it becomes equivalent to G-convergence. Treatments of G-convergence can be found elsewhere, including the books of Jikov et al. [1] and Dal Maso [2]. However, the treatment by Murat and Tartar has the advantage of being self-contained, elegant, compact, and quite general. As a result it remains, in our opinion, the best exposition of this basic material.

4. *A strange term coming from nowhere* by F. Murat and D. Cioranescu. This is a translation of a paper published in Nonlinear Partial Differential Equations and their Applications — College de France Seminar, H. Brezis, J.-L. Lions, and D. Cioranescu eds., Pitman, 1982, volume 2 pp. 98–138 and volume 3, pp. 154–178. The translation was arranged by D. Cioranescu with cooperation from F. Murat. It is reproduced in this form with the permission of Pitman. The focus of this work is somewhat different from the other chapters of this book. Attention is still on the macroscopic consequences of microstructures, and weak convergence still plays a fundamental role; however, in this work the

fine-scale boundary condition is of Dirichlet rather than Neumann or transmission type. There has been a lot of work on problems with similar boundary conditions but more general geometry, for example, Dal Maso and Garroni [3], and on problems involving Stokes flow, for example, Allaire [4] and Hornung et al. [5]. For work on structural optimization in problems of this type see Buttazzo and Dal Maso [6] and Sverak [7].

5. *Design of composite plates of extremal rigidity* by L. Gibiansky and A. Cherkaev. This is a translation of a Russian article first circulated as Ioffe Physico-Technical Institute Publication 914, Leningrad, 1984. The translation was done by Yury Gonorovsky and Leonid Gibiansky. This work provides an early application of homogenization to a problem of optimal design. Most prior work dealt with second-order scalar problems such as thermal conduction; this article deals instead with plate theory (and, by isomorphism, 2D elasticity). For subsequent related work see Kohn and Strang [8], Allaire and Kohn [9, 10], and especially the book of Bendsoe [11] and the review paper by Rozvany et al. [12] which have extensive bibliographies.

6. *Calculus of variations and homogenization* by F. Murat and L. Tartar. This is a translation of a paper published in *Les Methodes de l'Homogeneisation: Theorie et Applications en Physique,* Eyrolles, 1985, pp. 319–369. The translation was done by Anca-Maria Toader with cooperation from the authors. This work is reproduced in this form with the permission of Eyrolles. It presents a very complete treatment of optimal design problems in the setting of scalar second-order problems, and structures made from two isotropic materials. Such a treatment was made possible by the solution of the associated "G-closure problem" a few years before. The exposition of Murat and Tartar emphasizes the role of optimality conditions. For related work we refer once again to the book by Bendsoe [11], and also the article of Kohn and Strang [8].

7. *Effective characteristics of composite materials and the optimal design of structural elements* by K.A. Lurie and A. V. Cherkaev. This is a translation of the Russian article published in *Uspekhi Mekhaniki* 9:2, 1986, 3–81. However a portion of the Russian version has been omitted to avoid overlap with the paper by Gibiansky and Cherkaev, (Chapter 5 this volume.) The translation was done by Natalia Alexeeva with cooperation from the authors. The article is reproduced in this form with the permission of Uspekhi Mekhaniki. The paper presents a comprehensive review of work by the Russian community on homogenization methods applied to structural optimization and can be viewed as a theoretical introduction to optimal design problems illustrated by

a number of examples. The approach developed here is strongly influenced by advances in control theory (see the book by Lurie [13]) as well as by practical optimization problems. The paper is supplemented by an appendix describing early (1972) progress by Lurie and Simkina. That work in Russia was approximately contemporary with work by Murat and Tartar on similar issues in France in the early 1970s, including the first chapter of this book.

8. *Microstructures of composites of extremal rigidity and exact bounds on the associated energy density* by L.V. Gibiansky and A.V. Cherkaev. This is a translation of the Russian article first circulated as Ioffe Physico-Technical Institute Publication 1115, Leningrad, 1987. The translation was done by Yury Gonorovsky with cooperation from the authors. This work is a straight continuation of the problem discussed above, in Gibiansky and Cherkaev, this volume The bounds considered in this chapter by Gibiansky and Cherkaev concern the rigidity or compliance of a two-component elastic composite in three space dimensions; however, the paper reflects a subtle shift of emphasis. The mathematical community gradually realized during the 1980s that bounds on effective moduli are of broad interest in mechanics, beyond their value for relaxing problems of structural optimization. Here the translation method is applied for proving such bounds — based on the use of lower semicontinuous quadratic forms.

We give a highly selective list of more recent related work. Applications of the "compensated compactness" or the "translation method" to optimal bounds on effective conductivity include work by Lurie and Cherkaev [14, 16] and Tartar [15]. Bounds for other effective properties, including elasticity electromagnetic behavior, are discussed, for example, in Francfort and Murat [17], Gibiansky and Milton [18], Cherkaev and Gibiansky [19, 20], Milton [21, 22, 23], and Gibiansky and Torquato[24]. Polycrystal problems have been addressed by Nesi and Milton [25] and by Avellaneda et al. [26]. Other optimal energy bounds will be found in Milton and Kohn [27], Kohn and Lipton [28], and Allaire and Kohn [29]. Work on structural optimization using such bounds has been done by Allaire et al. [30] and Cherkaev and Palais [31]. Related methods have been used for the analysis of coherent phase transitions in crystalline solids; see, for example, the work of Ball and James [32, 33] concerning single crystals, and Bhattacharya and Kohn [34] concerning polycrystals. Some representative books from the mechanics literature on bounds and effective medium theories for composite materials are those by Christensen [35] and Nemat-Nasser and Hori [36]. These lists are representative, not complete; we apologize in advance to those whose work has not been mentioned.

We considered, at one point, preparing a comprehensive review of mathematical work related to the articles in this volume. But this seemed a

Herculean task, so we abandoned it. One can take pleasure in the idea that the subject has become difficult to review: it means the foundation represented by this book is occupied by a rich and multifaceted scientific edifice. The references given here should permit interested readers to explore the literature on their own.

We express our gratitude to the many people who have helped assemble this volume: the authors, who devoted much energy to proof-reading and improving the translations; the translators, who hopefully found their tasks interesting, because they certainly didn't get paid very much; and Edwin Beschler and his colleagues at Birkhäuser-Boston, who have shown unlimited patience as the target date for completion of the volume was reset time and again.

The subject is far from completion. Much remains to be understood, in many different areas, concerning effective moduli, structural optimization, multiwell variational problems, and coherent phase transitions. Some specific areas of current interest include bounds for multicomponent composites; bone remodeling; polycrystal plasticity, and "practical" suboptimal design. We hope this volume will accelerate progress by helping fill in the foundations of the field.

References

[1] Jikov, V.V., Kozlov, S.M., and Oleinik, O.A. (1994), *Homogenization of Differential Operators and Integral Functionals*, Springer-Verlag.

[2] Dal Maso, G. (1993), *An Introduction to Γ-Convergence*, Birkhäuser Boston.

[3] Dal Maso, G. and Garroni, A. (1994), New results on the asymptotic behavior of Dirichlet problems in perforated domains, *Math. Meth. Appl. Sci.* **3**, 373–407.

[4] Allaire, G. (1991), Homogenization of the Navier-Stokes equations in open sets perforated with tiny holes I. Abstract framework, a volume distribution of holes, *Arch. Rat. Mech. Anal.* **113**, 209–259.

[5] Hornung, U. et al. (in press), *Homogenization and Porous Media*, Springer-Verlag.

[6] Butazzo, G. and Dal Maso, G. (1991), Shape optimization for Dirichlet problems: relaxed formulation and optimality conditions, *Appl. Math. Optim.* **23**, 17–49.

[7] Sverak, V. (1993), On optimal shape design, *J. Math. Pures Appl.* **72**, 537–551.

[8] Kohn, R.V. and Strang, G. (1986), Optimal design and relaxation of variational problems, *Comm. Pure Appl. Math.* **39**, pp. 113–137, 139–182, 353–377.

[9] Allaire, G. and Kohn, R.V. (1993), Explicit optimal bounds on the elastic energy of a two-phase composite in two space dimensions, *Quart. Appl. Math.* **51**, 675–699.

[10] Allaire, G. and Kohn, R.V. (1993), Optimal design for minimum weight and compliance in plane stress,*Euro. J. Mechanics A/Solids* **12**, 839–878.

[11] Bendsoe, M. P. (1995), *Optimization of Structural Topology, Shape, and Material,* Springer, Berlin, New York.

[12] Rozvany G.I.N., Bendsoe M.P., and Kirsch U. (1994), Layout optimization of structures, *Appl. Mech. Rev.,* vol. 48, No 2, pp. 41–119; Addendum: *Appl. Mech. Rev.,* Vol. 49, No 1, p. 54.

[13] Lurie, K. (1993). *Applied Optimal Control,* Plenum, New York.

[14] Lurie, K.A. and Cherkaev, A. V. (1984), Exact estimates of conductivity of mixtures composed of two isotropical media taken in prescribed proportion, *Proceedings of Royal Soc. of Edinburgh,* **A**, 99(P1-2), 71-87. First version: *Phys.-Tech. Inst. Acad. Sci. USSR,* Preprint 783, 1982, (in Russian).

[15] Tartar, L. (1985), Estimations fines des coefficients homogeneises, Ennio de Giorgi's Colloquium, P. Kree,ed., *Pitman Research Notes in Math.* 125, 168–187.

[16] Lurie, K. and Cherkaev A. (1986), Exact estimates of conductivity of a binary mixture of isotropic compounds, *Proceed. Roy. Soc. Edinburgh, sect.A,* **104** (P1-2), p. 21–38. First version: Phys.-Tech. Inst. Acad. Sci. USSR, Preprint N.894, 1984 (in Russian).

[17] Francfort, G. and Murat, F. (1991), Homogenization and optimal bounds in linear elasticity, *Arch. Rational Mech. Anal.* **94**, pp. 301–307.

[18] Gibiansky, L.V., and Milton, G. W. (1993), On the effective viscoelastic moduli of two-phase media: I. Rigorous bounds on the complex bulk modulus, *Proc. R. Soc. Lond. A,* **440**, 163–188.

[19] Cherkaev A.V., and Gibiansky, L.V. (1992), The exact coupled bounds for effective tensors of electrical and magnetic properties of two-component two-dimensional composites, *Proceedings of Royal Society of Edinburgh* 122A, pp. 93–125.

[20] Cherkaev, A.V. and Gibiansky, L.V. (1993) Coupled estimates for the bulk and shear moduli of a two-dimensional isotropic elastic composite, *J. Mech. Phys. Solids* **41**, 937–980.

[21] Milton, G.W. (1990), On characterizing the set of possible effective tensors of composites: the variational method and the translation method, *Comm. Pure Appl. Math.* **43**, 63–125.

[22] Milton, G.W. (1990b), A brief review of the translation method for bounding effective elastic tensors of composites, Continuum Models and Discrete Systems, vol. 1, G. A. Maugin, ed., 60–74.

[23] Milton, G.W. Effective moduli of composites: exact results and bounds. In preparation

[24] Gibiansky, L.V. and S. Torquato, S. (1996), Rigorous link between the conductivity and elastic moduli of fiber-reinforced composite materials, *Trans. Roy. Soc. London* **A 452**, 253–283.

[25] Nesi, V. and Milton, G.W. (1991), Polycrystalline configurations that maximize electrical resistivity, *J. Mech. Phys. Solids* **39**, 525–542.

[26] M. Avellaneda, A.V. Cherkaev, L.V. Gibiansky, G.W. Milton, M. Rudelson (1996) A complete characterization of the possible bulk and shear moduli of planar polycrystals. J. Mech. Phys. Solids, in press.

[27] Milton, G.W. and Kohn, R.V. (1988) Variational bounds on the effective moduli of anisotropic composites, *J. Mech. Phys. Solids*, vol. 36, (1988), 597-629.

[28] Kohn, R.V. and Lipton, R. Optimal bounds for the effective energy of a mixture of isotropic, incompressible, elastic materials, *Arch. Rational Mech. Anal.*, 102, (1988), 331-350.

[29] Allaire, G. and Kohn, R.V. (1993) Optimal bounds on the effective behavior of a mixture of two well-ordered elastic materials, Quart. Appl. Math. 51, 643-674.

[30] Allaire, G., Bonnetier, E., Francfort, G., and Jouve, F. (1996) Shape optimization by the homogenization method, Numer. Math., in press.

[31] Cherkaev, A.V. and Palais, R. (1995) Optimal design of three-dimensional elastic structures. in:*Proceedings of the First World Congress of Structural and Multidisciplinary Optimization, Olhoff, N. and Rozvany, G. eds.*, Pergamon , 201-206

[32] Ball, J.M. and James, R.D. (1987) Fine phase mixtures as minimizers of energy, Arch. Rat. Mech. Anal. 100, 13-52.

[33] Ball, J.M. and James, R.D. (1992) Proposed experimental tests of a theory of fine microstructure and the two well problem, Phil. Trans. Roy. Soc. London 338A, 389-450.

[34] Bhattacharya, K. and Kohn, R.V. (1996) Elastic energy minimization and the recoverable strains of polycrystalline shape-memory materials, Arch. Rat. Mech. Anal., in press.

[35] Christensen, R.M. (1979) *Mechanics of Composite Materials* (Wiley-Interscience, New-York).

[36] Nemat-Nasser, S. and Hori, M. (1993) *Micromechanics: Overall Properties of Heterogeneous Materials*, North Holland.

On the Control of Coefficients
in Partial Differential Equations[*]

François Murat and Luc Tartar

1. Introduction

I. A few problems of the search for an optimal domain can be formulated as follows. Let Ω be an open bounded set of $\mathbb{R}^N$ ($N = 2$ or 3 in general). The problem is to find a subset Ω_1 of Ω such that if

$$a(x) = \begin{cases} \alpha & \text{on } \Omega_1, \\ \beta & \text{on } \Omega \setminus \Omega_1, \end{cases} \tag{1}$$

and u is the solution of the Dirichlet problem

$$\begin{cases} -\sum_i \dfrac{\partial}{\partial x_i}\left(a(x)\dfrac{\partial u}{\partial x_i}\right) = f & \text{in } \Omega, \\ u = 0 & \text{on the boundary } \partial\Omega \text{ of } \Omega, \end{cases} \tag{2}$$

the function

$$J(a) = \int_\Omega F(x, u(x))\, dx \tag{3}$$

reaches its mimimum.

II. If regularity properties are assumed on Ω, one can prove the existence of an optimal solution; see D. Chenais [1].

If no condition is imposed, then, depending on the cost function J, it may happen that there is no optimal solution; see F. Murat [2].

The aim of this note is to study this phenomenon.

---

[*] This article is a translation of an article originally written in French entitled *Problèmes de contrôle des coefficients dans des équations aux dérivées partielles*, published in *Control Theory, Numerical Methods and Computer Systems Modelling (Proceedings IRIA 1974)*, A. Bensoussan and J.-L. Lions, eds., Lecture Notes in Economics and Mathematical Systems, 107, Springer-Verlag, (1975), 420–426. The original article was signed only by L. Tartar who delivered the lecture in which he presented joint work that we were performing at that time. We prefer to present its English translation with our both names.

A. V. Cherkaev, R. Kohn (eds.), *Topics in the Mathematical Modelling of Composite Materials*, Modern Birkhäuser Classics, https://doi.org/10.1007/978-3-319-97184-1_1

2. The limit problem

I. Let us consider a minimizing sequence $a_n(x)$ of the preceding problem, that is to say, a sequence such that $J(a_n)$ decreases to the infimum of $J(a)$.

If f is a function in $L^2(\Omega)$, the solutions u_n belong to a bounded set of $H_0^1(\Omega)$, according to classical results for the Dirichlet problem. This implies that there exists a subsequence of the sequence u_n which converges weakly in $H_0^1(\Omega)$ to a function $u \in H_0^1(\Omega)$.

II. Unfortunately, the limit u is not, in general, a solution of a problem of type (2) with a taking only the values α and β (nor with a taking values in the interval $[\alpha, \beta]$).

But fortunately, one has some information about this function u, see S. Spagnolo [3], F. Murat and L. Tartar [4], namely,

III. There exists a subsequence a_m of a_n and a set of functions $a_{ij}(x)$, $i, j = 1, \ldots, N$, which are independent of f, satisfying

$$a_{ij}(x) = a_{ji}(x) \quad \text{a.e. in } \Omega, \tag{4}$$

$$\alpha \sum_i \xi_i^2 \le \sum_{i,j} a_{ij}(x)\xi_i\,\xi_j \le \beta \sum_i \xi_i^2 \quad \text{a.e. in } \Omega,\ \forall \xi_i \in \mathbb{R}^N, \tag{5}$$

such that the solution u_m converges (in $H_0^1(\Omega)$ weakly) to the solution u of the Dirichlet problem

$$\begin{cases} -\sum_{i,j} \dfrac{\partial}{\partial x_i}\left(a_{ij}(x)\dfrac{\partial u}{\partial x_j}\right) = f \quad \text{in } \Omega, \\ u = 0 \quad \text{on } \partial\Omega. \end{cases} \tag{6}$$

The determination of the coefficients $a_{ij}(x)$ of the limit problem (6) in terms of the sequence a_n is a local problem and in general it is not known how to solve it explicitly (except in the $N = 1$).

IV. In addition to (4) and (5), the matrix $a_{ij}(x)$ has the following property: if $\lambda_1(x) \le \lambda_2(x) \le \ldots \le \lambda_N(x)$ denote the eigenvalues of the matrix $a_{ij}(x)$, then

$$\frac{\alpha\beta}{\alpha + \beta - \lambda_N(x)} \le \lambda_1(x) \quad \text{a.e. in } \Omega. \tag{7}$$

Moreover one has the following convergences

$$
\begin{cases}
\dfrac{\partial u_m}{\partial x_i} \rightharpoonup \dfrac{\partial u}{\partial x_i} & \text{weakly in } L^2(\Omega), \quad i = 1, \ldots, N, \\[2ex]
a_m \dfrac{\partial u_m}{\partial x_i} \rightharpoonup \displaystyle\sum_j a_{ij} \dfrac{\partial u}{\partial x_i} & \text{weakly in } L^2(\Omega), \quad i = 1, \ldots, N, \\[2ex]
a_m \displaystyle\sum_i \left(\dfrac{\partial u_m}{\partial x_i}\right)^2 \rightharpoonup \displaystyle\sum_{ij} a_{ij} \dfrac{\partial u}{\partial x_i} \dfrac{\partial u}{\partial x_j} & \text{weakly } \star \text{ in } \mathcal{M}(\Omega),
\end{cases}
\tag{8}
$$

where $\mathcal{M}(\Omega)$ denotes the space of measures on Ω.

V. The inverse problem, that is, the characterization of the set of matrices $a_{ij}(x)$ that can be obtained by the above procedure, is not completely solved except for the cases $N = 1$ or 2.

In the case $N = 1$ one has the following result: if

$$
\frac{1}{a_m} \rightharpoonup \frac{1}{a} \quad \text{weakly } \star \text{ in } L^\infty(\Omega),
$$

then $a(x)$ is the coefficient that appears in the limit problem. All functions a with $a(x) \in [\alpha, \beta]$ can be obtained.

In the case $N = 2$ any matrix $a_{ij}(x)$ satisfying (4), (5), and (7) can be obtained as matrix of coefficients in the limit problem.

In the case $N > 2$ it is not known if all the matrices satisfying (4), (5), and (7) can be obtained by passing to the limit.[1] Nevertheless, we can obtain for instance the matrices a_{ij} whose eigenvalues are such that

$$
\alpha \le \lambda_1(x) \le \lambda_2(x) = \lambda_3(x) = \ldots = \lambda_N(x) \le \beta
\tag{9}
$$

with

$$
\frac{\alpha\beta}{\alpha + \beta - \lambda_N(x)} = \lambda_1(x).
\tag{10}
$$

Another example of admissible matrices are the matrices $a_{ij}(x) = a(x)\delta_{ij}$ with $a(x) \in [\alpha, \beta]$.

---

[1] It has been shown after the publication of the original of the present article that not all matrices satisfying (4), (5), and (7) can be obtained by passing to the limit: see Theorem 1 of L. Tartar, *Estimations fines de coefficients homogénéisés*, in *Ennio De Giorgi Colloquium*, P. Kree ed., Research Notes in Mathematics, 125, Pitman, (1985), 168–187, and Proposition 5.1 of E. Cabib and G. Dal Maso, *On a class of optimum problems in structural design*, J. Optim. Th. Appl., **56** (1988), 39–65.

3. Physical interpretation

I. Let two dielectric materials with dielectric permittivities α and β be given. One would like to fill the domain Ω by these two materials in order to minimize a certain function of the electrostatic potential $u(x)$ in the presence of a distribution of charge $\dfrac{1}{4\pi} f(x)$.

II. The electrostatic potential $u(x)$ is given by equation (2) (if the boundary of Ω is grounded), where $a(x)$ takes the values α or β in the regions occupied by one material or the other, and

$$E_i = \frac{\partial u}{\partial x_i} \text{ is the } i\text{th component of the electric field,}$$

$$D_i = aE_i \text{ is the } i\text{th component of the polarization field, and}$$

$$\frac{1}{4\pi} a \sum_i \left(\frac{\partial u}{\partial x_i} \right)^2 = \frac{E \cdot D}{4\pi} \text{ is the local energy.}$$

III. The case where there is no optimal solution corresponds to the situation in which, to minimize the function J, one has to intimately mix the materials in order to form a very heterogeneous material. Physically this material would rather be an alloy formed by the two materials. From the microscopic standpoint, it is still a mixture of two materials but, from the macroscopic one, it becomes a material with properties that are very different from the initial ones.

IV. This limit material is in general anisotropic and has a dielectric permittivity tensor a_{ij} which is not necessarily of the form $a\delta_{ij}$ (this form corresponds to an isotropic material).

It is clear that the determination of a_{ij} in the neighborhood of a point only depends on the manner in which the two materials are distributed in this neighborhood.

The mathematical convergences given in (8) indicate that macroscopically, the values of the electric field, the polarization field, and the energy converge to the corresponding quantities for the limit material.

Let us finally mention that the matrices satisfying (9) and (10) correspond to materials laminated in the direction of the eigenvector corresponding to the eigenvalue λ_1; that is to say, they are locally made of layers of material α and material β, perpendicular to the eigenvector, and the respective proportions of the materials determine the value of λ_1.

4. Necessary conditions of optimality

I. Suppose that there exists an optimal solution $a_0(x)$ taking only the values α and β, and corresponding to $u_0(x)$. Thus,

$$J(a) \geq J(a_0), \quad \text{for any } a \text{ satisfying (1).} \tag{11}$$

Suppose further that $F(x, u)$ is differentiable in u and introduce the function $w_0(x)$ solution of

$$\begin{cases} -\sum_i \frac{\partial}{\partial x_i}\left(a_0(x)\frac{\partial w_0}{\partial x_i}\right) = \frac{\partial F}{\partial u}(x, u_0) & \text{in } \Omega, \\ w_0 = 0 \quad \text{on } \partial\Omega. \end{cases} \tag{12}$$

II. If a_{ij} is admissible (i.e., is obtained as a limit), one can define

$$J(a_{ij}) = \int_\Omega F(x, u) \, dx$$

where u is the solution of (6).

Then, if F is smooth enough (i.e. if $u \mapsto \int_\Omega F(x, u) \, dx$ is continuous on $L^2(\Omega)$), one has

$$J(a_{ij}) \geq J(a_0) = J(a_0\delta_{ij}). \tag{13}$$

III. The application $a_{ij} \mapsto u$ is differentiable as well as the application $u \mapsto \int_\Omega F(x, u)dx$. Therefore, if a_0 is optimal, one deduces that the derivative of J in any admissible direction of variation is nonnegative.

Let δa_{ij} be an admissible direction of variation, that is to say,

$$\exists a_{ij}^n \text{ admissible}, \exists t_n > 0 \text{ such that } \lim_{t_n \to 0} \frac{a_{ij}^n - a_0\delta_{ij}}{t_n} = \delta a_{ij}.$$

Then a classical computation shows that

$$\sum_{ij} \int_\Omega \delta a_{ij} \frac{\partial u_0}{\partial x_i} \frac{\partial w_0}{\partial x_i} \, dx \leq 0. \tag{14}$$

IV. In view of (9) and (10) one knows some admissible directions of variation:

— in the region where $a_0(x) = \alpha$, the δa_{ij} with one eigenvalue equal to $\dfrac{\alpha}{\beta}$ and the other ones equal to $+1$; and

— in the region where $a_0(x) = \beta$, the δa_{ij} with one eigenvalue equal to $-\dfrac{\beta}{\alpha}$ and the other ones equal to -1.

In the region where $a_0(x) = \alpha$ and where $e(x)$ is a unit eigenvector corresponding to the eigenvalue $\dfrac{\alpha}{\beta}$, condition (14) can be rewritten as:

$$(\mathrm{grad}\ u_0, \mathrm{grad}\ w_0) + \left(\frac{\alpha}{\beta} - 1\right)(\mathrm{grad}\ u_0, e)\ (\mathrm{grad}\ w_0, e) \leq 0.$$

Let us choose $e(x)$ which maximizes the left hand side of this inequality. If the vectors $\mathrm{grad}\ u_0(x)$ and $\mathrm{grad}\ w_0(x)$ are not zero, the maximum is reached when $e(x)$ is orthogonal to the bisector of the angle they form. Let $\theta(x)$ be the measure of this angle. One derives the condition:

$$\begin{cases} \text{in the region where } a_0(x) = \alpha, \text{ one has} \\[2mm] \text{either } |\mathrm{grad}\ u_0(x)|\,|\mathrm{grad}\ w_0(x)| = 0 \text{ or } \cos\theta(x) \leq -\dfrac{\beta - \alpha}{\beta + \alpha}. \end{cases} \quad (15)$$

Similarly,

$$\begin{cases} \text{in the region where } a_0(x) = \beta, \text{ one has} \\[2mm] \text{either } |\mathrm{grad}\ u_0(x)|\,|\mathrm{grad}\ w_0(x)| = 0 \text{ or } \cos\theta(x) \geq \dfrac{\beta - \alpha}{\beta + \alpha}. \end{cases} \quad (16)$$

V. In the case $N = 2$ one knows the characterization of the set of those a_{ij} where the solution of the minimization problem can be found. Consequently, one can give necessary conditions for a_{ij}^0 to be optimal.

Let w_0 denote the solution of

$$\begin{cases} -\displaystyle\sum_{ij} \frac{\partial}{\partial x_i}\left(a_{ij}\frac{\partial w_0}{\partial x_j}\right) = \frac{\partial F}{\partial u}(x, u_0) \quad \text{in } \Omega, \\[4mm] w_0 = 0 \quad \text{on } \partial\Omega. \end{cases} \quad (12')$$

One finds the same optimality condition (14).

In addition to the region where $a_{ij}^0(x) = \alpha\delta_{ij}$, which leads to condition (15), and to the region where $a_{ij}^0(x) = \beta\delta_{ij}$, which leads to condition (16), one has to examine the region where

$$\alpha < \frac{\alpha\beta}{\alpha + \beta - \lambda_2^0(x)} < \lambda_1^0(x) < \beta,$$

and the region where

$$\alpha < \frac{\alpha\beta}{\alpha + \beta\lambda_2^0(x)} = \lambda_1^0(x) < \lambda_2^0(x) < \beta.$$

In the first region, all the directions of variation are admissible and one obtains the condition

$$\begin{cases} \text{in the region where } \alpha < \dfrac{\alpha\beta}{\alpha + \beta - \lambda_2^0(x)} < \lambda_1^0(x) < \lambda_2^0(x) < \beta, \\ \text{one has } |\text{grad } u_0(x)|\, |\text{grad } w_0(x)| = 0. \end{cases} \qquad (17)$$

In the second region, let $\{e_1(x), e_2(x)\}$ be an orthonormal basis of eigenvectors corresponding to the eigenvalues $\lambda_1^0(x)$ and $\lambda_2^0(x)$. One easily verifies that the matrices

$$\pm \begin{pmatrix} 0 & 1 \\ 1 & 0 \end{pmatrix}$$

as well as the matrix

$$\begin{pmatrix} \dfrac{\alpha\beta}{(\alpha + \beta - \lambda_2^0(x))^2} & 0 \\ 0 & 1 \end{pmatrix}$$

are admissible directions of variation δa_{ij}. Using these three matrices in (14), one finally finds the following condition, where $\theta(x)$ is the measure of the angle between grad $u_0(x)$ and grad $w_0(x)$:

$$\begin{cases} \text{in the region where } \alpha < \dfrac{\alpha\beta}{\alpha + \beta - \lambda_2^0(x)} < \lambda_1^0(x) < \lambda_2^0(x) < \beta, \\ \text{one has either } |\text{grad } u_0(x)|\, |\text{grad } w_0(x)| = 0, \\ \text{or } \cos\theta = \dfrac{(\lambda_1^0(x))^2 - \alpha\beta}{(\lambda_1^0(x))^2 + \alpha\beta} \text{ and the bisector of the angle} \\ \text{made by grad } u_0(x) \text{ and grad } w_0(x) \text{ is an eigenvector } e_2(x) \\ \text{corresponding to the eigenvalue } \lambda_2^0(x). \end{cases} \qquad (18)$$

References

[1] D. Chenais, Un résultat d'existence dans un problème d'identification de domaine, *C. R. Acad. Sci. Paris* Série A, **276** (1973), 547–550.

D. Chenais, On the existence of a solution in a domain identification problem, *J. Math. Anal. Appl.* **52** (1975), 189–219.

[2] F. Murat, Un contre-exemple pour le problème du contrôle dans les coefficients, *C. R. Acad. Sci. Paris* Série A, **273** (1971), 708–711.

F. Murat, Théorèmes de non-existence pour des problèmes de contrôle dans les coefficients, *C. R. Acad. Sci. Paris* Série A, **274** (1972), 395–398.

F. Murat, Contre-exemples pour divers problèmes où le contrôle intervient dans les coefficients, *Ann. Mat. Pura Appl.* **12** (1977), 49–68.

[3] S. Spagnolo, Sul limite delle soluzioni di problemi de Cauchy relativi all'equazione del calore, *Ann. Sc. Norm. Sup. Pisa* **21** (1967), 657–699.

S. Spagnolo, Sulla convergenza di soluzioni di equazioni paraboliche ed ellitiche, *Ann. Sc. Norm. Sup. Pisa* **22** (1968), 571–597.

A. Marino and S. Spagnolo, Un tipo di approssimazione dell'operatore $\sum D_i(a_{ij}D_j)$ con operatori $\sum D_i(\beta D_i)$, *Ann. Sc. Norm. Sup. Pisa* **23** (1969), 657–673.

E. De Giorgi and S. Spagnolo, Sulla convergenza degli integrali dell'energia per operatori ellitici del secondo ordine, *Boll. Unione Mat. Ital.* **8** (1973), 391–411.

[4] F. Murat and L. Tartar, to appear.[2]

---

[2] This article was supposed to be a detailed version of the present note, but was never written. Our article *Calculus of Variations and Homogenization* published in the present volume represents a further stage of our work in this domain.

Estimations of Homogenized Coefficients

Luc Tartar

1. Introduction

The homogenization method gives the possibility of finding equations satisfied by macroscopic quantities from equations satisfied by the physical quantities and from information on the microscopic composition (or on the microscopic structure).

The mathematical method is essentially based on the study of solutions of partial differential equations (or of systems of PDEs) with highly oscillating coefficients. In general these solutions are very close, in a weak topology, to the solution of an homogenized equation with rather smooth coefficients called "homogenized or effective coefficients." Note that the homogenized equation can be very different from the original one.

When one has information about the microscopic structure, explicit formulae (which need more or less classical computations), may be obtained for the effective coefficients. This is the case, for instance, when the coefficients are periodic with a very small period. In the general case, the existence of explicit formulae is not known.

When having partial information about the microscopic structure, of statistical order for instance, one can only prove some inequalities satisfied by the homogenized coefficients. We are interested in this kind of result in this article.

2. The model problem

Let us consider the solution of an equation of the following type:

$$-\sum_{ij} \frac{\partial}{\partial x_i}\left(a_{ij}^{\varepsilon}(x)\frac{\partial u^{\varepsilon}}{\partial x_j}\right) = f^{\varepsilon} \quad \text{in } \Omega \subset \mathbb{R}^N, \tag{1}$$

where the coefficients a_{ij}^{ε}, which are measurable functions, satisfy a uniform ellipticity condition

$$\begin{cases} \text{(i). } \exists\, \alpha > 0, \quad \text{such that } \forall \lambda \in \mathbb{R}^N, \quad \sum_{ij} a_{ij}^{\varepsilon}\lambda_i\lambda_j \geq \alpha|\lambda|^2 \quad \text{a.e.,} \\[2mm] \text{(ii). } \exists\, M > 0, \quad \text{such that } |a_{ij}^{\varepsilon}(x)| \leq M \quad \text{a.e.} \end{cases} \tag{2}$$

If f^{ε} converges to f^0 and u^{ε} to u^0 (in topologies to be made precise), then u^0 satisfies an equation of the type:

$$-\sum_{ij} \frac{\partial}{\partial x_i}\left(q_{ij}(x)\frac{\partial u^0}{\partial x_j}\right) = f^0 \quad \text{in } \Omega \subset \mathbb{R}^N, \tag{3}$$

© Springer Nature Switzerland AG 2018

A. V. Cherkaev, R. Kohn (eds.), *Topics in the Mathematical Modelling of Composite Materials*, Modern Birkhäuser Classics, https://doi.org/10.1007/978-3-319-97184-1_2

where q_{ij} are called homogenized coefficients.

In the case where $a_{ij}^\varepsilon(x) = a_{ij}\left(\dfrac{x}{\varepsilon}\right)$, one can show that there exists an asymptotic expansion of the form

$$u^\varepsilon(x) = u^0(x) + \varepsilon u^1\left(x, \frac{x}{\varepsilon}\right) + \dots \qquad (4)$$

We have here a new type of convergence, called H-convergence (or convergence in the sense of homogenization). This convergence is denoted by

$$a^\varepsilon \xrightarrow{\ H\ } q$$

and is characterized by the following property.

Characteristic property of H-convergence.

Let $a^\varepsilon \xrightarrow{\ H\ } q$. Then, if u^ε satisfies

$$\frac{\partial u^\varepsilon}{\partial x_j} \rightharpoonup \frac{\partial u^0}{\partial x_j} \quad \text{weakly in } L^2(\Omega), \ j = 1, \dots, N, \qquad (5)$$

and

$$-\sum_{ij} \frac{\partial}{\partial x_i}\left(a_{ij}^\varepsilon(x)\frac{\partial u^\varepsilon}{\partial x_j}\right) \to f^0 \quad \text{strongly in } H^{-1}(\Omega), \qquad (6)$$

one has

$$\sum_j a_{ij}^\varepsilon(x)\frac{\partial u^\varepsilon}{\partial x_j} \rightharpoonup \sum_j q_{ij}\frac{\partial u^0}{\partial x_j} \quad \text{weakly in } L^2(\Omega), \ i = 1, \dots, N. \qquad (7)$$

Remark 1. In particular, u^0 is the solution of equation (3).

The fundamental result in this theory is the following compactness theorem.

Theorem 1. *Let a_{ij}^ε be a sequence of functions satisfying condition (2). Then, there exists a subsequence that H-converges to q_{ij} and moreover, q_{ij} satisfy (2.i) with the same α and in general (2.ii), with a different constant M'.*

There are two important situations (with several variants), where one knows explicit formulae for q_{ij}: the case where the coefficients are periodic and the case where the material is laminated. For these situations we have the following results:

Theorem 2.(Periodic case). *If the coefficients a_{ij}^ε are of the form $a_{ij}\left(\dfrac{x}{\varepsilon}\right)$ where a_{ij} are periodic, the whole sequence a^ε H-converges to q when ε*

tends to 0, whose entries q_{ij} are constant. They are determined by solving N partial differential equations and afterwards by calculating an average on the period as follows.

(i). For $k = 1, \ldots, N$, let w_k be the solution, determined up to an additive constant, of the system

$$
\begin{cases}
-\sum_{ij} \frac{\partial}{\partial y_i} \left(a_{ij}(y) \frac{\partial w_k}{\partial y_j} \right) = 0, \\[2mm]
w_k - y_k \quad periodic.
\end{cases}
$$

(ii). One computes

$$
q_{ik} = \text{the average on one period of } \sum_{j} \left(a_{ij}(y) \frac{\partial w_k}{\partial y_j} \right).
$$

Theorem 3. (Laminated material). *If the coefficients a_{ij}^ε are functions of x_1 only, then $a^\varepsilon \xrightarrow{H} q$ (with q_{ij} functions of x_1 only) is equivalent to:*

(i).

$$
\frac{1}{a_{11}^\varepsilon} \rightharpoonup \frac{1}{q_{11}} \quad \text{weakly} \star \text{ in } L^\infty(\Omega).
$$

(ii). For $i \neq 1$

$$
\begin{cases}
\dfrac{a_{i1}^\varepsilon}{a_{11}^\varepsilon} \rightharpoonup \dfrac{q_{i1}}{q_{11}} \quad \text{weakly} \star \text{ in } L^\infty(\Omega), \\[3mm]
\dfrac{a_{1i}^\varepsilon}{a_{11}^\varepsilon} \rightharpoonup \dfrac{q_{1i}}{q_{11}} \quad \text{weakly} \star \text{ in } L^\infty(\Omega).
\end{cases}
$$

(iii). For $i \neq 1$, $j \neq 1$

$$
a_{ij}^\varepsilon - \frac{a_{i1}^\varepsilon a_{1j}^\varepsilon}{a_{11}^\varepsilon} \rightharpoonup q_{ij} - \frac{q_{i1} q_{1j}}{q_{11}} \quad \text{weakly} \star \text{ in } L^\infty(\Omega).
$$

Remark 2. If $a_{ij}^\varepsilon = a_{ji}^\varepsilon$, $\forall i, j$, then $q_{ij} = q_{ji}$, $\forall i, j$. In this case, if there exists $\beta > 0$ such that $\forall \lambda$, $\sum_{ij} a_{ij}^\varepsilon \lambda_i \lambda_j \leq \beta |\lambda|^2$, a.e., then q_{ij} satisfy this inequality with the same constant β. It can also be shown that for this situation, H-convergence coincides with G-convergence; for details see [1].

Remark 3. H-convergence has a local character. Indeed, the coefficients q_{ij} are uniquely determined (up to a set of measure zero) and the knowledge of a_{ij}^ε on an open subset ω is sufficient for the characterization of q_{ij} on ω

It should be mentioned that H-convergence is different from the usuˑ weak convergences: one can easily construct two subsequences a^ε anˊ such that, for any continuous function F, $F(a_{ij}^\varepsilon)$ and $F(b_{ij}^\varepsilon)$ have the weak limit but a^ε and b^ε H-converge to different H-limits.

Remark 4. One can prove that equations (5), (6), and (7) imply that

$$\sum_{ij} a_{ij}^\varepsilon \frac{\partial u^\varepsilon}{\partial x_j} \frac{\partial u^\varepsilon}{\partial x_i} \rightharpoonup \sum_{ij} q_{ij} \frac{\partial u^0}{\partial x_j} \frac{\partial u^0}{\partial x_i} \quad \text{weakly in } L_{\text{loc}}^1. \tag{8}$$

Remark 5. Suppose that u^ε in equation (1) is the electrostatic potential. Then $E_\varepsilon = $ -grad u_ε is the electrical field and $D_\varepsilon = A_\varepsilon$ grad u_ε is the electrical induction field. The limit u_0 denotes the mean (macroscopic) potential, E_0 and D_0 denote the mean fields and are related by $D_0 = qE_0$, where q is the effective conductivity tensor and *is not the average of* A_ε. The density of electrostatic energy is $e_k = (E_\varepsilon, D_\varepsilon)$ and (8) says that the mean density is equal to $e_0 = (E_0, D_0)$.

This fact has to be underlined since for a function G, the mean value of $G(E_\varepsilon, D_\varepsilon)$ is not, in general, $G(E_0, D_0)$.

Remark 6. Let us point out that for proving the preceding results, the methods we employ are of the variational type and do not make use of any property specific to the second order operators.

3. Bounds

In this section we make use of the notations:
— if M and N are two matrices, then

$$M \leq N \iff (M\lambda, \lambda) \leq (N\lambda, \lambda), \ \forall \lambda \in \mathbb{R}^N.$$

— A_ε is the matrix of coefficients a_{ij}^ε, and Q that of coefficients q_{ij}, and we suppose that these matrices are symmetric.

H-convergence has interesting properties with respect to the ordering. Let us present some of them.

Theorem 4. *If* $A_\varepsilon \leq B_\varepsilon$ *(with* $A_\varepsilon^* = A_\varepsilon$*) and* $A_\varepsilon \xrightarrow{H} Q$, $B_\varepsilon \xrightarrow{H} R$, *then* $Q \leq R$.

Theorem 5. *Let* $A_\varepsilon^* = A_\varepsilon$ *with* $A_\varepsilon \xrightarrow{H} Q$. *If* $A_\varepsilon \rightharpoonup A_0$ *weakly* $\star$ *in* $(L^\infty)^{N^2}$ *and* $A_\varepsilon^{-1} \rightharpoonup B_0^{-1}$ *weakly* $\star$ *in* $(L^\infty)^{N^2}$, *then* $B_0 \leq Q \leq A_0$.

Remark 7. Even in the situation of explicit formulae from Theorems 2 and 3, these inequalities are as difficult to prove as in the general case.

Remark 8. Making use of explicit formulae, one can find cases where $Q - B_0$ and $A_0 - Q$ both have one eigenvalue equal to 0.

Remark 9. Going back to the interpretations from Remark 5, we can rewrite Theorem 5 as follows:

effective conductivity $\leq$ *mean conductivity,*

effective resistivity $\leq$ *mean resistivity.*

Remark 10. Let us apply Theorem 5 to the case

$$a_{ij}^{\varepsilon} = \delta_{ij} a\left(\frac{x}{\varepsilon}, \frac{y}{\varepsilon}\right),$$

where a, periodic of period 1 in x and y, is defined in the unit square by

$$a(x, y) = \begin{cases} \alpha & \dfrac{1}{3} \leq x, y \leq \dfrac{2}{3}, \\ 1 & \text{elsewhere.} \end{cases}$$

Then, for obvious reasons of symmetry, $q_{ij} = q\delta_{ij}$ and q satisfies

$$\frac{9\alpha}{8\alpha + 1} \leq q \leq \frac{8 + \alpha}{9}.$$

It is clear that this formula is not very precise when $\alpha \to 0$ or $\alpha \to \infty$.

Consider now the function b defined by

$$b(x, y) = \begin{cases} \alpha & \dfrac{1}{3} \leq x \leq \dfrac{2}{3}, \\ 1 & \text{elsewhere,} \end{cases}$$

and let us apply Theorem 3 in order to have the H-limit of $b\left(\dfrac{x}{\varepsilon}, \dfrac{y}{\varepsilon}\right)$. We get, also by using Theorem 4, other inequalities for q:

— if $\alpha \leq 1$, one has

$$\frac{2 + \alpha}{3} \leq q,$$

which improves the former inequality for $0 \leq \alpha \leq \dfrac{1}{4}$,

— if $\alpha \geq 1$, one has

$$q \leq \frac{3\alpha}{2\alpha + 1},$$

which is an improvement for $4 \leq \alpha$.

These last inequalities are clearly more precise than the preceding ones if α goes to 0 or to ∞.

To obtain the preceding inequalities we used the weak limit of A_{ε} (or of A_{ε}^{-1}), which is obtained by integrating A_{ε} on sets of positive measure and afterwards passing to the limit. More precise inequalities can be obtained by integrating on sets of dimension 1 or $N - 1$. For each direction of the hyperplane, we obtain two inequalities. In order to state the results, we need to introduce a notation. If we choose the direction $x_1 = 0$, we have to make use of the following convergence:

$$M_\varepsilon \overset{*1}{\rightharpoonup} M \iff \begin{cases} \dfrac{1}{m_{11}^\varepsilon} \rightharpoonup \dfrac{1}{m_{11}}, & \text{weakly } \star \text{ in } L^\infty \\[2mm] \dfrac{m_{i1}^\varepsilon}{m_{11}^\varepsilon} \rightharpoonup \dfrac{m_{i1}}{m_{i1}}, & \text{weakly } \star \text{ in } L^\infty \\[2mm] \dfrac{m_{1j}^\varepsilon}{m_{11}^\varepsilon} \rightharpoonup \dfrac{m_{1j}}{m_{i1}}, & \text{weakly } \star \text{ in } L^\infty \\[2mm] m_{ij}^\varepsilon - \dfrac{m_{i1}^\varepsilon m_{1j}^\varepsilon}{m_{11}^\varepsilon} \rightharpoonup m_{ij} - \dfrac{m_{i1} m_{1j}}{m_{i1}}, & \text{weakly } \star \text{ in } L^\infty. \end{cases}$$

Theorem 6. *Let* $\omega(x_2, \dots, x_n)$ *be a positive function of integral 1 and let* ρ *be the measure* $\delta_{x_1} \otimes \omega$. *If*

$$A_\varepsilon * \rho \overset{*1}{\longrightarrow} B_0,$$

then

$$Q * \rho \le B_0. \qquad\qquad \blacksquare$$

For an $(N-1)$-direction, we have

Theorem 7. *Let* $\omega(x_1)$ *be a positive function of integral 1 and let* ρ *be the measure* $\omega \otimes \delta_{x_2 = \dots = x_N}$. *If*

$$A_\varepsilon^{-1} * \rho \overset{*1}{\longrightarrow} C_0^{-1},$$

then

$$Q * \rho \ge C_0.$$

Remark 11. In the case where $A_\varepsilon = a_\varepsilon I$, the algorithm described in Theorem 6 consists of doing an arithmetic average in $x_2, \dots, x_N$ first, and then, similarly an harmonic average in x_1. The algorithm in Theorem 7 begins with the harmonic average in x_1.

Remark 12. It should be noted that the convergence $*1$ (see Theorem 3) is a very natural one. The mathematical object A is, in fact, an application that transforms a 1-differential form $E = \sum_i E_i \, dx_i$ into the $(N-1)$-form $D = \sum_i D_i \, dx_1 \dots \widehat{dx_i} \dots, dx_N$, and the terms $\dfrac{1}{a_{11}}$, $\dfrac{a_{i1}}{a_{11}}$, $a_{ij} - \dfrac{a_{i1} a_{1j}}{a_{11}}$ appear when writing the exterior product $E \wedge D = \sum_i E_i D_i \, dx$ by using $D_1, E_2, \dots, E_N$ (which are the only components making sense on the hyperplanes $x_1 = constant$).

Remark 13. If Theorems 6 and 7 are applied to the example treated in Remark 10, one gets the more precise inequalities

$$\frac{2+7\alpha}{3(2\alpha+1)} \leq q \leq \frac{3(2+\alpha)}{7+2\alpha}.$$

If $\alpha \to 0$ (or ∞), then one can see (from the formulae of Theorem 2 since in this case one can explicitly write down the solutions) that q goes to $\frac{2}{3}$ (respectively, $\frac{3}{2}$).

Proofs.

(a). We know how to construct, for each $\lambda \in \mathbb{R}^N$, a sequence of vector-valued functions v_ε such that

$$\begin{cases} v_\varepsilon \rightharpoonup \lambda & \text{weakly in } (L^2(\Omega))^N, \\ \text{curl } v_\varepsilon = 0, \\ \text{div } (A_\varepsilon v_\varepsilon) & \text{converges strongly in } H^{-1}(\Omega), \end{cases}$$

and hence,

$$\begin{cases} A_\varepsilon v_\varepsilon \rightharpoonup Q\lambda & \text{weakly in } (L^2(\Omega))^N, \\ (A_\varepsilon v_\varepsilon, v_\varepsilon) \rightharpoonup (Q\lambda, \lambda) & \text{weakly in } L^1_{\text{loc}}(\Omega). \end{cases}$$

This is the essential part of the information given by H-convergence and Theorem 1 (which we assume here). Roughly speaking, it is sufficient to solve $-\text{div} (A_\varepsilon \text{grad } u_\varepsilon - Q\lambda) = 0$ with $u_\varepsilon - (\lambda, x) \in H^1_0(\Omega)$ and set $v_\varepsilon = \text{grad } u_\varepsilon$. ∎

(b). (Theorem 4). Let v_ε be the preceding sequence associated with A_ε and let us construct as previously, a sequence w_ε associated with B_ε, both corresponding to the vector λ.

We pass to the limit in the inequality

$$(A_\varepsilon (v_\varepsilon - w_\varepsilon), (v_\varepsilon - w_\varepsilon)) \geq 0.$$

By construction, $(A_\varepsilon v_\varepsilon, v_\varepsilon)$ and $(A_\varepsilon v_\varepsilon, w_\varepsilon)$ converge to $(Q\lambda, \lambda)$ in L^1_{loc} weakly. On the other hand, $(A_\varepsilon w_\varepsilon, w_\varepsilon) \leq (B_\varepsilon w_\varepsilon, w_\varepsilon)$, which converges to $(R\lambda, \lambda)$ in L^1_{loc} weakly. Hence $(R\lambda, \lambda) \geq (Q\lambda, \lambda)$. ∎

(c). (Theorem 5). Let again v_ε be the sequence associated with A_ε and λ. Passing first to the limit in the inequality

$$(A_\varepsilon (v_\varepsilon + \mu), (v_\varepsilon + \mu)) \geq 0,$$

one has

$$(Q\lambda, \ \lambda) + 2(Q\lambda, \ \mu) + (A_0\mu, \ \mu) \geq 0,$$

which implies

$$A_0 \geq Q.$$

Passing afterwards to the limit in the inequality

$$(A_\varepsilon^{-1}(A_\varepsilon \ v_\varepsilon + \mu), \ (A_\varepsilon \ v_\varepsilon + \mu) \geq 0,$$

one gets

$$(Q\lambda, \ \lambda) + 2(\lambda, \ \mu) + (B_0^{-1}\mu, \ \mu) \geq 0,$$

which yields

$$B_0 \leq Q. \qquad \blacksquare$$

(d). (Theorem 6). One makes use of the inequality

$$\rho * (A_\varepsilon \ (v_\varepsilon + w_\varepsilon), \ v_\varepsilon + w_\varepsilon) \geq 0,$$

where we choose $w_\varepsilon(x_1)$ such that curl $w_\varepsilon = 0$ (hence $w_{\varepsilon 2}, \ldots, w_{\varepsilon N}$ are constants) and $w_\varepsilon \rightharpoonup \mu$ in $L^\infty(\Omega)$ weakly $\star$. We have

$$\rho * (A_\varepsilon \ v_\varepsilon, \ v_\varepsilon) \rightharpoonup \rho * (Q\lambda, \ \lambda) \quad \text{weakly in } L^1_{\text{loc}}(\Omega),$$
$$\rho * (A_\varepsilon \ v_\varepsilon, \ w_\varepsilon) \rightharpoonup \rho * (Q\lambda, \ \mu) \quad \text{weakly in } L^2(\Omega).$$

Let us now choose w_ε in order to minimize the limit of $\rho * (A_\varepsilon \ w_\varepsilon, \ w_\varepsilon) = ((\rho * (A_\varepsilon) \ w_\varepsilon, \ w_\varepsilon)$ (since w_ε depends on x_1 only). The smallest limit is $(B_0\mu, \ \mu)$ as a consequence of the following lemma.

Lemma 1. *Let* $M_\varepsilon \overset{*1}{\longrightarrow} M_0$ *with* $M_\varepsilon \geq 0$. *If* $w_\varepsilon \rightharpoonup \mu$ *in* $L^\infty(\Omega)$ *weakly* $\star$ *with* $w_{\varepsilon 2}, \ldots, w_{\varepsilon N}$ *constants, and* $(M_\varepsilon \ w_\varepsilon, \ w_\varepsilon) \rightharpoonup \ell$ *in* $L^\infty(\Omega)$ *weakly* $\star$, *then necessarily* $\ell \geq (M_0\mu, \ \mu)$ *and moreover, there exists a sequence such that* $\ell = (M_0\mu, \ \mu)$. *(The best sequence is obtained by writing* $(M_\varepsilon w_\varepsilon)_1 = \mu_1$, $w_{\varepsilon j} \to \mu_j$ *for* $j > 1$.)*

In view of this lemma,

$$\Big((\rho * Q)\lambda, \ \lambda\Big) + 2\Big((\rho * Q)\lambda, \ \mu\Big) + (B_0\mu, \ \mu) \geq 0, \quad \forall \lambda, \mu,$$

and thus

$$\rho * Q \leq B_0. \qquad \blacksquare$$

(e). (Theorem 7). One makes use of the inequality

$$\rho * (A_\varepsilon^{-1}(A_\varepsilon \ v_\varepsilon + w_\varepsilon), \ A_\varepsilon \ v_\varepsilon + w_\varepsilon) \geq 0$$

where we choose $w_\varepsilon(x_2, \ldots, x_N)$ such that div $w_\varepsilon = 0$ (hence $w_{\varepsilon 2}, \ldots, w_{\varepsilon N}$ are constants) and $w_\varepsilon \rightharpoonup \mu$ in $L^\infty(\Omega)$ weakly $\star$. We have

$$\rho * (A_\varepsilon \, v_\varepsilon, \, v_\varepsilon) \rightharpoonup \rho * (Q\lambda, \, \lambda) \qquad \text{weakly in } L^1_{\text{loc}}(\Omega),$$
$$\rho * (v_\varepsilon, \, w_\varepsilon) \rightharpoonup \rho * (\lambda, \, \mu) = (\lambda, \, \mu) \quad \text{weakly in } L^2(\Omega).$$

Let us now choose w_ε in order to minimize the limit of $\rho * (A_\varepsilon^{-1} \, w_\varepsilon, \, w_\varepsilon) = \left(\left(\rho * (A_\varepsilon^{-1}) \right) w_\varepsilon, \, w_\varepsilon \right)$ (since w_ε do not depend on x_1). The smallest limit is $(C_0^{-1}\mu, \, \mu)$. It follows that

$$\left((\rho * Q)\lambda, \, \lambda \right) + 2(\lambda, \, \mu) + (C_0^{-1}\mu, \, \mu) \geq 0, \quad \forall \lambda, \mu,$$

hence

$$\rho * Q \geq C_0. \qquad \blacksquare$$

In the preceding proofs we used the fact that if v_ε and w_ε are two sequences such that

$$\begin{cases} v_\varepsilon \rightharpoonup v \quad \text{weakly in } (L^2(\Omega))^N, \\ \text{curl } v_\varepsilon = 0 \end{cases}$$

and

$$\begin{cases} w_\varepsilon \rightharpoonup w \quad \text{weakly in } (L^2(\Omega))^N, \\ \text{div } w_\varepsilon \qquad \text{converges strongly in } H^{-1}(\Omega), \end{cases}$$

then

$$(v_\varepsilon, \, w_\varepsilon) \rightharpoonup (v, \, w) \quad \text{in } \mathcal{D}'(\Omega).$$

This result, which can easily be obtained by writing $v_\varepsilon = \text{grad } w_\varepsilon$ and integrating by parts, is the first example of the application of a new method: *compensated compactness*. Using other examples, one can derive other inequalities on Q but, unfortunately, they are not very explicit.

For obtaining other inequalities, let us consider first a quadratic form B defined on the space of dimension $2N^2$ formed by pairs of matrices $(V, \, W)$ satisfying the condition

$$\begin{cases} B(V, W) \geq 0 \quad \text{if } \exists \xi \in \mathbb{R}^N, \ \xi \neq 0, \ \exists \lambda \in \mathbb{R}^N \quad \text{such that} \\ V_{ij} = \lambda_j \xi_i, \ \forall i, j \ \text{ and } \ \sum_i w_{ij}\xi_i = 0, \ \forall j. \end{cases} \qquad (9)$$

If further, $L(V, W)$ is a linear form in V, W, one defines

$$F(A) = \sup_V \left\{ B(V, \, AV) + L(V, \, AV) \right\}. \qquad (10)$$

We have the result:

Theorem 8. *Let F be defined by* equation (10) *with B satisfying* (9). *If*

$$A_\varepsilon \xrightarrow{H} Q,$$

$$F(A_\varepsilon) \rightharpoonup \overline{F} \quad weakly \star in \ L^\infty,$$

then

$$F(Q) \leq \overline{F}.$$

Proof. For any given matrix V, one can find a sequence V_ε such that

$$\begin{cases} V_\varepsilon \rightharpoonup V \quad \text{weakly in } (L^2(\Omega))^{N^2}, \\ \text{curl } (V_\varepsilon \lambda) = 0, \\ \text{div } (A_\varepsilon V_\varepsilon \lambda) \text{ converges strongly in } H^{-1}(\Omega), \forall \lambda, \end{cases}$$

and hence

$$A_\varepsilon V_\varepsilon \rightharpoonup QV \quad \text{weakly in } (L^2(\Omega))^{N^2}.$$

Compensated compactness implies that if $B(V_\varepsilon, A_\varepsilon V_\varepsilon)$ converges vaguely to a measure μ_0, then $\mu_0 \geq B(V, QV)$. One has

$$F(A_\varepsilon) \geq B(V_\varepsilon, A_\varepsilon V_\varepsilon) + L(V_\varepsilon, A_\varepsilon V_\varepsilon),$$

where the right-hand side term converges vaguely to $\mu_0 + L(V, QV)$. But

$$\mu_0 + L(V, QV) \geq B(V, QV) + L(V, QV),$$

whence

$$\overline{F} \geq B(V, QV) + L(V, QV), \ \forall V$$

and thus

$$\overline{F} \geq F(Q). \qquad \blacksquare$$

Remark 14. Theorem 5 corresponds to the case where B and L are, respectively, of the form $-(V\lambda, W\lambda)$, $\lambda \in \mathbb{R}^N$ and $a(V\lambda, \mu) + b(V\lambda, \mu)$ with $a, b \in \mathbb{R}$, $\lambda, \mu \in \mathbb{R}^N$.

Remark 15. Theorem 8 is valid without any hypothesis of symmetry of A_ε. However, in the nonsymmetric case even the analogous form of Theorem 5 gives a complicated result as in the following. Suppose that

$$\begin{cases} A_\varepsilon \left(\dfrac{A_\varepsilon + A_\varepsilon^\star}{2} \right)^{-1} A_\varepsilon^\star \rightharpoonup B_0, \\ \left(\dfrac{A_\varepsilon + A_\varepsilon^\star}{2} \right)^{-1} A_\varepsilon^\star \rightharpoonup C_0, \\ \left(\dfrac{A_\varepsilon + A_\varepsilon^\star}{2} \right)^{-1} \rightharpoonup D_0, \end{cases}$$

then for all $w, z \in \mathbb{R}^N$ we have the inequality

$$(B_0 w, \; w) + (C_0 w, \; z) + (w, \; C_0 z) + (D_0 z, \; z)$$
$$\geq \left(\left(\frac{Q + Q^\star}{2} \right)^{-1} (Q^\star w + z), \; Q^\star w + z \right).$$

Remark 16. An important problem is to characterize the values of Q when all the weak limits of the functions $G(A_\varepsilon)$ are known.

For example, consider a mixture of two isotropic materials with conductivities α and β and proportions $1 - \theta$ and θ. The question is: what are the possible conductivities of the homogenized material? More generally, what can be said about the homogenized coefficients when some statistical properties of the repartition of materials are known?

Remark 17. The purpose of the preceding was to expose techniques and results and it is important to notice that all are valid for other variational situations.

A courageous reader could prove inequalities of the same type for the case of the system of elasticity. It would be better to first refine the method in order to obtain a characterization for the model case.

4. Comments

Homogenization in a general framework (i.e., not restricted to the periodical case) was studied, in the case of symmetrical operators of order 2 under the name of G-convergence, by Spagnolo and De Giorgi (this approach is exposed for instance, in Spagnolo [1]) and in the general case by Murat and Tartar (we refer the reader to [2], [3] and to the notes [4], when published!). The "compensated compactness" is developed by Murat and Tartar in [5] and [6].

The other aspect of the homogenization, the periodic case, can be found in Babuška [7] and Bensoussan et al. [8].

5. References

[1] S. Spagnolo, Convergence in energy for elliptic operators, in: *Numerical Solutions of Partial Differential Equations III*, B. Hubbard, ed., Academic Press (1976), 469–498.

[2] L. Tartar, Quelques remarques sur l'homogénéisation, in *Functional Analysis and Numerical Analysis*, Procceedings of the Japan-France Seminar 1976, H. Fujita, ed., Japan Society for the Promotion of Science, (1978), 469–482.

[3] F. Murat, Thèse, Université Pierre et Marie Curie, Paris (1977).

[4] L. Tartar, Cours Peccot. Collège de France, Paris (1977).

[5] F. Murat, Compacité par compensation, *Ann. Scu. Norm. Sup. Pisa* **5** (1978), 489–507.

[6] L. Tartar, Une nouvelle méthode de résolution d'équations aux dérivées partielles non linéaires, in: *Journées d'Analyse non linéaire.* Lecture Notes in Mathematics, (665), Springer Verlag (1978), 228–241.

[7] I. Babuška, Homogenization and its applications, in: *Numerical Solutions of Partial Differential Equations III*, B. Hubbard, ed., Academic Press (1976), 89–116.

[8] A. Bensoussan, J. L. Lions, and G. Papanicolaou, *Asymptotic Methods in Periodic Structures*, North Holland, Amsterdam (1978).

H-Convergence

François Murat and Luc Tartar

Foreword to the English Translation

The article is the translation of notes originally written in French that were intended as a first draft for a joint book which has yet to be written. These notes presented part of the material that Luc Tartar taught in his *Cours Peccot* at the Collège de France in March 1977 and were also based on a series of lectures given by François Murat at Algiers University in March 1978. They were subsequently reproduced by mimeograph in the *Séminaire d'Analyse Fonctionnelle et Numérique de l'Université d'Alger 1977/78* under the signature of only François Murat. We have chosen to return to our original project by cosigning the present translation.

We would like to note that a small change in the definition of the set $M(\alpha, \beta, \Omega)$, which is introduced and used in the following, would result in an improvement of the presentation of these notes. Indeed, define $M'(\alpha, \gamma, \Omega)$ as the set of those matrices $A \in [L^\infty(\Omega)]^{N^2}$ which are such that $(A(x)\lambda, \lambda) \geq \alpha \mid \lambda \mid^2$ and $((A)^{-1}(x)\lambda, \lambda) \geq \gamma \mid \lambda \mid^2$ for any λ in $\mathbf{R}^N$ and a.e. x in Ω. A proof similar to that presented hereafter implies that the H-limit of a sequence of matrices of $M'(\alpha, \gamma, \Omega)$ also belongs to $M'(\alpha, \gamma, \Omega)$, whereas the H-limit of a sequence of matrices of $M(\alpha, \beta, \Omega)$ only belongs to $M(\alpha, (\beta^2/\alpha), \Omega)$ when the matrices are not symmetric.

1 Notation

Ω is an open subset of $\mathbf{R}^N$.
$\omega \subset\subset \Omega$ denotes a bounded open subset ω of Ω such that $\overline{\omega} \subset \Omega$.
$\alpha, \beta, \alpha', \beta'$ are strictly positive real numbers satisfying

$$0 < \alpha < \beta < +\infty \,,$$

$$0 < \alpha' < \beta' < +\infty.$$

$(\cdot, \cdot)$ and $|\cdot|$ respectively denote the euclidean inner product and norm on $\mathbf{R}^N$.
$(e_1, \ldots, e_N)$ is the canonical basis of $\mathbf{R}^N$.
$E = \{\epsilon = 1/n : n \in \mathbf{Z}^+ - \{0\}\}$.
$E', E'', \ldots$ are infinite subsets of E (subsequences).

$$M(\alpha, \beta, \Omega) = \{A \in [L^\infty(\Omega)]^{N^2} : (A(x)\lambda, \lambda) \geq \alpha \mid \lambda \mid^2, \mid A(x)\lambda \mid \leq \beta \mid \lambda \mid$$
$$\text{for any } \lambda \in \mathbf{R}^N \text{ and a.e. } x \text{ in } \Omega\}.$$

© Springer Nature Switzerland AG 2018
A. V. Cherkaev, R. Kohn (eds.), *Topics in the Mathematical Modelling of Composite Materials*, Modern Birkhäuser Classics, https://doi.org/10.1007/978-3-319-97184-1_3

If A is an element of $M(\alpha, \beta, \Omega)$ and u is an element of $H_0^1(\Omega)(= W_0^{1,2}(\Omega))$,

$$-div\,(A\,grad\,u) = -\sum_{i=1}^N \frac{\partial}{\partial x_i}(\sum_{j=1}^N A_{ij}\frac{\partial u}{\partial x_j}).$$

2   Introductory Remarks

Let $A^\epsilon, \epsilon \in E$, be a sequence of elements of $M(\alpha, \beta, \Omega)$. Then, for any ϵ, any bounded open set Ω, and any f in $H^{-1}(\Omega)$, there exists a unique solution of

$$\begin{cases} -div\,(A^\epsilon grad\,u^\epsilon) = f & \text{in } \Omega, \\ u^\epsilon \in H_0^1(\Omega). \end{cases}$$

Furthermore one has

$$\alpha \parallel u^\epsilon \parallel_{H_0^1(\Omega)} \leq \parallel f \parallel_{H^{-1}(\Omega)},$$

which implies the existence of a subsequence E' such that, for ϵ in E',

$$u^\epsilon \rightharpoonup u^0 \text{ weakly in } H_0^1(\Omega).$$

The following question is raised: does u^0 satisfy an equation of the same type as that satisfied by u^ϵ?

Whenever the matrices A^ϵ converge almost everywhere to a matrix A^0, A^ϵ converges to A^0 in $[L^p(\Omega)]^{N^2}$ for any finite p, and the weak limit of $A^\epsilon grad\,u^\epsilon$ in $[L^2(\Omega)]^N$ is $A^0 grad\,u^0$ (for ϵ in E'). Therefore u^0 is the solution of

$$\begin{cases} -div\,(A^0 grad\,u^0) = f & \text{in } \Omega, \\ u^0 \in H_0^1(\Omega). \end{cases}$$

Note that the uniqueness of u^0 is ensured because the pointwise limit A^ϵ of A^0 belongs to $M(\alpha, \beta, \Omega)$.

In the absence of pointwise convergence of the matrices A^ϵ the setting is drastically different, as illustrated by the one-dimensional case.

3   The One-Dimensional Case

Set $\Omega = (0,1)$, take f in $L^2(\Omega)$ and A^ϵ in $M(\alpha, \beta, \Omega)$, which is here just $M(\alpha, \beta, \Omega) = \{A^\epsilon \in L^\infty(\Omega) : \alpha \leq A^\epsilon(x) \leq \beta \text{ a.e. in } \Omega\}$.

Define u^ϵ as the unique solution of

$$\begin{cases} -\frac{d}{dx}(A^\epsilon \frac{du^\epsilon}{dx}) = f & \text{in } \Omega, \\ u^\epsilon \in H_0^1(\Omega). \end{cases}$$

Since $\alpha \parallel u^\epsilon \parallel_{H_0^1(\Omega)} \leq \parallel f \parallel_{H^{-1}(\Omega)}$, a subsequence E' of E is such that

$$u^\epsilon \rightharpoonup u^0 \quad \text{weakly in } H_0^1(\Omega) \,, \; \epsilon \in E'.$$

Set $\xi^\epsilon = A^\epsilon \dfrac{du^\epsilon}{dx}$. The function ξ^ϵ is bounded in $H^1(\Omega)$ because

$$\| \xi^\epsilon \|_{L^2(\Omega)} \le \frac{\beta}{\alpha} \| f \|_{H^{-1}(\Omega)} \quad \text{and} \quad \frac{d\xi^\epsilon}{dx} = -f \quad \text{in } \Omega.$$

Hence a subsequence E'' of E' is such that

$$\xi^\epsilon \to \xi^0 \quad \text{strongly in } L^2(\Omega), \; \epsilon \in E''.$$

Since A^ϵ belongs to $M(\alpha, \beta, \Omega)$,

$$\frac{1}{\beta} \le \frac{1}{A^\epsilon(x)} \le \frac{1}{\alpha} \quad \text{a.e. in } \Omega,$$

and a subsequence E''' of E is such that

$$\frac{1}{A^\epsilon} \rightharpoonup \frac{1}{A^0} \quad \text{weak-* in } L^\infty(\Omega), \epsilon \in E'''.$$

Furthermore A^0 belongs to $M(\alpha, \beta, \Omega)$.

The limit of each side of the equality

$$\frac{1}{A^\epsilon} \xi^\epsilon = \frac{du^\epsilon}{dx}, \; \epsilon \in E''',$$

is computable and it yields

$$\frac{1}{A^0} \xi^0 = \frac{du^0}{dx}.$$

Since $\dfrac{d\xi^0}{dx} = -f$, u^0 is a solution of

$$\begin{cases} -\dfrac{d}{dx}(A^0 \dfrac{du^0}{dx}) = f \quad \text{in } \Omega, \\ u^0 \in H_0^1(\Omega), \end{cases}$$

and it is unique because A^0 belongs to $M(\alpha, \beta, \Omega)$.

Note that if B^0 is the weak-* limit in $L^\infty(\Omega)$ of A^ϵ for a subsequence E'''' of E''', then A^0 is generally different from B^0 as easily seen upon consideration of the following example:

$$\begin{cases} A^\epsilon(x) = \alpha \;, & k\epsilon \le x < (k + \dfrac{1}{2})\epsilon, \\ A^\epsilon(x) = \beta \;, & (k + \dfrac{1}{2})\epsilon \le x < (k + 1)\epsilon, \end{cases}$$

with $k \in \mathbf{Z}^+$, in which case

$$\begin{cases} \dfrac{1}{A^0} = \dfrac{1}{2}(\dfrac{1}{\alpha} + \dfrac{1}{\beta}), \\ B^0 = \dfrac{1}{2}(\alpha + \beta). \end{cases}$$

The reader should, however, refrain from drawing the hasty conclusion that weak-* convergence in $[L^\infty(\Omega)]^{N^2}$ of the inverse matrices $(A^\epsilon)^{-1}$ of A^ϵ is the key to the understanding of the problem in the N-dimensional case. Consider, for example, the following setting.

4   Layering

A sequence $A^\epsilon, \epsilon \in E$, of elements of $M(\alpha, \beta, \Omega)$ such that $A^\epsilon(x) = A^\epsilon(x_1)$ is investigated. Since it satisfies

$$A_{11}^\epsilon(x) = (A^\epsilon(x)e_1, e_1) \geq \alpha \mid e_1 \mid^2 = \alpha,$$

a subsequence E' of E is such that

$$\begin{cases} \dfrac{1}{A_{11}^\epsilon} \rightharpoonup \dfrac{1}{A_{11}^0}, \\[2ex] \dfrac{A_{i1}^\epsilon}{A_{11}^\epsilon} \rightharpoonup \dfrac{A_{i1}^0}{A_{11}^0}, \quad i > 1, \\[2ex] \dfrac{A_{1j}^\epsilon}{A_{11}^\epsilon} \rightharpoonup \dfrac{A_{1j}^0}{A_{11}^0}, \quad j > 1, \\[2ex] A_{ij}^\epsilon - \dfrac{A_{i1}^\epsilon A_{1j}^\epsilon}{A_{11}^\epsilon} \rightharpoonup A_{ij}^0 - \dfrac{A_{i1}^0 A_{1j}^0}{A_{11}^0}, \quad i > 1, j > 1, \end{cases} \qquad (1)$$

for $\epsilon \in E'$. The convergences in equation (1) are to be understood as weak-* convergences in $L^\infty(\Omega)$.

If Ω is bounded and f is an element of $L^2(\Omega)$, the solution u^ϵ of

$$\begin{cases} -div\,(A^\epsilon grad\, u^\epsilon) = f \quad \text{in } \Omega, \\ u^\epsilon \in H_0^1(\Omega), \end{cases}$$

is such that, for a subsequence E'' of E',

$$u^\epsilon \rightharpoonup u^0 \quad \text{weakly in } H_0^1(\Omega), \epsilon \in E''.$$

Let us prove that u^0 is the solution of

$$\begin{cases} -div\,(A^0 grad\,u^0) = f & \text{in } \Omega, \\ \\ u^0 \in H_0^1(\Omega)\,, \end{cases} \tag{2}$$

with A^0 defined through (1).

Let $\omega = \prod\limits_{i=1}^{N} (a_i, b_i)$ be a rectangle such that $\omega \subset \Omega$. Set $\omega' = \prod\limits_{i=2}^{N} (a_i, b_i)$ and

$$\xi_i^\epsilon = \sum_{j=1}^{N} A_{ij}^\epsilon \frac{\partial u^\epsilon}{\partial x_j}, \quad 1 \le i \le N.$$

Each of the ξ_i^ϵ's is bounded in $L^2(\omega)$ and

$$-\frac{\partial \xi_1^\epsilon}{\partial x_1} = f + \sum_{i=2}^{N} \frac{\partial \xi_i^\epsilon}{\partial x_i}\,.$$

Thus ξ_1^ϵ is bounded in $H^1((a_1, b_1); H^{-1}(\omega'))$.

The identity mapping from $L^2(\omega')$ into $H^{-1}(\omega')$ is compact, which implies, by virtue of Aubin's compactness lemma, that

$$\Xi = \{\xi \in L^2((a_1, b_1)); L^2(\omega')) : \frac{\partial \xi}{\partial x_1} \in L^2((a_1, b_1)); H^{-1}(\omega'))\}$$

is compactly embedded in $L^2((a_1, b_1); H^{-1}(\omega'))$. Thus, at the expense of extracting a subsequence E''' of E'', we are at liberty to assume that

$$\begin{cases} \xi_i^\epsilon \rightharpoonup \xi_i^0 & \text{weakly in } L^2(\omega)\,, \\ \\ \xi_1^\epsilon \to \xi_1^0 & \text{strongly in } L^2((a_1, b_1)); H^{-1}(\omega'))\,, \\ \\ u^\epsilon \rightharpoonup u^0 & \text{strongly in } L^2(\omega)\,, \end{cases} \tag{3}$$

for ϵ in E'''.

But A_{ij}^ϵ is a function of x_1 and only x_1, thus

$$\frac{\partial u^\epsilon}{\partial x_1} + \sum_{j=2}^{N} \frac{\partial}{\partial x_j}(\frac{A_{1j}^\epsilon}{A_{11}^\epsilon} u^\epsilon) = \frac{1}{A_{11}^\epsilon} \xi_1^\epsilon,$$

$$\xi_i^\epsilon = \frac{A_{i1}^\epsilon}{A_{11}^\epsilon} \xi_1^\epsilon + \sum_{j=2}^{N} \frac{\partial}{\partial x_j}((A_{ij}^\epsilon - \frac{A_{i1}^\epsilon A_{1j}^\epsilon}{A_{11}^\epsilon})u^\epsilon), \quad i > 1.$$

The limit of every single term in the preceding equalities is immediately computable upon recalling equations (1) and (3). For example, if φ is an arbitrary element of $C_0^\infty(\omega)$,

$$\int_\omega \frac{1}{A_{11}^\epsilon} \xi_1^\epsilon \varphi dx = < \xi_1^\epsilon, \frac{1}{A_{11}^\epsilon} \varphi > ,$$

where $< , >$ stands for the duality bracket between $L^2((a_1, b_1); H^{-1}(\omega'))$ and $L^2((a_1, b_1); H_0^1(\omega'))$. We finally obtain

$$\xi_i^0 = \sum_{j=1}^N A_{ij}^0 \frac{\partial u^0}{\partial x_j}, \quad i \geq 1,$$

which yields equation (2) because $-div\, \xi^0 = f$ in ω , and $\omega \subset \Omega$ is arbitrary.

5   Definition of the H-Convergence

Definition 1 *A sequence* $A^\epsilon, \epsilon \in E$, *of elements of* $M(\alpha, \beta, \Omega)$ $H-converges$ *to an element* A^0 *of* $M(\alpha', \beta', \Omega)$ $(A^\epsilon \xrightarrow{H} A^0)$ *if and only if, for any* $\omega \subset\subset \Omega$ *and any* f *in* $H^{-1}(\omega)$, *the solution* u^ϵ *of*

$$\begin{cases} -div\, (A^\epsilon grad\, u^\epsilon) = f & in\ \omega, \\ \\ u^\epsilon \in H_0^1(\omega) , \end{cases} \tag{4}$$

is such that

$$\begin{cases} u^\epsilon \rightharpoonup u^0 & weakly\ in\ H_0^1(\omega) , \\ \\ A^\epsilon grad\, u^\epsilon \rightharpoonup A^0 grad\, u^0 & weakly\ in\ [L^2(\omega)]^N , \end{cases} \tag{5}$$

for $\epsilon \in E$, *where* u^0 *is the solution of*

$$\begin{cases} -div\, (A^0 grad\, u^0) = f & in\ \omega , \\ \\ u^0 \in H_0^1(\omega). \end{cases}$$

Remarks

1. According to the results obtained in Sections 2, 3, and 4, the following results hold true:

    (i) If A^ϵ converges to A^0 a.e. in Ω, then $A^\epsilon \xrightarrow{H} A^0$.

    (ii) If $N = 1$, $A^\epsilon \xrightarrow{H} A^0$ if and only if $\frac{1}{A^\epsilon} \rightharpoonup \frac{1}{A^0}$ weak-* in $L^\infty(\Omega)$, as easily seen upon approximation in $H^{-1}(\Omega)$ of f by functions of $L^2(\Omega)$ (see Section 3).

    (iii) If $A^\epsilon(x) = A^\epsilon(x_1)$, and if $A^\epsilon \xrightarrow{H} A^0$, equation (1) is satisfied. Conversely if (1) is satisfied, then it can be shown that A^0 is coercive and Section 4 implies that $A^\epsilon \xrightarrow{H} A^0$.

2. If equation (4) is interpreted as the equation for the electrostatic potential u^ϵ, A^ϵ as the tensor of dielectric permittivity , $E^\epsilon = grad\, u^\epsilon$ as the electric field, and $D^\epsilon = A^\epsilon grad\, u^\epsilon$ as the polarization field, then convergence (5) is a statement about the weak convergence of the fields E^ϵ and D^ϵ. It is shown later on that the electrostatic energy $e^\epsilon = (D^\epsilon, E^\epsilon) = (A^\epsilon grad\, u^\epsilon, grad\, u^\epsilon)$ is also a weakly converging quantity.

3. The concept of *H*-convergence generalizes that of *G*convergence introduced by Spagnolo (see, for example, Spagnolo [5] and De Giorgi and Spagnolo [2]). Furthermore, the theory of periodic homogenization, as developed in A. Bensoussan et al. [1], may be construed as a systematic study of the *H*-convergence in a periodic framework. The latter reference offers a thorough bibliography as well as a wealth of open problems.

6 Locality

In essence, *H*-convergence amounts to a statement of convergence of the inverse operators $[-div\,(A^\epsilon grad\,)]^{-1}$, which are bounded linear mappings from $H^{-1}(\Omega)$ into $H_0^1(\Omega)$, when both spaces $H^{-1}(\Omega)$ and $H_0^1(\Omega)$ are endowed with their weak topologies. The underlying topology satisfies the property of uniqueness of the *H*-limit, and the *H*-limit is local as demonstrated in the following proposition:

Proposition 1 *(i) A sequence A^ϵ, $\epsilon \in E$, of elements of $M(\alpha, \beta, \Omega)$ has at most one H-limit.*

(ii) Let A^ϵ and B^ϵ, $\epsilon \in E$, be two sequences in $M(\alpha, \beta, \Omega)$ that satisfy

$$\begin{cases} A^\epsilon \xrightarrow{H} A^0 \, , \\[2mm] B^\epsilon \xrightarrow{H} B^0 \, . \end{cases}$$

and are such that $A^\epsilon = B^\epsilon$ on an open set $\omega \subset \Omega$. Then $A^0 = B^0$ on ω.

Proof:

Let A^0 be an *H*-limit of A^ϵ, $\epsilon \in E$. Consider $\omega \subset\subset \omega_1 \subset \Omega$, $\varphi \in C_0^\infty(\omega_1)$ with $\varphi = 1$ on ω, and define, for any λ in $\mathbf{R}^N$,

$$f_\lambda = -div\,(A^0(x)grad\,((\lambda, x)\varphi(x))).$$

Then u_λ^ϵ, defined as the solution of

$$\begin{cases} -div\,(A^\epsilon grad\,u_\lambda^\epsilon) = f_\lambda \quad in\;\omega_1, \\[2mm] u_\lambda^\epsilon \in H_0^1(\omega_1)\,, \end{cases}$$

for $\epsilon \in E$ and $\epsilon = 0$, is such that

$$\begin{cases} u_\lambda^0(x) = (\lambda, x)\varphi(x), \\[2mm] u_\lambda^\epsilon \rightharpoonup u_\lambda^0 \quad weakly\;in\;H_0^1(\omega_1), \\[2mm] A^\epsilon grad\,u_\lambda^\epsilon \rightharpoonup A^0 grad\,u_\lambda^0 \quad weakly\;in\;[L^2(\omega_1)]^N. \end{cases}$$

If B^0 is another H-limit for A^ϵ, then

$$A^\epsilon grad\,u_\lambda^\epsilon \rightharpoonup B^0 grad\,u_\lambda^0 \quad weakly\;in\;[L^2(\omega_1)]^N.$$

Thus $A^0 grad\,u_\lambda^0 = B^0 grad\,u_\lambda^0$ and, since $grad\,u_\lambda^0 = \lambda$ in ω, $A^0 = B^0$ in ω, which proves (i). The proof of (ii) is immediate in view of (i) together with the definition of H-convergence. ∎

7 Two Fundamental Lemmata

Lemma 1 *Let Ω be an open subset of $\mathbf{R}^N$ and $\xi^\epsilon, v^\epsilon, \epsilon \in E$, be such that*

$$\begin{cases} \xi^\epsilon \in [L^2(\Omega)]^N\,, \\ \xi^\epsilon \rightharpoonup \xi^0 \quad weakly\;in\;[L^2(\Omega)]^N\,, \\ div\,\xi^\epsilon \to div\,\xi^0 \quad strongly\;in\;H^{-1}(\Omega), \end{cases}$$

$$\begin{cases} v^\epsilon \in H^1(\Omega)\,, \\ v^\epsilon \rightharpoonup v^0 \quad weakly\;in\;H^1(\Omega). \end{cases}$$

Then

$$(\xi^\epsilon, grad\,v^\epsilon) \rightharpoonup (\xi^0, grad\,v^0) \quad weakly\text{-}*\;in\;\mathcal{D}'(\Omega).$$

Remarks

1. The product $(\xi^\epsilon, grad\,v^\epsilon)$ is that of two weakly and not strongly converging sequences; thus it is a miracle that the limit of the product should be equal to the product of the limits. This phenomenon is known as compensated compactness (see Murat [4] and Tartar [7]).

2. The product $(\xi^\epsilon, grad\,v^\epsilon)$ is bounded in $L^1(\Omega)$ independently of ϵ. Thus it actually converges vaguely to a measure. However, it does not in general converge weakly in $L^1(\Omega)$ (see Murat [6] for a counterexample).

Proof of Lemma 1:

Let φ be an element of $C_0^\infty(\Omega)$. Then

$$\int_\Omega (\xi^\epsilon, grad\, v^\epsilon)\varphi dx = - < div\, \xi^\epsilon, \varphi v^\epsilon >_{H^{-1}(\Omega),H_0^1(\Omega)} - \int_\Omega (\xi^\epsilon, grad\, \varphi)v^\epsilon dx.$$

Passing to the limit in each term of the right-hand side is easy (use Rellich's theorem in the second term). Integration by parts of the resulting expression yields the desired result.   ∎

Lemma 2 *Let Ω be an open subset of $\mathbf{R}^N$. Let A^ϵ belong to $M(\alpha,\beta,\Omega)$ for $\epsilon \in E$. Assume that, for $\epsilon \in E$,*

$$\begin{cases} u^\epsilon \in H^1(\Omega), \\ u^\epsilon \rightharpoonup u^0 \quad weakly\ in\ H^1(\Omega), \\ \xi^\epsilon = A^\epsilon grad\, u^\epsilon \rightharpoonup \xi^0 \quad weakly\ in\ [L^2(\Omega)]^N, \\ -div\,(A^\epsilon grad\, u^\epsilon) \to -div\, \xi^0 \quad strongly\ in\ H^{-1}(\Omega), \end{cases} \quad (6)$$

$$\begin{cases} v^\epsilon \in H^1(\Omega), \\ v^\epsilon \rightharpoonup v^0 \quad weakly\ in\ H^1(\Omega), \\ \eta^\epsilon = {}^t A^\epsilon grad\, v^\epsilon \rightharpoonup \eta^0 \quad weakly\ in\ [L^2(\Omega)]^N, \\ -div\,({}^t A^\epsilon grad\, v^\epsilon) \to -div\, \eta^0 \quad strongly\ in\ H^{-1}(\Omega). \end{cases} \quad (7)$$

Then

$$(\xi^0, grad\, v^0) = (grad\, u^0, \eta^0) \quad a.e.\ in\ \Omega. \quad (8)$$

Proof:

The proof is immediate upon observing that

$$(\xi^\epsilon, grad\, v^\epsilon) = (A^\epsilon grad\, u^\epsilon, grad\, v^\epsilon) = (grad\, u^\epsilon, {}^t A^\epsilon grad\, v^\epsilon) = (grad\, u^\epsilon, \eta^\epsilon),$$

and through application of Lemma 1.

Note that equality (8) is a pointwise equality, which is a much stronger statement than an integral equality.   ∎

8   Irrelevance of the Boundary Conditions. Convergence of the Energy

Proposition 2 *If $A^\epsilon, \epsilon \in E$, belongs to $M(\alpha,\beta,\Omega)$ and H-converges to A^0 which belongs to $M(\alpha',\beta',\Omega)$, then ${}^t A^\epsilon, \epsilon \in E, H$-converges to ${}^t A^0$.*

Proof:

Let $\omega \subset\subset \Omega$ and g be an element of $H^1(\omega)$. Let v^ϵ be the solution of

$$\begin{cases} -div\,({}^tA^\epsilon grad\,v^\epsilon) = g & \text{in } \omega, \\ v^\epsilon \in H^1_0(\omega)\,, \end{cases}$$

for $\epsilon \in E$. Our task is to show that

$$\begin{cases} v^\epsilon \rightharpoonup v^0 & \text{weakly in } H^1_0(\omega), \\ {}^tA^\epsilon grad\,v^\epsilon \rightharpoonup {}^tA^0 grad\,v^0 & \text{weakly in } [L^2(\omega)]^N\,, \end{cases}$$

for $\epsilon \in E$, where v^0 is the solution of

$$\begin{cases} -div\,({}^tA^0 grad\,v^0) = g & \text{in } \omega, \\ v^0 \in H^1_0(\omega)\,. \end{cases}$$

Because $v^\epsilon, \epsilon \in E$ is bounded in $H^1_0(\omega)$, a subsequence E' of E is such that

$$\begin{cases} v^\epsilon \rightharpoonup v & \text{weakly in } H^1_0(\omega), \\ {}^tA^\epsilon grad\,v^\epsilon \rightharpoonup \eta & \text{weakly in } [L^2(\omega)]^N\,, \end{cases}$$

for ϵ in E'. Furthermore, $-div\,\eta = g$ in ω.

For any f in $H^{-1}(\omega)$, u^ϵ defined as the solution of

$$\begin{cases} -div\,(A^\epsilon grad\,u^\epsilon) = f & \text{in } \omega, \\ u^\epsilon \in H^1_0(\omega)\,, \end{cases}$$

for $\epsilon \in E$ and $\epsilon = 0$, is such that

$$\begin{cases} u^\epsilon \rightharpoonup u^0 & \text{weakly in } H^1_0(\omega), \\ A^\epsilon grad\,u^\epsilon \rightharpoonup A^0 grad\,u^0 & \text{weakly in } [L^2(\omega)]^N\,, \end{cases}$$

for $\epsilon \in E$, because A^ϵ H-converges to A^0 for $\epsilon \in E$.

Application of Lemma 2 yields

$$(A^0 grad\,u^0, grad\,v) = (grad\,u^0, \eta) \quad \text{a.e. in } \omega.$$

As f spans $H^{-1}(\omega)$, u^0 spans $H^1_0(\omega)$; thus, if $\omega_1 \subset\subset \omega$, $grad\,u^0$ can be taken to be any $\lambda \in \mathbf{R}^N$ on ω_1 and we obtain

$$(A^0\lambda, grad\,v) = (\lambda, \eta) \quad \text{a.e. in } \omega_1 \text{ and for any } \lambda \in \mathbf{R}^N,$$

which implies that

$$\eta = {}^t A^0 \, grad \, v \qquad \text{a.e. in } \omega.$$

Since $-div \, \eta = g$, we conclude that $v = v^0$ and $\eta = {}^t A^0 grad \, v^0$.

Since ${}^t A^0$ is unique, v^0 is unique and the whole sequence $\epsilon \in E$ (and not only the subsequence $\epsilon \in E'$) is found to converge. ∎

Theorem 1 *Assume that $A^\epsilon, \epsilon \in E$, belongs to $M(\alpha, \beta, \Omega)$ and H-converges to $A^0 \in M(\alpha', \beta', \Omega)$. Assume that*

$$\begin{cases} u^\epsilon \in H^1(\Omega) \,, \\[2mm] f^\epsilon \in H^{-1}(\Omega) \,, \\[2mm] -div \, (A^\epsilon grad \, u^\epsilon) = f^\epsilon \quad in \, \Omega \,, \\[2mm] u^\epsilon \rightharpoonup u^0 \quad weakly \, in \, H^1(\Omega) \,, \\[2mm] f^\epsilon \to f^0 \quad strongly \, in \, H^{-1}(\Omega) \,, \end{cases}$$

for $\epsilon \in E$. Then

$$A^\epsilon grad \, u^\epsilon \rightharpoonup A^0 grad \, u^0 \qquad weakly \, in \, [L^2(\Omega)]^N \,,$$

$$(A^\epsilon grad \, u^\epsilon, grad \, u^\epsilon) \rightharpoonup (A^0 grad \, u^0, grad \, u^0) \qquad weakly\text{-}* \, in \, \mathcal{D}'(\Omega).$$

The proof of Theorem 1 is analogous to that of Proposition 2: it merely uses Proposition 2 and Lemmata 1 and 2.

It can be further proved, with the help of Meyers' regularity theorem (see Meyers [3]), that the energy $(A^\epsilon grad \, u^\epsilon, grad \, u^\epsilon)$ actually converges weakly in $L^1_{loc}(\Omega)$.

9 Sequential Compactness of $M(\alpha, \beta, \Omega)$ for the Topology Induced by H-convergence

The notion of H-convergence finds its *raison d'être* in the following theorem.

Theorem 2 *Let $A^\epsilon, \epsilon \in E$ belong to $M(\alpha, \beta, \Omega)$. There exists a subsequence E' of E and a matrix A^0 in $M(\alpha, \frac{\beta^2}{\alpha}, \Omega)$ such that A^ϵ H-converges to A^0 for $\epsilon \in E'$.*

Proof:

The proof of Theorem 2 consists of the following steps.

Step 1:

Proposition 3 *Let F and G be two Banach spaces, with F separable and G reflexive. Let $T^\epsilon, \epsilon \in E$ be elements of $\mathcal{L}(F, G)$ satisfying*

$$\| T^\epsilon \|_{\mathcal{L}} \leq C .$$

Then there exist a subsequence E' of E and an element T^0 of $\mathcal{L}(F, G)$ such that, for any element f of F, $T^\epsilon f \to T^0 f$ weakly in $G, \epsilon \in E'$.

Proof:

Take X to be a countable dense subset of F. A diagonal process ensures the existence of a subsequence E' of E such that $T^\epsilon x$ has a weak limit in G denoted by $T^0 x$ for $\epsilon \in E'$ and $x \in X$.

Fix f in F and g' in G' and approximate f by elements $x \in X$. This allows one to prove that $< T^\epsilon f, g' >_{G,G'}$ is a Cauchy sequence for $\epsilon \in E'$. Denote the corresponding limit by $< T^0 f, g' >_{G,G'}$; then T^0 is linear and bounded. Specifically,

$$\| T^0 f \|_G \leq \liminf_{\epsilon \in E'} \| T^\epsilon f \|_G \leq C \| f \|_F.$$

■

Step 2:

Proposition 4 *Let V be a reflexive separable Banach space and $T^\epsilon, \epsilon \in E$ be elements of $\mathcal{L}(V, V')$ such that*

$$\begin{cases} \| T^\epsilon \|_{\mathcal{L}} & \leq \ \beta , \\[2mm] < T^\epsilon v, v >_{V',V} & \geq \ \alpha \| v \|_V^2 , \ v \in V. \end{cases}$$

Then there exist a subsequence E' of E and an element T^0 in $\mathcal{L}(V, V')$ such that

$$\begin{cases} \| T^0 \|_{\mathcal{L}} & \leq \ \beta^2/\alpha , \\[2mm] < T^0 v, v >_{V',V} & \geq \ \alpha \| v \|_V^2 , \ v \in V, \end{cases}$$

which satisfy for any f in V',

$$(T^\epsilon)^{-1} f \to (T^0)^{-1} f \quad weakly \ in \ V, \epsilon \in E'.$$

Proof:

By virtue of Lax–Milgram's lemma, T^ϵ has an inverse $(T^\epsilon)^{-1}$ that satisfies $\|\|\,(T^\epsilon)^{-1}\,\|\|_{\mathcal{L}} \leq 1/\alpha$. Application of Proposition 3 yields a subsequence E' of E and an element S in $\mathcal{L}(V',V)$ such that, for any element f of V',

$$(T^\epsilon)^{-1}f \rightharpoonup Sf \quad \text{weakly in } V, \ \epsilon \in E'.$$

Since

$$< (T^\epsilon)^{-1}f, f >_{V,V'} \ = \ < (T^\epsilon)^{-1}f, \, T^\epsilon(T^\epsilon)^{-1}f >_{V,V'}$$

$$\geq \alpha \,\|\,(T^\epsilon)^{-1}f\,\|_V^2 \ \geq \ \frac{\alpha}{\beta^2}\,\|\|\,T^\epsilon\,\|\|_{\mathcal{L}}^2 \,\|\,(T^\epsilon)^{-1}f\,\|_V^2 \geq \frac{\alpha}{\beta^2}\,\|\,f\,\|_{V'}^2,$$

we obtain

$$< Sf, f >_{V,V'} \geq \frac{\alpha}{\beta^2}\,\|\,f\,\|_{V'}^2.$$

Thus S, being coercive, is invertible. Denote by $T^0 \in \mathcal{L}(V,V')$ its inverse. It satisfies, for any element v of V,

$$\frac{\alpha}{\beta^2}\,\|\,T^0v\,\|_V^2 \ \leq \ < ST^0v, T^0v >_{V,V'} \ \leq \|\,v\,\|_V \,\|\,T^0v\,\|_{V'}.$$

Hence $\|\|\,T^0\,\|\|_{\mathcal{L}} \leq \beta^2/\alpha$.

Since

$$\alpha\,\|\,(T^\epsilon)^{-1}f\,\|_V^2 \ \leq \ < T^\epsilon(T^\epsilon)^{-1}f, (T^\epsilon)^{-1}f >_{V',V}$$

$$= \ < f, (T^\epsilon)^{-1}f >_{V',V},$$

the sequential weak lower semicontinuity of $\|\ \|_V$ implies

$$\alpha\,\|\,Sf\,\|_V^2 \leq \ < f, Sf >_{V',V},$$

and the choice of $f = Tv^0, v \in V$, finally yields

$$\alpha\,\|\,v\,\|_V^2 \leq \ < T^0v, v >_{V',V}.$$

∎

Step 3:

For the remainder of the proof of Theorem 2 it will be assumed that Ω is bounded. If such was not the case the argument would be applied to $\Omega \cap \{x \in \mathbf{R}^N : |\,x\,| \leq p\}$ with $p \in \mathbf{Z}^+$ and a diagonalization argument would permit us to conclude.

We propose to manufacture a sequence of test functions to be later inserted into Lemma 2. To this effect a bounded open set Ω' of $\mathbf{R}^N$ with

$\Omega \subset \Omega'$ is considered. We define B^ϵ to be an element of $M(\alpha, \beta, \Omega')$ such that

$$B^\epsilon = {}^t A^\epsilon \text{ in } \Omega.$$

(Take for example $B^\epsilon = \alpha I$ in $\Omega' \setminus \Omega$.)

Set

$$\mathcal{B}^\epsilon = -div\,(B^\epsilon grad\,) \in \mathcal{L}(H_0^1(\Omega'); H^{-1}(\Omega')).$$

Proposition 4 implies the existence of a subsequence E' of E and of an element $\mathcal{B}^0 \in \mathcal{L}(H_0^1(\Omega'); H^{-1}(\Omega'))$ such that, for any element g in $H^{-1}(\Omega')$,

$$(\mathcal{B}^\epsilon)^{-1} g \rightharpoonup (\mathcal{B}^0)^{-1} g \text{ weakly in } H_0^1(\Omega),$$

when $\epsilon \in E'$. Let φ be an element of $C_0^\infty(\Omega')$ such that $\varphi = 1$ on Ω and, for any $i \in \{1, \ldots, N\}$, set

$$g_i = \mathcal{B}^0(x_i \varphi(x)) \in H^{-1}(\Omega').$$

Define $v_i^\epsilon, \epsilon \in E', i \in \{1, \ldots, N\}$, as

$$v_i^\epsilon = (\mathcal{B}^\epsilon)^{-1} g_i.$$

The restriction of v_i^ϵ to Ω belongs to $H^1(\Omega)$ and satisfies

$$\begin{cases} v_i^\epsilon \rightharpoonup x_i \quad \text{weakly in } H^1(\Omega), \\ -div\,({}^t A^\epsilon grad\, v_i^\epsilon) = g_i \quad \text{in } \Omega. \end{cases}$$

At the possible expense of the extraction of a subsequence E'' of E', we are at liberty to further assume that

$$ {}^t A^\epsilon grad\, v_i^\epsilon \rightharpoonup \eta_i \quad \text{weakly in } [L^2(\Omega)]^N,$$

when $\epsilon \in E'', i \in \{1, \ldots, N\}$.

Note that $-div\,\eta_i = g_i$ in Ω and that the functions $v_i^\epsilon, \epsilon \in E''$ satisfy equation (7) in Lemma 2 with $v^0 = x_i$.

We now define a matrix $A^0 \in [L^2(\Omega)]^{N^2}$ by

$$(A^0)_{ij} = (\eta_i)_j \in L^2(\Omega), \quad i, j \in \{1, \ldots, N\}.$$

The matrices $A^\epsilon, \epsilon \in E''$, are shown to H-converge to A^0.

Step 4:

Let $\omega \subset\subset \Omega$. Define the isomorphism $\mathcal{A}^\epsilon$ by

$$\mathcal{A}^\epsilon = -div\,(A^\epsilon grad\,) \in \mathcal{L}(H_0^1(\omega); H^{-1}(\omega)),$$

and set

$$\mathcal{C}^\epsilon = A^\epsilon grad\,((A^\epsilon)^{-1}) \in \mathcal{L}(H^{-1}(\omega); [L^2(\omega)]^N).$$

Then, for any element f in $H^{-1}(\omega)$,

$$\| \mathcal{C}^\epsilon f \|_{[L^2(\omega)]^N} \le \beta \, \| (A^\epsilon)^{-1} f \|_{H^1_0(\omega)} \le \frac{\beta}{\alpha} \, \| f \|_{H^{-1}(\omega)} \, .$$

Direct applications of Proposition 3 to $\mathcal{C}^\epsilon$ and of Proposition 4 to A^ϵ, imply the existence of a subsequence E_ω of E'', of $\mathcal{C}^0 \in \mathcal{L}(H^{-1}(\omega); [L^2(\omega)]^N)$, and of an isomorphism $A^0 \in \mathcal{L}(H^1_0(\omega); H^{-1}(\omega))$, such that, for any element f in $H^{-1}(\omega)$,

$$\begin{cases} (A^\epsilon)^{-1} f \rightharpoonup (A^0)^{-1} f & \text{weakly in } H^1_0(\omega)\,, \\[2mm] \mathcal{C}^\epsilon f \rightharpoonup \mathcal{C}^0 f & \text{weakly in } [L^2(\omega)]^N. \end{cases}$$

Note that E_ω depends upon the choice of ω.

The sequence $u^\epsilon = (A^\epsilon)^{-1} f$, $\epsilon \in E''$, satisfies

$$\begin{cases} u^\epsilon \rightharpoonup u^0 = (A^0)^{-1} f & \text{weakly in } H^1_0(\omega), \\[2mm] A^\epsilon grad\,u^\epsilon \rightharpoonup \mathcal{C}^0 f = \xi^0 & \text{weakly in } [L^2(\omega)]^N, \\[2mm] -div\,(A^\epsilon grad\,u^\epsilon) = f & \text{in } \omega, \end{cases}$$

which is precisely equation (6) of Lemma 2.

Thus application of Lemma 2 to u^ϵ and v^ϵ_i yields, for $i \in \{1, \ldots, N\}$,

$$(\xi^0, grad\,x_i) = (grad\,u^0, \eta_i) \quad \text{a.e. in } \omega\,,$$

which, in view of the definition of A^0, is precisely

$$\mathcal{C}^0 f = \xi^0 = A^0 grad\,u^0 \quad \text{a.e. in } \omega.$$

Step 5:

The matrix A^0, which is by definition an element of $[L^2(\omega)]^{N^2}$, is such that $A^0 grad\,u^0$ belongs to $[L^2(\omega)]^N$ for any u^0 in $H^1_0(\omega)$. We prove that A^0 belongs to $M(\alpha, \frac{\beta^2}{\alpha}, \omega)$. Indeed, application of Lemma 1 to $A^\epsilon grad\,u^\epsilon$ and u^ϵ, $\epsilon \in E_\omega$ yields

$$(A^\epsilon grad\,u^\epsilon, grad\,u^\epsilon) \rightharpoonup (A^0 grad\,u^0, grad\,u^0) \quad \text{weakly-* in } \mathcal{D}'(\omega).$$

Let φ be an arbitrary nonnegative element of $\mathcal{C}^\infty_0(\omega)$. The inequality

$$\int_\omega \varphi\,(A^\epsilon grad\,u^\epsilon, grad\,u^\epsilon)\,dx \ge \alpha \int_\omega \varphi\,|\,grad\,u^\epsilon\,|^2\,dx$$

implies

$$\int_\omega \varphi \left(A^0 grad\, u^0, grad\, u^0 \right) dx \geq \alpha \int_\omega \varphi \mid grad\, u^0 \mid^2 dx.$$

Since the preceding result holds true for any u^0 in $H_0^1(\omega)$, taking $u^0(x) = (\lambda, x)$ on the support of φ yields

$$(A^0(x)\lambda, \lambda) \geq \alpha \mid \lambda \mid^2, \quad \lambda \in \mathbf{R}^N, \text{ a.e. } x \in \omega.$$

On the other hand, for any $\mu \in \mathbf{R}^N$ (see the beginning of the proof of Proposition 4),

$$((A^\epsilon)^{-1}(x)\mu, \mu) \geq \frac{\alpha}{\beta^2} \mid \mu \mid^2, \quad \text{a.e. } x \in \omega.$$

Let φ be an arbitrary nonnegative element of $C_0^\infty(\omega)$. The inequality

$$\int_\omega \varphi \left(grad\, u^\epsilon, A^\epsilon grad\, u^\epsilon \right) dx \geq \frac{\alpha}{\beta^2} \int_\omega \varphi \mid A^\epsilon grad\, u^\epsilon \mid^2 dx$$

implies

$$\int_\omega \varphi \left(grad\, u^0, A^0 grad\, u^0 \right) dx \geq \frac{\alpha}{\beta^2} \int_\omega \varphi \mid A^0 grad\, u^0 \mid^2 dx.$$

Since the preceding result holds true for any u^0 in $H_0^1(\omega)$, taking $u^0(x) = (\lambda, x)$ on the support of φ yields

$$(\lambda, A^0(x)\lambda) \geq \frac{\alpha}{\beta^2} \mid A^0(x)\lambda \mid^2, \quad \lambda \in \mathbf{R}^N, \text{ a.e. } x \in \omega,$$

and thus

$$\mid A^0(x)\lambda \mid \leq \frac{\beta^2}{\alpha} \mid \lambda \mid.$$

We have proved that A^0 belongs to $M(\alpha, \frac{\beta^2}{\alpha}, \omega)$.

Step 6:

Because $A^0 \in M(\alpha, \frac{\beta^2}{\alpha}, \omega)$, the limit u^0 of $u^\epsilon, \epsilon \in E_\omega$ is uniquely defined, independently of E_ω, through

$$\begin{cases} -div\, (A^0 grad\, u^0) = f \quad \text{in } \omega, \\ \\ u^0 \in H_0^1(\omega). \end{cases}$$

Thus there is no need to extract E_ω from E'' and the sequences u^ϵ and $A^\epsilon grad\, u^\epsilon$ converge for $\epsilon \in E''$. But E'' is independent of ω. Thus $A^\epsilon, \epsilon \in E''$, H-converges to A^0. ∎

10   Definition of the Corrector Matrix P^ϵ

Let $A^\epsilon, \epsilon \in E$, be a sequence of elements of $M(\alpha, \beta, \Omega)$ that H-converges to $A^0 \in M(\alpha, \beta', \Omega)$. Consider $\omega \subset\subset \Omega$, $\lambda \in \mathbf{R}^N$, and $\epsilon \in E$ and define w_λ^ϵ such that

$$
\begin{cases}
w_\lambda^\epsilon \in H^1(\omega), \\[2mm]
w_\lambda^\epsilon \rightharpoonup (\lambda, x) \quad \text{weakly in } H^1(\omega), \\[2mm]
-div\,(A^\epsilon\,grad\,w_\lambda^\epsilon) \to -div\,(A^0\lambda) \quad \text{strongly in } H^{-1}(\omega).
\end{cases}
\tag{9}
$$

The existence of w_λ^ϵ is readily asserted upon solving

$$
\begin{cases}
-div\,(A^\epsilon\,grad\,w_\lambda^\epsilon) = -div\,(A^0\,grad\,((\lambda, x)\varphi(x))) \quad \text{in } \omega_1, \\[2mm]
w_\lambda^\epsilon \in H_0^1(\omega_1),
\end{cases}
$$

with $\omega \subset\subset \omega_1 \subset\subset \Omega$ and φ an element of $C_0^\infty(\omega_1)$ such that $\varphi = 1$ on ω.

Definition 2 *Let $A^\epsilon, \epsilon \in E$ be a sequence of elements of $M(\alpha, \beta, \Omega)$ that H-converges to $A^0 \in M(\alpha, \beta', \Omega)$. The corrector matrix $P^\epsilon \in [L^2(\omega)]^{N^2}$ is defined by*

$$
P^\epsilon \lambda = grad\,w_\lambda^\epsilon, \quad \lambda \in \mathbf{R}^N, \quad \epsilon \in E,
\tag{10}
$$

where the sequence w_λ^ϵ satisfies (9).

Remarks

1. It can easily be shown from equation (9) that the matrix P^ϵ is "unique" to the extent that if P^ϵ and $\tilde{P}^\epsilon$, $\epsilon \in E$, are two such sequences, then

$$
P^\epsilon - \tilde{P}^\epsilon \to 0 \quad \text{strongly in } [L_{loc}^2(\omega)]^{N^2}.
$$

2. The sequence P^ϵ is bounded in $[L^2(\omega)]^{N^2}$ independently of ϵ. Bounds for this sequence in $[L^q(\omega)]^{N^2}$, $q > 2$ can be achieved through application of Meyers' regularity result (see Meyers [3]).

3. In the case of layers where $A^\epsilon(x) = A^\epsilon(x_1)$ (see Step 4), the functions w_λ^ϵ are of the form

$$
w_\lambda^\epsilon(x) = (\lambda, x) + z_\lambda^\epsilon(x_1),
$$

and it is easily proved that P^ϵ can be defined by

$$
\begin{cases}
P_{11}^\epsilon = \dfrac{A_{11}^0}{A_{11}^\epsilon}, \\[3mm]
P_{1j}^\epsilon = \dfrac{A_{1j}^0 - A_{1j}^\epsilon}{A_{11}^\epsilon}, \quad j > 1, \\[3mm]
P_{ii}^\epsilon = 1, \quad i > 1, \\[3mm]
P_{ij}^\epsilon = 0, \quad i, j > 1, \ i \neq j.
\end{cases}
\tag{11}
$$

4. Note that the previous remark immediately demonstrates that a sequence Q^ϵ associated with ${}^t A^\epsilon$ through Definition 2 does not generally coincide with ${}^t P^\epsilon$. Indeed, in the case of layers, both P^ϵ and Q^ϵ given by (11) have nonzero terms only on the diagonal and in the first line.

Proposition 5 *Let P^ϵ be the sequence of corrector matrices defined through Definition 2. Then, as $\epsilon \in E$,*

$$
P^\epsilon \ \longrightarrow \ I \quad \text{weakly in } [L^2(\omega)]^{N^2},
$$

$$
A^\epsilon P^\epsilon \ \longrightarrow \ A^0 \quad \text{weakly in } [L^2(\omega)]^{N^2},
$$

$$
{}^t P^\epsilon A^\epsilon P^\epsilon \ \longrightarrow \ A^0 \quad \text{weakly-* in } [\mathcal{D}'(\omega)]^{N^2}.
$$

Proof:

The sequence P^ϵ is bounded in $[L^2(\omega)]^{N^2}$. If φ is an arbitrary element of $[\mathcal{C}_0^\infty(\omega)]^N$, that is, if

$$
\varphi = \sum_{i=1}^N \varphi_i e_i, \quad \varphi_i \in \mathcal{C}_0^\infty(\omega),
$$

one has

$$
\begin{cases}
\displaystyle \int_\omega P^\epsilon \varphi \, dx = \int_\omega \sum_{i=1}^N \varphi_i P^\epsilon e_i \, dx = \int_\omega \sum_{i=1}^N \varphi_i \, grad \, w_{e_i}^\epsilon \, dx \\[4mm]
\displaystyle \longrightarrow \int_\omega \sum_{i=1}^N \varphi_i e_i dx = \int_\omega \varphi \, dx.
\end{cases}
$$

Thus P^ϵ converges weakly to I in $[L^2(\omega)]^N$. The remaining statements of convergence are obtained in a similar way with the help of Theorem 1 and Lemma 1. ∎

11 Strong Approximation of $grad\, u^\epsilon$. Correctors

Theorem 3 *Assume that $A^\epsilon, \epsilon \in E$ belongs to $M(\alpha, \beta, \Omega)$ and H-converges to $A^0 \in M(\alpha, \beta', \Omega)$. Assume that*

$$
\begin{cases}
u^\epsilon \in H^1(\omega), \\[2mm]
f^\epsilon \in H^{-1}(\omega), \\[2mm]
-div\,(A^\epsilon\, grad\, u^\epsilon) = f^\epsilon \quad in\ \omega, \\[2mm]
u^\epsilon \rightharpoonup u^0 \quad weakly\ in\ H^1(\omega), \\[2mm]
f^\epsilon \to f^0 \quad strongly\ in\ H^{-1}(\omega),
\end{cases}
\tag{12}
$$

where ω is such that $\omega \subset\subset \Omega$. Let P^ϵ be the corrector matrix introduced in Definition 2. Then one has for $\epsilon \in E$:

$$
\begin{cases}
grad\, u^\epsilon = P^\epsilon\, grad\, u^0 + z^\epsilon, \\[2mm]
z^\epsilon \to 0 \quad strongly\ in\ [L^1_{loc}(\omega)]^N.
\end{cases}
\tag{13}
$$

Further, if

$$
\begin{cases}
P^\epsilon \in [L^q(\omega)]^{N^2}, \quad \|P^\epsilon\|_{[L^q(\omega)]^{N^2}} \le C, \quad 2 \le q \le +\infty, \\[2mm]
grad\, u^0 \in [L^p(\omega)]^N, \quad 2 \le p < +\infty,
\end{cases}
\tag{14}
$$

then

$$
z^\epsilon \to 0 \quad strongly\ in\ [L^r_{loc}(\omega)]^N,
\tag{15}
$$

with

$$
\frac{1}{r} = max(\frac{1}{2}, \frac{1}{p} + \frac{1}{q}).
$$

Finally, if

$$
\int_\omega (A^\epsilon\, grad\, u^\epsilon,\, grad\, u^\epsilon)\, dx \to \int_\omega (A^0\, grad\, u^0,\, grad\, u^0)\, dx,
\tag{16}
$$

then

$$
z^\epsilon \to 0 \quad strongly\ in\ [L^r(\omega)]^N.
\tag{17}
$$

Remarks

1. Theorem 3 provides a "good" approximation for $grad\, u^\epsilon$ in the strong topology of L^1_{loc}, L^r_{loc}, or even L^r. Such an approximation is a useful tool in the study of the limit of non linear functions of $grad\, u^\epsilon$.

2. When u^0 is more regular, that is, when $u^0 \in H^2(\omega)$, Theorem 3 immediately implies that

$$u^\epsilon = u^0 + \sum_{i=1}^{N} (w^\epsilon_{e_i} - x_i)\frac{\partial u^0}{\partial x_i} + r^\epsilon \quad \text{with} \quad r^\epsilon \to 0 \quad \text{strongly in } W^{1,1}_{loc}(\omega).$$

The term $\sum_{i=1}^{N} (w^\epsilon_{e_i} - x_i)\frac{\partial u^0}{\partial x_i}$ may be seen as a correcting term. In the case where $A^\epsilon(x) = A(x/\epsilon)$ with A a periodic matrix, it is precisely the term of order ϵ in the asymptotic expansion for u^ϵ (see Bensoussan et al. [1]).

3. In the absence of any hypothesis on the behavior of u^ϵ near the boundary of ω (note the absence of any kind of boundary condition on u^ϵ in (12)) the estimates (13) and (17) on $grad\, u^\epsilon - P^\epsilon grad\, u^0$ are only local estimates. Assumption (16) alleviates this latter obstacle; it is met in particular when u^ϵ is the solution of an homogeneous Dirichlet boundary value problem.

4. An approximation of $grad\, u^\epsilon$ by $P^\epsilon grad\, u$ in the strong topology of $[L^2_{loc}(\omega)]^N$ is obtained as soon as the corrector matrix P^ϵ is bounded in $[L^q(\omega)]^{N^2}$ with q large enough. Since $grad\, u^\epsilon$ is bounded in $[L^2(\omega)]^N$, such an approximation may be deemed "natural." It is unfortunately not available in general. The most pleasant setting is, of course, the case where $q = +\infty$.

5. The case where $p = +\infty$ in (14) also results in the statements (15) and (17), but its proof requires Meyers' regularity theorem to be done.

Proof of Theorem 3:

The proof consists of two steps.

Step 1:

Proposition 6 *In the setting of Theorem 3, the following convergence holds true for any φ in $[C_0^\infty(\omega)]^N$, ϕ in $C_0^\infty(\omega)$ and $\epsilon \in E$.*

$$
\begin{cases}
\displaystyle \int_\omega \phi\left(A^\epsilon(grad\,u^\epsilon - P^\epsilon \varphi),\ (grad\,u^\epsilon - P^\epsilon\ \varphi)\right)dx \\[4mm]
\displaystyle \to \int_\omega \phi\left(A^0(grad\,u^0 - \varphi),\ (grad\,u^0 - \varphi)\right)dx.
\end{cases}
\tag{18}
$$

Proof:

Set $\varphi = \sum\limits_{i=1}^{N} \varphi_i e_i$, $\varphi_i \in C_0^\infty(\omega)$. Then

$$
\int_\omega \phi\left(A^\epsilon(grad\,u^\epsilon - P^\epsilon \varphi),\ (grad\,u^\epsilon - P^\epsilon\ \varphi)\right)dx
$$

$$
= \int_\omega \phi\left(A^\epsilon\,grad\,u^\epsilon,\ grad\,u^\epsilon\right)dx + \sum_{j=1}^{N}\int_\omega \phi\left(A^\epsilon\,grad\,u^\epsilon, P^\epsilon e_j\right)\varphi_j\,dx
$$

$$
+ \sum_{i=1}^{N}\int_\omega \phi\left(A^\epsilon P^\epsilon e_i,\ grad\,u^\epsilon\right)\varphi_i\,dx + \sum_{i,j=1}^{N}\int_\omega \phi\left(A^\epsilon P^\epsilon e_i, P^\epsilon e_j\right)\varphi_i\,\varphi_j\,dx.
$$

Each term in the preceding equality passes to the limit, with the help of Theorem 1 for the first one, Lemma 1 together with the definition of P^ϵ for the second and third ones, and Proposition 5 for the last one. This proves (18).

Whenever assumption (16) is satisfied, the choice $\phi = 1$ is licit because the first term passes to the limit as well as the other terms that contain at least one φ_i which has compact support.

Step 2:

If u^0 belongs to $C_0^\infty(\omega)$, the first step permits us to conclude upon setting $\varphi = grad\,u^0$. Otherwise an approximation process is required. The regularity hypothesis (14) is assumed with no loss of generality since (13) is recovered from (15) if $p = q = 2$.

Let δ be an arbitrary (small) positive number. Choose $\varphi \in [C_0^\infty(\omega)]^N$ such that

$$
\| grad\,u^0 - \varphi \|_{[L^p(\omega)]^N} \le \delta,
$$

which is possible since $p < +\infty$. Then

$$
\| P^\epsilon\,grad\,u^0 - P^\epsilon \varphi \|_{[L^s(\omega)]^N} \le C\delta, \quad \text{if } \frac{1}{s} = \frac{1}{p} + \frac{1}{q}.
$$

Take $\omega_1 \subset\subset \omega$ and $\phi \in C_0^\infty(\omega)$, $\phi = 1$ on ω_1, $0 \le \phi \le 1$ in ω. Proposition 6 then yields

$$\limsup_{\epsilon \in E} \alpha \| \operatorname{grad} u^\epsilon - P^\epsilon \varphi \|^2_{[L^2(\omega_1)]^N}$$

$$\leq \limsup_{\epsilon \in E} \int_\omega \phi \left(A^\epsilon (\operatorname{grad} u^\epsilon - P^\epsilon \varphi), (\operatorname{grad} u^\epsilon - P^\epsilon \varphi) \right) dx$$

$$= \int_\omega \phi \left(A^0 (\operatorname{grad} u^0 - \varphi), (\operatorname{grad} u^0 - \varphi) \right) dx$$

$$\leq \beta' \| \operatorname{grad} u^0 - \varphi \|^2_{[L^2(\omega)]^N} \leq C \delta^2.$$

The two results we have obtained imply that

$$z^\epsilon = \operatorname{grad} u^\epsilon - P^\epsilon \operatorname{grad} u = (\operatorname{grad} u^\epsilon - P^\epsilon \varphi) - (P^\epsilon \operatorname{grad} u^0 - P^\epsilon \varphi)$$

satisfies

$$\limsup_{\epsilon \in E} \| z^\epsilon \|_{L^r[(\omega_1)]^N} \leq C \delta,$$

with $r = \min(2, s)$. Letting δ tend to 0 yields (15).

When assumption (16) holds, the proof remains valid with the choice $\phi = 1$ and $\omega_1 = \omega$, and (17) is thus established. ∎

We conclude with a straightforward application of Theorem 3.

Proposition 7 *Consider, in the setting of Theorem 3, a sequence $a^\epsilon, \epsilon \in E$, with*

$$\begin{cases} a^\epsilon \in [L^\infty(\omega)]^N, \quad \| a^\epsilon \|_{[L^\infty(\omega)]^N} \leq C, \\ {}^t P^\epsilon a^\epsilon \rightharpoonup a^0 \quad \text{weakly in } [L^2(\omega)]^N. \end{cases}$$

Then

$$(a^\epsilon, \operatorname{grad} u^\epsilon) \rightharpoonup (a^0, \operatorname{grad} u^0) \quad \text{weakly in } L^2(\omega).$$

The proof is immediate upon recalling (13). The same idea also permits, at the expense of a few technicalities, handling the case where u^ϵ converges weakly in $H^1(\Omega)$ and satisfies an equation of the type:

$$-\operatorname{div} \left(A^\epsilon \operatorname{grad} u^\epsilon + b^\epsilon u^\epsilon + c^\epsilon \right) + (d^\epsilon, \operatorname{grad} u^\epsilon) + e^\epsilon u^\epsilon = f^\epsilon \quad \text{in } \Omega.$$

References

[1] A. Bensoussan, J.-L. Lions, and G. Papanicolaou, *Asymptotic methods in periodic structures*, North-Holland, Amsterdam, 1978.

[2] E. De Giorgi and S. Spagnolo, Sulla convergenza degli integrali dell' energia per operatori ellittici del secondo ordine, *Boll. UMI* **8** (1973), 391–411.

[3] N. G. Meyers, An L^p-estimate for the gradient of solutions of second order elliptic equations, *Ann. Sc. Norm. Sup. Pisa* **17** (1963), 189–206.

[4] F. Murat, Compacité par compensation, *Ann. Sc. Norm. Sup. Pisa* **5** (1978), 489–507.

[5] S. Spagnolo, Sulla convergenza di soluzioni di equazioni paraboliche ed ellittiche, *Ann. Sc. Norm. Sup. Pisa* **22** (1968), 571–597.

References Added at the Time of the Translation

[6] F. Murat, "Compacité par compensation II," in: *Proceedings of the international meeting on recent methods in non linear analysis (Rome, May 1978)*,E. De Giorgi, E. Magenes and U. Mosco, eds., Pitagora Editrice, Bologna, 1979, 245–256.

[7] L. Tartar, "Compensated compactness and applications to partial differential equations.," in: *Non linear analysis and mechanics, Heriot-Watt Symposium, Volume IV*, R.J. Knops, ed., Research Notes in Mathematics, **39**, Pitman, Boston, 1979, 136–212.

A Strange Term Coming from Nowhere

Doina Cioranescu and François Murat

Introduction

Let Ω be a bounded open set in $\mathbb{R}^N$ and let us perforate it by holes: we obtain an open set Ω^ε. Consider the Dirichlet problem in the domain Ω^ε. The general questions with which we are concerned are the following. Do the solutions u^ε converge to a limit u when the parameter ε tends to zero? If this limit exists, can it be characterized?

Several situations can occur depending on the behavior of the holes. *The first situation* is when each compact set K included in Ω is absorbed by the domains Ω^ε (i.e., $K \subset \Omega^\varepsilon$ if ε is small enough); then the functions u^ε converge to the solution of the Dirichlet problem in Ω. *The second situation* is when the characteristic function of the set of holes converges (weakly-$\star$ in $L^\infty(\Omega)$) to a function that is strictly positive. In this case the limit u is necessarily zero. These two situations are classical and there are a great number of papers dealing with them, see for instance, Courant and Hilbert [7], Chap.VI-2.6, Nečas [16], Chap.3-6, and Rauch and Taylor [18].

One can also divide Ω in several regions, where one of these two situations takes place but *other situations are possible*. We may consider, for instance, the case where the domain Ω is perforated by an increasing number of holes, regularly distributed but smaller and smaller as ε tends to zero. This is the type of situation that we study here. Roughly speaking, three cases can occur in this situation: either the holes, in spite of their number, are too small and u^ε converges to the solution of the Dirichlet problem in Ω (as in the first situation), or the holes are too big and then u^ε converges to zero (as in the second situation). Between these two cases there is a third one where the holes have a critical size depending on their number and distribution, and where the limit of u^ε is the solution of a Dirichlet problem in Ω with another operator which is the sum of the initial one and of an additional ("strange") term, that comes in from the holes.

In this article, we also study the behavior of solutions of variational inequalities with highly oscillating obstacles. These inequalities are associated with the preceding Dirichlet problems through the definition of the obstacles. The obstacles we consider here are equal to a given function ψ in Ω^ε and to zero inside the holes (note that the solutions of the variational inequalities are defined in the whole of Ω).

The method we use is the so-called energy method introduced by Tartar [22], [23] for studying homogenization problems. It consists of con-

© Springer Nature Switzerland AG 2018
A. V. Cherkaev, R. Kohn (eds.), *Topics in the Mathematical Modelling of Composite Materials*, Modern Birkhäuser Classics, https://doi.org/10.1007/978-3-319-97184-1_4

structing suitable test functions that are used in the equations as well as in variational inequalities. It is a quite general method allowing us to treat Dirichlet problems in a domain with small holes for operators of arbitrary order. On the other hand, the use of truncations when treating variational inequalities with highly oscillating obstacles constrains us to deal only with variational inequalities with second order operators.

This paper is divided into sections. In Section 1 we assume the existence of special test functions and we show how to use them to derive at the limit the "strange" term for Dirichlet problems. In Section 2 we give several examples where these test functions are explicitly built up. The two main examples are related to small holes with a precise diameter, distributed either in the whole space or on a hyperplane. Section 3 is devoted to a further study of the general abstract framework introduced in Section 1. We investigate the limit of the energy of a sequence of functions which vanish on the holes. We also show that the solutions u^ε of Dirichlet problems in Ω^ε can be written (up to a remainder) as the product of the solution of the limit problem by the special test functions: this is the corrector type result. Finally, in Section 4 we study in the same abstract framework the convergence of the solutions of variational inequalities with obstacles that are equal to zero inside the holes and to a given function ψ in Ω^ε.

Most of the results presented here are not new, however, the presentation is original and, it is hoped, clear and easy to understand. It also points out the connection that exists between Dirichlet problems in domains with small holes and the class of variational inequalities previously presented.

The Dirichlet problem in a domain with small holes has been studied by Hrouslov [12], [13], Marcenko and Hrouslov [15], Rauch and Taylor [18], and more recently by Papanicolaou and Varadhan [17], Sanchez-Palencia [21] and Cioranescu [3], [4]. The variational inequalities with highly oscillating obstacles we consider here were first studied by Carbone and Colombini [2]. These examples were then included in a more general presentation by De Giorgi, Dal Maso and Longo [11], Dal Maso and Longo [10], Dal Maso [8], [9] and Attouch and Picard [1].

To conclude this introduction, let us remark that this paper only deals with Dirichlet boundary conditions, and that the results concerning other boundary conditions are completely different. The Neumann boundary condition was studied by Hrouslov [14], the torsion boundary condition by Rauch and Taylor [19]. These two cases were also studied by Cioranescu and Saint Jean Paulin [5], [6] and Saint Jean Paulin [20].

1. Dirichlet problems in perforated domains. Abstract framework

Let Ω be a bounded open set of $\mathbb{R}^N$. Consider for every ε, where ε takes its values in a sequence of positive numbers which tends to zero, some closed subsets T_i^ε, $1 \le i \le n(\varepsilon)$, which are the "holes." The domain Ω^ε is defined by removing the holes T_i^ε from Ω, that is,

$$\Omega^\varepsilon = \Omega - \bigcup_{i=1}^{n(\varepsilon)} T_i^\varepsilon.$$

Let $f \in L^2(\Omega)$ and consider the Dirichlet problem in Ω^ε: Find u^ε such that

$$\begin{cases} -\Delta u^\varepsilon = f & \text{in } \mathcal{D}'(\Omega^\varepsilon), \\ u^\varepsilon \in H_0^1(\Omega^\varepsilon). \end{cases} \tag{1.1}$$

The equivalent variational formulation is: find u^ε such that

$$\begin{cases} \displaystyle\int_{\Omega^\varepsilon} \operatorname{grad} u^\varepsilon \operatorname{grad} v \, dx = \int_{\Omega^\varepsilon} fv \, dx, & \forall v \in H_0^1(\Omega^\varepsilon), \\ u^\varepsilon \in H_0^1(\Omega^\varepsilon). \end{cases} \tag{1.2}$$

Let us denote by $\widetilde{u^\varepsilon}$ the extension by zero of u^ε, to the whole of Ω, that is,

$$\widetilde{u^\varepsilon} = \begin{cases} u^\varepsilon & \text{in } \Omega^\varepsilon, \\ 0 & \text{in the holes } T_i^\varepsilon, \quad 1 \le i \le n(\varepsilon). \end{cases}$$

Then $\widetilde{u^\varepsilon}$ belongs to $H_0^1(\Omega)$ and taking $v = u^\varepsilon$ in (1.2), one has

$$\int_\Omega |\operatorname{grad} \widetilde{u^\varepsilon}|^2 \, dx = \int_{\Omega^\varepsilon} |\operatorname{grad} u^\varepsilon|^2 \, dx = \int_{\Omega^\varepsilon} fu^\varepsilon \, dx = \int_\Omega f\widetilde{u^\varepsilon} \, dx$$
$$\le \|f\|_{L^2(\Omega)} \|\widetilde{u^\varepsilon}\|_{L^2(\Omega)}.$$

By the Poincaré inequality for the domain Ω, there exists $\alpha > 0$ such that

$$\alpha \|v\|_{L^2(\Omega)} \le \|\operatorname{grad} v\|_{(L^2(\Omega))^N}, \quad \forall v \in H_0^1(\Omega),$$

so that one derives

$$\|\widetilde{u^\varepsilon}\|_{H_0^1(\Omega)} = \|\operatorname{grad} \widetilde{u^\varepsilon}\|_{(L^2(\Omega))^N} \le \frac{1}{\alpha} \|f\|_{L^2(\Omega)}. \tag{1.3}$$

Consequently, by passing to a subsequence, still denoted by u^ε, we may assume that

$$\widetilde{u^\varepsilon} \rightharpoonup u \quad \text{weakly in } H_0^1(\Omega).$$

The questions are then the following ones: Can the limit u be identified? Is u the solution of some equation? We would like, of course, to give a positive answer to these questions. We would also like the limit equation to be analogous to that satisfied by u^ε but, if possible, with an additional term and this where the charm of the problem lies! For this very reason, we shall assume the following

Hypotheses on the holes. We suppose that there exist a sequence of functions w_ε and a distribution μ such that

(H.1) $w^\varepsilon \in H^1(\Omega)$,

(H.2) $w^\varepsilon = 0$ on the holes T_i^ε, $1 \le i \le n(\varepsilon)$,

(H.3) $w^\varepsilon \rightharpoonup 1$ weakly in $H^1(\Omega)$,

(H.4) $\mu \in W^{-1,\infty}(\Omega)$,

(H.5) $\begin{cases} \text{for every sequence } v^\varepsilon \text{ such that } v^\varepsilon = 0 \quad \text{on } T_i^\varepsilon, \ 1 \le i \le n(\varepsilon), \\[4pt] \text{satisfying } v^\varepsilon \rightharpoonup v \text{ weakly in } H^1(\Omega) \text{ with } v \in H^1(\Omega), \text{ one has} \\[4pt] < -\Delta w^\varepsilon, \varphi v^\varepsilon >_{H^{-1}(\Omega), H_0^1(\Omega)} \to < \mu, \varphi v >_{H^{-1}(\Omega), H_0^1(\Omega)}, \\[4pt] \text{for all } \varphi \in \mathcal{D}(\Omega). \end{cases}$

These hypotheses especially (H5), seem rather unusual. They are discussed at the end of this section. We show in Section 2 that actually they are fulfilled in some model examples.

For the moment, let us use these hypotheses in order to prove some results.

Proposition 1.1. *If w^ε and μ satisfy hypotheses* (H.1) *to* (H.5), *one has*

$$< \mu, \varphi >= \lim_{\varepsilon \to 0} \int_\Omega |\mathrm{grad}\, w^\varepsilon|^2 \, \varphi \, dx, \quad \forall \varphi \in \mathcal{D}(\Omega). \tag{1.4}$$

It follows immediately that μ, the limit in the distribution sense of $|\mathrm{grad}\, w^\varepsilon|^2$, is a positive measure.

Proof. Taking $v^\varepsilon = w^\varepsilon$ and $v = 1$ in (H.5), which is allowed, one has

$$< -\Delta w^\varepsilon, \varphi w^\varepsilon >_{H^{-1}(\Omega), H_0^1(\Omega)} \to < \mu, \varphi w >_{H^{-1}(\Omega), H_0^1(\Omega)},$$

for all $\varphi \in \mathcal{D}(\Omega)$. But

$$< -\Delta w^\varepsilon, \varphi w^\varepsilon >_{H^{-1}(\Omega), H_0^1(\Omega)} = \int_\Omega \mathrm{grad}\, w^\varepsilon \, \mathrm{grad}\, (\varphi w^\varepsilon) \, dx$$

$$= \int_\Omega |\mathrm{grad}\, w^\varepsilon|^2 \varphi \, dx + \int_\Omega w^\varepsilon \mathrm{grad}\, w^\varepsilon \mathrm{grad}\, \varphi \, dx.$$

When passing to the limit in ε, the last term tends to zero. Indeed, by (H.3), grad $w^\varepsilon \rightharpoonup 0$ weakly in $(L^2(\Omega))^N$ and $w^\varepsilon \to 1$ strongly in $L^2_{loc}(\Omega)$ by the Rellich–Kondrashov theorem. ∎

Theorem 1.2. *Under hypotheses (H.1) to (H. 5), the solutions u^ε of (1.1) satisfy*

$$\widetilde{u^\varepsilon} \rightharpoonup u \quad \text{weakly in } H_0^1(\Omega),$$

where u is the unique solution of

$$\begin{cases} -\Delta u + \mu u = f & \text{in } \mathcal{D}'(\Omega), \\ u \in H_0^1(\Omega). \end{cases} \tag{1.5}$$

The limit u is thus the solution of a Dirichlet problem for the operator $-\Delta + \mu$, which contains a "strange term" when compared to the initial operator $-\Delta$.

Proof. As previously proved in equation (1.3), $\|\widetilde{u^\varepsilon}\|_{H_0^1(\Omega)}$ is bounded independently of ε. By passing to a subsequence, still denoted by $\widetilde{u^\varepsilon}$, there exists $u \in H_0^1(\Omega)$ such that

$$\widetilde{u^\varepsilon} \rightharpoonup u \quad \text{weakly in } H_0^1(\Omega) \text{ and strongly in } L^2(\Omega).$$

It remains to identify this limit u.

By hypotheses (H.1) and (H.2), for any $\varphi \in \mathcal{D}(\Omega)$, the function $w^\varepsilon \varphi$ belongs to $H_0^1(\Omega^\varepsilon)$ as $\varphi = 0$ on $\partial\Omega$ and $w^\varepsilon = 0$ on the holes.[1] Therefore one can take $v = w^\varepsilon \varphi$ in the variational formulation (1. 2). Thus

$$\int_\Omega \varphi \,\text{grad}\, \widetilde{u^\varepsilon} \,\text{grad}\, w^\varepsilon \, dx + \int_\Omega w^\varepsilon \text{grad}\, \widetilde{u^\varepsilon} \,\text{grad}\, \varphi \, dx = \int_\Omega f w^\varepsilon \varphi \, dx.$$

It is a simple matter to pass to the limit in the second and third terms of this equality as (see (H. 3)) $w^\varepsilon \to 1$ strongly in $L^2_{loc}(\Omega)$. One gets

$$\int_\Omega w^\varepsilon \text{grad}\, \widetilde{u^\varepsilon} \,\text{grad}\, \varphi \, dx \to \int_\Omega \text{grad}\, u \,\text{grad}\, \varphi \, dx,$$

$$\int_\Omega f w^\varepsilon \varphi \, dx \to \int_\Omega f\varphi \, dx.$$

---

[1] To be entirely correct, one should state (H. 2) in the following way. Let $Q^\varepsilon = \mathbb{R}^N - \cup_{i=1}^{n(\varepsilon)} T_i^\varepsilon$. The function w^ε is the extension by 0 in the holes T_i^ε of a function belonging to $H_0^1(Q^\varepsilon)$. This formulation does not require any regularity on the boundaries $\partial\Omega$ and ∂T_i^ε, $1 \le i \le n(\varepsilon)$, and what happens at infinity does not enter as Ω is bounded. One should also reformulate the assumption $v^\varepsilon = 0$ on the holes T_i^ε" in (H.5), as follows. For every $\varphi \in \mathcal{D}(\Omega)$, φv^ε is the extension by 0 in the holes T_i^ε of a function belonging to $H_0^1(\Omega^\varepsilon)$.

As for the first term, integrating by parts yields

$$\begin{cases} \int_\Omega \varphi \; \text{grad} \; \widetilde{u^\varepsilon} \; \text{grad} \; w^\varepsilon \; dx = < -\Delta w^\varepsilon, \varphi \widetilde{u^\varepsilon} >_{H^{-1}(\Omega),H_0^1(\Omega)} \\ \\ \qquad\qquad\qquad\qquad - \int_\Omega \widetilde{u^\varepsilon} \; \text{grad} \; \varphi \; \text{grad} \; w^\varepsilon \; dx. \end{cases}$$

We easily pass to the limit in the last two terms. For the first one, by using hypothesis (H. 5) (which was done for this very purpose!), we get

$$< -\Delta w^\varepsilon, \varphi \widetilde{u^\varepsilon} >_{H^{-1}(\Omega),H_0^1(\Omega)} \to < \mu, \varphi u >_{H^{-1}(\Omega),H_0^1(\Omega)} .$$

For the second one, by making use of (H. 3) and of the strong convergence of $\widetilde{u^\varepsilon}$ in $L^2(\Omega)$, we get

$$\int_\Omega \widetilde{u^\varepsilon} \; \text{grad} \; \varphi \; \text{grad} \; w^\varepsilon \; dx \to \int_\Omega u \; \text{grad} \; \varphi \; \text{grad} \; 1 \; dx = 0.$$

Summing up these convergence results, it follows that u satisfies

$$\int_\Omega \text{grad} \; u \; \text{grad} \; \varphi \; dx + < \mu u, \varphi > = \int_\Omega f\varphi \; dx, \quad \forall \varphi \in \mathcal{D}(\Omega).$$

The product μu of $\mu \in W^{-1,\infty}(\Omega)$ by $u \in H_0^1(\Omega)$ belongs to $H^{-1}(\Omega)$. The function u is therefore the solution of equation (1.5) and is unique because μ is nonnegative (see equation (1.4)).

The uniqueness of the limit implies that the whole sequence u^ε converges to u, and this completes the proof of Theorem 1.1. ∎

Remark 1.3. Theorem 1.2 is in fact a *local* result (with respect to Ω) as the boundary condition on $\partial\Omega$ does not play any role. In order to prove this result, consider functions $\underline{u}^\varepsilon$ satisfying

$$\begin{cases} -\Delta \underline{u}^\varepsilon = f & \text{in } \mathcal{D}'(\Omega^\varepsilon), \\ \underline{u}^\varepsilon = 0 & \text{on } \partial T_i^\varepsilon, \; 1 \le i \le n(\varepsilon), \end{cases}$$

(without any condition on $\partial\Omega$). If we assume that $\widetilde{\underline{u}^\varepsilon}$, the extensions by zero in the holes of $\underline{u}^\varepsilon$, are such that

$$\widetilde{\underline{u}^\varepsilon} \rightharpoonup \underline{u} \quad \text{weakly in } H^1(\Omega),$$

one can perform the same proof as before to obtain a result similar to the result of Theorem 1.2. Indeed in the proof of Theorem 1.2, we multiplied equation (1.1) by φw^ε, with $\varphi \in \mathcal{D}(\Omega)$, and so what happens on $\partial\Omega$ has no influence on the result. Consequently, we deduce that if hypotheses (H.1) to (H.5) are fulfilled, then $\underline{u}$ satisfies

$$-\Delta\underline{u} + \mu\underline{u} = f \quad \text{in } \mathcal{D}'(\Omega).$$

Comments on hypotheses (H.1) to (H.5)

As we already said (in footnote 1), hypotheses (H.1) to (H.5) can be reformulated without assuming any regularity on the boundaries $\partial\Omega$ and ∂T_i^ε, $1 \leq i \leq n(\varepsilon)$. We also proved (Proposition 1.1) that these hypotheses imply that μ is a non negative measure. We can now give some further consequences of hypotheses (H1) to (H5).

Remark 1.4. If there exist some w^ε and μ satisfying hypotheses (H.1) to (H.5), they are "quasi-unique." To prove this, let us consider $\{w^\varepsilon, \mu\}$ and $\{\underline{w}^\varepsilon, \underline{\mu}\}$ satisfying hypotheses (H.1) to (H.5). By taking $v^\varepsilon = w^\varepsilon - \underline{w}^\varepsilon$ in (H.5), one has

$$\begin{cases} \mu = \underline{\mu}, \\ w^\varepsilon - \underline{w}^\varepsilon \to 0 \quad \text{strongly in } H^1_{\text{loc}}(\Omega). \end{cases}$$

Remark 1.5. If there exist w^ε satisfying hypotheses (H.1), (H.2), and (H.3) and if, furthermore,

$$w^\varepsilon \to 1 \quad \text{strongly in } H^1_{\text{loc}}(\Omega),$$

hypotheses (H.4) and (H.5) are satisfied with $\mu = 0$ and there is no additional term in (1.5). Therefore, if one wants to have $\mu \neq 0$, one has to take care to have weak (and not strong) convergence in (H.3).

Remark 1.6. Notice that (H.5) does not imply the strong convergence in $H^{-1}(\Omega)$ of $-\Delta w^\varepsilon$ to μ. Indeed, we supposed in (H.5) that the term $< -\Delta w^\varepsilon, \varphi v^\varepsilon >_{H^{-1}(\Omega),H^1_0(\Omega)}$ has a limit *not for any* sequence φv^ε that converges weakly in $H^1_0(\Omega)$, but only for some very particular ones, that is, those whose elements vanish on the holes.

In fact hypothesis (H.5) could be replaced by the following (quite close) one.

$$\begin{cases} \text{There exist two sequences } \mu_\varepsilon, \gamma_\varepsilon \in H^{-1}(\Omega) \text{ such that} \\ \mu_\varepsilon \to \mu \text{ strongly in } H^{-1}(\Omega), \ \gamma_\varepsilon \rightharpoonup \mu \text{ weakly in } H^{-1}(\Omega) \text{ and} \\ < \gamma_\varepsilon, v_\varepsilon >= 0, \ \forall v^\varepsilon \in H^1_0(\Omega) \text{ with } v^\varepsilon = 0 \text{ on } T_i^\varepsilon, \ 1 \leq i \leq n(\varepsilon). \end{cases} \quad \text{(H.5)}'$$

Clearly (H.5)′ implies (H.5). We have chosen to assume (H.5) because it naturally appears in the proof of Theorem 1.2. In return, hypothesis (H.5)′ is much more eloquent: it asserts that $-\Delta w^\varepsilon$ is the difference between two sequences having the same limit (this is quite normal since, due to (H.3), $-\Delta w^\varepsilon \rightharpoonup 0$ weakly in $H^{-1}(\Omega)$). But one of these sequences, namely γ_ε, does not play any role in the Dirichlet problem in Ω_ε, since the last condition in (H.5)′ means that γ_ε is "invisible" for the elements of $H_0^1(\Omega)$ vanishing on the holes. On the other hand, the sequence μ_ε behaves in a better way than expected: it converges strongly instead of weakly.

To complete this remark, let us point out that in the following examples we check that (H.5)′ is satisfied: as previously mentioned, this implies that (H.5) is satisfied.

Remark 1.7. Hypotheses (H.1), (H.2), and (H.3), which seem innocent, are in reality quite restrictive. The fact that there exists a sequence of functions vanishing on the holes and weakly converging to 1 in $H^1(\Omega)$, implies some conditions on the asymptotic behavior of the holes. For instance, in the one-dimensional case, these hypotheses imply that the holes accumulate on the boundary $\partial\Omega$ as $\varepsilon \to 0$: if there exists a sequence of holes whose "centers" x_i^ε converge to $x^0 \in \Omega$, necessarily $\lim_{\varepsilon \to 0} w^\varepsilon(x^0) = 0$ because of the Sobolev embedding $H^1(\mathbb{R}) \subset C^{0,1/2}(\mathbb{R})$. But this is in contradiction with hypothesis (H.3) which asserts that $\lim_{\varepsilon \to 0} w^\varepsilon(x) = 1$ for all $x \in \Omega$. This is why we drop the case $N = 1$ in Section 2, when looking to examples.

Generally speaking, hypotheses (H.1), (H.2) and (H.3) imply that the holes are *not too big*.

Remark 1.8. Hypothesis (H.3) can be weakened by assuming only $\mu \in H^{-1}(\Omega)$. If sufficient for proving Theorem 1.2, this new hypothesis introduces some difficulties in the proof of the uniqueness of the solution of (1.5). Namely, the product μv for $v \in H_0^1(\Omega)$ is not more in $H^{-1}(\Omega)$; it is only a distribution. One can get rid of this difficulty by proving that if v^ε and v are such that

$$\begin{cases} v^\varepsilon \rightharpoonup v & \text{weakly in } H_0^1(\Omega), \\ v^\varepsilon = 0 & \text{on } T_i^\varepsilon, \ 1 \le i \le n(\varepsilon), \end{cases}$$

then $v \in L^2(\mu)$ (i.e., $|v|^2$ is measurable with respect to the measure μ). The proof makes use of truncations and of techniques similar to those we use in Section 3.

Generalizations

It is easy to establish the analogue of Theorem 1.2 in the case of Dirichlet problems (1.1) written for the operator $-\operatorname{div}(A \operatorname{grad})$ in place of $-\Delta$, where A is a coercive matrix, not necessarily symmetric and belonging to

$(L^\infty(\Omega))^{N^2}$. Following along the lines of the proof of Theorem 1.2 with this operator instead of $-\Delta$, one sees immediately that only the transposed matrix ${}^tA$ appears in the proof. Accordingly, it it sufficient to replace in hypothesis (H.5) the line

$$< -\Delta w^\varepsilon, \varphi v^\varepsilon >_{H^{-1}(\Omega), H_0^1(\Omega)} \to < \mu, \varphi v >_{H^{-1}(\Omega), H_0^1(\Omega)}$$

by

$$< -\mathrm{div}({}^tA\ \mathrm{grad}\ w^\varepsilon), \varphi v^\varepsilon >_{H^{-1}(\Omega), H_0^1(\Omega)} \to < \mu, \varphi v >_{H^{-1}(\Omega), H_0^1(\Omega)},$$

to get what we call hypothesis (H.5)$_A$. It is obvious that the test functions w^ε and the measure μ related to an operator $-\mathrm{div}(A\ \mathrm{grad})$ are in general not identical to those related to $-\Delta$, just consider the case $A = 2$ Id.

It should be remarked that among the arguments used in the proofs, none is specific to second order operators: neither truncation nor maximum principle were used until now. Consequently, we may apply our method to Dirichlet problems for higher order operators to obtain results similar to Theorem 1.2.

Remark 1.9. Set $H = H_0^1(\Omega)$, $\mathcal{K} = \mathcal{D}(\Omega)$ and introduce the space $H^\varepsilon = \{v \in H_0^1(\Omega) : v = 0 \text{ on } T_i^\varepsilon, 1 \le i \le n(\varepsilon)\}$. Consider the operator β^ε defined on $\mathcal{K}$ by

$$\beta^\varepsilon \phi = (1 - w^\varepsilon)\phi.$$

Theorem 1.2 is then a consequence of the abstract results obtained by Cioranescu [3]. We adopted here a more "concrete" point of view, which traces back to the method introduced by Tartar [23] for homogenization problems. Another "abstract" presentation (which in our "concrete" framework is, roughly speaking, equivalent to consider the product by w^ε as a projection of H on H^ε) was used by Hrouslov [12], [13], and by Marcenko and Hrouslov [15], where many examples and applications can be found.

2. Examples

Hypotheses (H.1) to (H.5) are "abstract" and the theory of Section 1, even if interesting, is empty if some examples of domains Ω^ε satisfying them are not given. The aim of this section is precisely to give such examples.

In the sequel we suppose $N \ge 2$. As shown in Remark 1.7, in the one-dimensional case the imbedding $H^1(\mathbb{R}) \subset C^{0,1/2}(\mathbb{R})$ prevents us from finding examples of domains satisfying hypotheses (H.1) to (H.5) unless the holes concentrate on the boundary of Ω (which is not what we want to have!).

Model example 2.1. Spherical holes periodically distributed in volume

For each value of ε, we cover $\mathbb{R}^n (n \geq 2)$ by cubes P_i^ε of size 2ε. From each cube we remove the ball T_i^ε of radius $a^\varepsilon (a^\varepsilon > 0)$ centered at the very center of the cube. In this way, $\mathbb{R}^N$ is perforated by spherical identical holes. Set

$$Q^\varepsilon = \mathbb{R}^N \backslash \bigcup T_i^\varepsilon,$$

$$\Omega_\varepsilon = \Omega \cap Q^\varepsilon = \Omega \backslash \bigcup_{i=1}^{n(\varepsilon)} T_i^\varepsilon.$$

This means that we removed from Ω small balls of radius a^ε whose centers are the nodes of a lattice in $\mathbb{R}^N$ with cell size 2ε (see Figure 1).

The question is now: can we choose a^ε in such a way that there exist w^ε and μ satisfying hypotheses (H.1) to (H.5) of Section 1? We show that the answer is yes by constructing these functions explicitly.

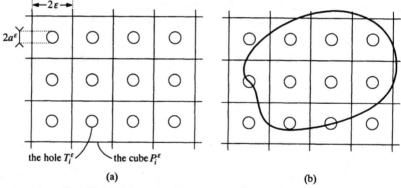

(a)   (b)

Figure 1.   Small balls periodically distributed in $\mathbb{R}^n$.

Let us define w^ε on each cube P_i^ε by setting (see Figure 2)

$$\begin{cases} w^\varepsilon = 0 & \text{in } T_i^\varepsilon, \\ \Delta w^\varepsilon = 0 & \text{in } B_i^\varepsilon - T_i^\varepsilon, \\ w^\varepsilon = 1 & \text{in } P_i^\varepsilon - B_i^\varepsilon, \\ w^\varepsilon \text{ continuous at the interfaces,} \end{cases} \qquad (2.1)$$

where $B_i^\varepsilon \subset P_i^\varepsilon$ is the ball of radius ε concentric with T_i^ε. This defines w^ε in the all of $\mathbb{R}^N$, hence in Ω. The function w^ε is periodic with period $]-\varepsilon, +\varepsilon[^N$ but can not be written in the form $w(\frac{x}{\varepsilon})$, except in the case where $a^\varepsilon = C\varepsilon$, $0 < C < 1$.

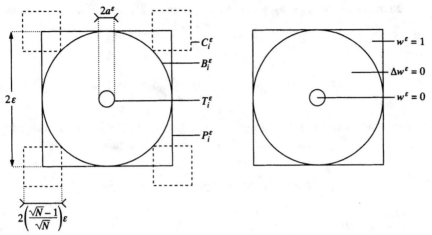

Figure 2. Definition of W^ε.

Now it is easy to compute w^ε in polar coordinates in the annulus $B_i^\varepsilon - T_i^\varepsilon$. One has

$$
\begin{cases}
w^\varepsilon = \dfrac{\log a^\varepsilon - \log r}{\log a^\varepsilon - \log \varepsilon} & \text{if } N = 2, \\[2mm]
w^\varepsilon = \dfrac{(a^\varepsilon)^{-(N-2)} - r^{-(N-2)}}{(a^\varepsilon)^{-(N-2)} - \varepsilon^{-(N-2)}} & \text{if } N \geq 3.
\end{cases}
\tag{2.2}
$$

Define μ by

$$
\begin{cases}
\mu = \dfrac{\pi}{2}\dfrac{1}{C_0} & \text{if } N = 2, \\[2mm]
\mu = \dfrac{S_N(N-2)}{2N} C_0^{N-2} & \text{if } N \geq 3,
\end{cases}
\tag{2.3}
$$

where S_N is the surface of the unit sphere in $\mathbb{R}^N$ and C_0 is a constant, $C_0 > 0$.

Theorem 2.2. *If we choose*

$$
\begin{cases}
a^\varepsilon = \exp\left(-\dfrac{C_0}{\varepsilon^2}\right) & \text{if } N = 2, \\[2mm]
a^\varepsilon = C_0\, \varepsilon^{N/N-2} & \text{if } N \geq 3,
\end{cases}
\tag{2.4}
$$

then w^ε and μ defined by (2.1) and (2.3) satisfy hypotheses (H.1) to (H.5).

Proof. The function w^ε is continuous, piecewise smooth, and vanishes on the holes T_i^ε, $1 \leq i \leq n(\varepsilon)$. Therefore, hypotheses (H.1) and (H.2) are satisfied for any choice of $a^\varepsilon < \varepsilon$. Furthermore, μ is a constant and hypothesis (H.4) is also satisfied .

In order to check (H.3), we compute $\|\text{grad } w^\varepsilon\|^2_{(L^2(\Omega))^N}$. As the number $n(\varepsilon)$ of cubes P_i^ε covering Ω is about $\dfrac{\text{meas } \Omega}{(2\varepsilon)^N}$, we obtain

$$\|\text{grad } w^\varepsilon\|^2_{(L^2(\Omega))^N} \sim \frac{\text{meas } \Omega}{(2\varepsilon)^N} \int_{P_i^\varepsilon} |\text{grad } w^\varepsilon|^2 \, dx,$$

where, thanks to (2.2), we can compute explicitly the right-hand side. For instance, for $N \geq 3$ (the computation for $N = 2$ is similar),

$$\int_{P_i^\varepsilon} |\text{grad } w^\varepsilon|^2 \, dx = \frac{S_N(N-2)}{(a^\varepsilon)^{-(N-2)} - \varepsilon^{-(N-2)}}.$$

If $a^\varepsilon = C_0 \, \varepsilon^{N/(N-2)} (N \geq 3)$ we have

$$\|\text{grad } w^\varepsilon\|^2_{(L^2(\Omega))^N} \to \frac{(\text{meas } \Omega) \, S_N(N-2)}{2^N} C_0^{N-2}$$

and, since $0 \leq w^\varepsilon \leq 1$, w^ε is bounded in $H^1(\Omega)$. Hence, by taking a subsequence still denoted ε, we have

$$w^\varepsilon \rightharpoonup w \quad \text{weakly in } H^1(\Omega) \text{ and strongly in } L^2(\Omega).$$

To identify this w, remark that $w^\varepsilon = 1$ in the small squares C_i^ε of size $\dfrac{(2(\sqrt{N} - 1)}{\sqrt{N})}\varepsilon$ centered at the vertices of the cube P_i^ε (see Figure 2). If χ_C^ε denotes the characteristic function of C_i^ε, then

$$w^\varepsilon \, \chi_C^\varepsilon = \chi_C^\varepsilon.$$

But χ_C^ε tends to $\left[\dfrac{\sqrt{N} - 1}{\sqrt{N}}\right]$ weakly-$\star$ in $L^\infty(\Omega)$, so that if we pass to the limit in the preceding identity, we have $w = 1$. Thus (H.3) is satisfied.

It is more delicate to verify hypothesis (H.5). As a matter of fact, we will prove that (H.5)′ is satisfied, which by Remark 1.6 implies (H.5).

From the definition (2.1) of w^ε, it is clear that Δw^ε vanishes everywhere except on the spheres ∂B_i^ε and ∂T_i^ε; that is,

$$-\Delta w^\varepsilon = \mu^\varepsilon - \gamma^\varepsilon,$$

where μ^ε denote the measures supported by ∂B_i^ε and γ^ε those supported by ∂T_i^ε. Let us prove that

$$\mu^\varepsilon \to \mu \quad \text{strongly in } H^{-1}(\Omega), \tag{2.5}$$

which shows that (H.5)' is fulfilled, since γ^ε only changes the boundaries of the holes. Indeed,

$$< \gamma^\varepsilon, v^\varepsilon >= 0, \quad \forall v^\varepsilon \in H_0^1(\Omega) \text{ such that } v^\varepsilon = 0 \text{ on } T_i^\varepsilon, 1 \leq i \leq n(\varepsilon).$$

In view of (2.2), one can compute explicitly μ^ε. To this end denote by δ_i^ε the Dirac mass supported by the sphere ∂B_i^ε; that is,

$$< \delta_i^\varepsilon, \varphi >= \int_{\partial B_i^\varepsilon} \varphi(s)\, ds, \quad \forall \varphi \in \mathcal{D}(\mathbb{R}^N). \tag{2.6}$$

Then for $N \geq 3$ we have

$$\mu^\varepsilon = \sum_{i=1}^{n(\varepsilon)} \frac{\partial w^\varepsilon}{\partial n}\bigg|_{\partial B_i^\varepsilon} \delta_i^\varepsilon = \frac{(N-2)C_0^{N-2}}{1 - C_0^{N-2}\varepsilon^2} \sum_{i=1}^{n(\varepsilon)} \varepsilon \delta_i^\varepsilon \quad \text{in } \mathbb{R}^N,$$

and for $N = 2$,

$$\mu^\varepsilon = \frac{1}{C_0 + \varepsilon^2 \log \varepsilon} \sum_{i=1}^{n(\varepsilon)} \varepsilon \delta_i^\varepsilon \quad \text{in } \mathbb{R}^2.$$

Using Lemma 2.3 below, one deduces (2.5). This completes the proof of Theorem 2.2. ∎

Lemma 2.3. *Let $N \geq 2$. Assume that the centers of the spheres ∂B_i^ε, $1 \leq i \leq n(\varepsilon)$, of radius ε are the nodes of a lattice in $\mathbb{R}^N$ with cell size 2ε. If δ_i^ε are the Dirac masses supported by ∂B_i^ε, then*

$$\sum_i \varepsilon \delta_i^\varepsilon \to \frac{S_N}{2^N} \quad \text{strongly in } W_{\text{loc}}^{-1,\infty}(\mathbb{R}^N).$$

Proof. Introduce the auxiliary function q^ε defined as follows.

$$\begin{cases} -\Delta q^\varepsilon = N & \text{in } B_i^\varepsilon, \\ \dfrac{\partial q^\varepsilon}{\partial n} = \varepsilon & \text{on } \partial B_i^\varepsilon. \end{cases}$$

If the following compatibility condition,

$$\int_{B_i^\varepsilon} -\Delta q^\varepsilon\, 1\, dx = - \int_{\partial B_i^\varepsilon} \frac{\partial q^\varepsilon}{\partial n}\, 1\, ds,$$

is satisfied (and it is the very case here), the preceding problem has a unique solution up to an additive constant. One can explicitly compute this solution which in polar coordinates satisfies

$$\frac{\partial q^\varepsilon}{\partial r} = r \quad \text{in } B_i^\varepsilon.$$

We fix the constant by requiring $q^\varepsilon = 0$ on ∂B_i^ε. Then we extend q^ε by 0 in $P_i^\varepsilon - B_i^\varepsilon$ and so the function q^ε is defined in all of $\mathbb{R}^N$.

Since $|\text{grad } q^\varepsilon| \leq \varepsilon$ a.e., we have

$$q^\varepsilon \to 0 \quad \text{strongly in } W^{1,\infty}_{\text{loc}}(\mathbb{R}^N).$$

Let χ_B^ε denote the characteristic function of the balls B_i^ε of radius ε. Then

$$-\Delta q^\varepsilon = N\chi_B^\varepsilon + \sum_{i=1}^{n(\varepsilon)} \varepsilon \delta_i^\varepsilon,$$

where δ_i^ε is defined by (2.6). But

$$-\Delta q^\varepsilon \to 0 \quad \text{strongly in } W^{-1,\infty}(\mathbb{R}^N)$$

and

$$\chi_B^\varepsilon \to \frac{1}{N}\frac{S_N}{2^N} \quad \text{weakly-} \star \text{ in } L^\infty(\mathbb{R}^N) \text{ and hence strongly in } W^{-1,\infty}_{\text{loc}}(\mathbb{R}^N).$$

The use of these convergences in the preceding expression of $-\Delta q^\varepsilon$ gives the claimed result and completes the proof of the lemma. ∎

Remark 2.4. In this model example of small spherical holes periodically distributed in volume in $\mathbb{R}^N$, we constructed w^ε and μ satisfying hypotheses that are stronger than (H.1) to (H.5). Indeed, w^ε and μ also satisfy

(H.0) $0 \leq w^\varepsilon \leq 1$ a.e. in Ω,

(H.1)″ $w^\varepsilon \in W^{1,\infty}(\Omega)$,

(H.5)″ $\begin{cases} \text{for every sequence } v^\varepsilon \text{ such that } v^\varepsilon = 0 \text{ on } T_i^\varepsilon, \ 1 \leq i \leq n(\varepsilon), \\ \text{satisfying } v^\varepsilon \to v \text{ weakly in } H^1(\Omega) \text{ with } v \in H^1(\Omega), \text{ one has} \\ < -\Delta w^\varepsilon, \varphi v^\varepsilon >_{W^{-1,\infty}(\Omega), W_0^{1,1}(\Omega)} \to < \mu, \varphi v >_{W^{-1,\infty}(\Omega), W_0^{1,1}(\Omega)}, \\ \text{for all } \varphi \in H_0^1(\Omega). \end{cases}$

Let us make some comments on these additional hypotheses and on the way one can check that they are satisfied.

Hypothesis (H.0) is a consequence of the maximum principle (and, of course, of the explicit construction of w^ε). We make use of this hypothesis in Section 4 in the study of variational inequalities with highly oscillating obstacles.

Let us point out that hypothesis (H.1)$''$ is in fact an "individual" regularity assumption on w^ε. It can even be proven that if w^ε is bounded in $W^{1,p}(\Omega)$ with $p > 2$, one would necessarily have $\mu = 0$. Hypothesis (H.1)$''$ enables us to write down expressions that would not make sense with the sole assumption $w^\varepsilon \in H^1(\Omega)$. For instance, we can consider $< \Delta w^\varepsilon, \varphi w^\varepsilon >$ for $\varphi \in H_0^1(\Omega)$ and $v^\varepsilon \in H^1(\Omega)$. Hypothesis (H.1)$''$ needs some smoothness of the boundaries of the holes ∂T_i^ε to be satisfied.

The difference between (H.5)$''$ and (H.5) lies in the fact that we have taken here $\varphi \in H_0^1(\Omega)$ and not in $\mathcal{D}(\Omega)$. We use (H.5)$''$ in order to weaken the regularity assumptions on the function ψ when studying variational inequalities with obstacles oscillating between 0 and ψ. In the examples, the proof of this hypothesis essentially depends on the choice of the sets ∂B_i^ε on which the measures μ^ε are concentrated. In our model example, (H.5)$''$ follows directly from Lemma 2.3.

We also emphasize that the fact that hypotheses (H.0), (H.1)$''$ and (H.5)$''$ are satisfied is not a particular case restricted to the preceding model example only. They are general hypotheses that are satisfied by all the following examples.   ∎

Remark 2.5. Theorem 2.2 states the following: If the holes of Ω^ε are balls of radius a^ε as defined by (2.4) with centers located at the nodes of a lattice in $\mathbb{R}^N$ with cell size 2ε, then an additional term, namely the constant μ defined by (2.3), appears in the limit equation of the Dirichlet problems formulated over Ω^ε.

It may be asked what happens when the holes are "smaller" or "bigger" than the "critical" value of a^ε defined by (2.4).

Let us consider first the case where the holes are balls whose centers are the nodes of a lattice in $\mathbb{R}^N$ with cell size 2ε, the radius $\widehat{a}^\varepsilon$ of which tends faster to zero than the a^ε defined by (2.4). The explicit computation done in the proof of Theorem 2.2 shows that if w^ε is still defined by (2.1), one has

$$\|\text{grad } w^\varepsilon\|^2_{(L^2(\Omega))^N} \to 0$$

and not, as before, to a strictly positive constant. This implies that w^ε converges *strongly* to 1 in $H^1(\Omega)$ and, as we saw in Remark 1.5, hypotheses (H.1) to (H.5) are then satisfied with $\mu = 0$. In short, there is no more "strange term" in the limit equation if the holes are "smaller" than those defined by (2.4).

D. Cioranescu and F. Murat

In return, if we consider "bigger" holes, that is, holes whose radius $\widehat{a}^\varepsilon$ tends to zero slowly with respect to a^ε given by (2.4), one can prove that the limit u of the solutions $\widetilde{u^\varepsilon}$ of the Dirichlet problems in Ω^ε is always zero. The proof consists of bounding from below the first eigenvalue of the Dirichlet problem in Ω^ε by an explicit quantity that tends to infinity if $\widehat{a}^\varepsilon$ tends to zero slowly with respect to a^ε (see, for instance, Rauch and Taylor[18], p. 44).

Let us treat some other examples.

Example 2.6. Nonspherical holes periodically distributed in volume in $\mathbb{R}^N$

Let us introduce a version of the model example by now considering holes that are not balls. More precisely, suppose that $\mathbb{R}^N$ is covered by cubes P_i^ε of size 2ε. Assume that the hole $\overline{T}_i^\varepsilon$ is a closed set included in the ball of radius a^ε given by (2.4) and centered at the center of the cube P_i^ε. The domain Ω^ε is obtained by removing from Ω the holes $\overline{T}_i^\varepsilon, 1 \le i \le n(\varepsilon)$. We have thus perforated $\mathbb{R}^N$ by holes that are not balls as in the previous example (see Figure 3).

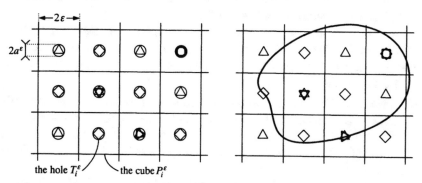

Figure 3. Non-spherical holes distributed in volume in $\mathbb{R}^N$.

It should be mentioned that for a fixed ε, the holes $\overline{T}_i^\varepsilon$ cannot be deduced from one another by translations, since they can have different forms, but they still have a size smaller than or equal to a^ε. Notice also that we did not make any regularity assumption on the boundaries of the holes.

In this case $\overline{w}^\varepsilon$ is defined in the following way. In the cube P_i^ε, $\overline{w}^\varepsilon$ satisfies

$$\begin{cases} \overline{w}^\varepsilon = 0 & \text{in } \overline{T}_i^\varepsilon, \\ \Delta \overline{w}^\varepsilon = 0 & \text{in } B_i^\varepsilon - \overline{T}_i^\varepsilon, \\ \overline{w}^\varepsilon = 1 & \text{in } P_i^\varepsilon - B_i^\varepsilon, \\ \overline{w}^\varepsilon \in H^1(P_i^\varepsilon), \end{cases} \qquad (2.7)$$

where B_i^ε still denotes the ball of radius ε included in P_i^ε (compare with the definition (2.1) of w^ε in the case of spherical holes).

We have

$$\|\text{grad } \overline{w}^\varepsilon\|^2_{(L^2(B_i^\varepsilon))^N} = \inf \|\text{grad } v\|^2_{(L^2(B_i^\varepsilon))^N} \leq \|\text{grad } w^\varepsilon\|^2_{(L^2(B_i^\varepsilon))^N},$$

where the infimum is taken over the functions $v \in H^1(B_i^\varepsilon)$, with $v = 0$ on $\overline{T}_i^\varepsilon$ and $v = 1$ on ∂B_i^ε. It follows that if a^ε is given by (2.4), $\overline{w}^\varepsilon$ is bounded in $H^1(\Omega)$. Therefore $|\text{grad } \overline{w}^\varepsilon|^2$ is bounded in $L^1(\Omega)$ and, extracting a subsequence still denoted by setting ε, we can define a positive measure $\overline{\mu}$ by

$$< \overline{\mu}, \varphi >= \lim_{\varepsilon \to 0} \int_\Omega |\text{grad } \overline{w}^\varepsilon|^2 \, \varphi \, dx, \quad \forall \varphi \in \mathcal{D}(\Omega). \tag{2.8}$$

We can now formulate the following result:

Theorem 2.7. *The functions $\overline{w}^\varepsilon$ and the measure $\overline{\mu}$ defined, respectively, by (2.7) and (2.8) satisfy hypotheses (H.1) to (H.5).*

Proof. It is obvious that (H.1), (H.2), and (H.3) are satisfied. In order to check (H.4) and (H.5), remark that

$$-\Delta \overline{w}^\varepsilon = \overline{\mu}^\varepsilon - \overline{\gamma}^\varepsilon,$$

where $\overline{\gamma}^\varepsilon$ is a measure supported by the boundaries of the holes $\partial \overline{T}_i^\varepsilon$ and thus does not play any role in $\mathcal{D}(\Omega^\varepsilon)$ (see Remark 1.6). The measure $\overline{\mu}^\varepsilon$ is defined by

$$\overline{\mu}^\varepsilon = \sum_{i=1}^{n(\varepsilon)} \frac{\partial \overline{w}^\varepsilon}{\partial n} \delta_i^\varepsilon,$$

where δ_i^ε is the Dirac mass supported by ∂B_i^ε, see equation (2.6).

By the maximum principle (see Figure 4), one has

$$0 \leq \frac{\partial \overline{w}^\varepsilon}{\partial n} \leq \frac{\partial w^\varepsilon}{\partial n} \quad \text{on } \partial B_i^\varepsilon$$

and (H.4) and (H.5) now follow from the following compactness result. ∎

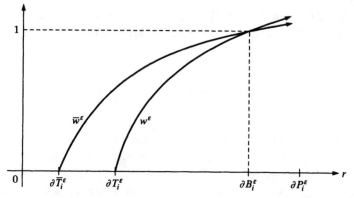

Figure 4. Comparison between w^ε and $\overline{w}^\varepsilon$, $\frac{\partial w}{\partial n}$ and $\frac{\partial \overline{w^\varepsilon}}{\partial n}$.

Lemma 2.8. *Let μ^ε be a sequence of nonnegative measures that strongly converge in $H^{-1}(\Omega)$. Let $\overline{\mu}^\varepsilon$ be a sequence of measures such that*

$$0 \le \overline{\mu}^\varepsilon \le \mu^\varepsilon.$$

Then $\overline{\mu}^\varepsilon$ belong to $H^{-1}(\Omega)$ and are bounded independently of ε in this space. Moreover, the sequence $\overline{\mu}^\varepsilon$ is relatively compact in the strong topology of $H^{-1}(\Omega)$.

Proof. Let $\varphi \in \mathcal{D}(\Omega), \varphi \ge 0$. We have

$$0 \le\; < \overline{\mu}^\varepsilon, \varphi > \;\le\; < \mu^\varepsilon, \varphi > \;\le\; \|\mu^\varepsilon\|_{H^{-1}(\Omega)} \|\varphi\|_{H^1_0(\Omega)}.$$

Writing $v = v^+ - v^-$ for any $v \in H^1_0(\Omega)$ and regularizing, one proves that μ^ε belongs to $H^{-1}(\Omega)$ and is bounded in this space.

Assume now that for a subsequence (still denoted $\overline{\mu}^\varepsilon$) we have

$$\overline{\mu}^\varepsilon \rightharpoonup \overline{\mu} \quad \text{weakly in } H^{-1}(\Omega).$$

Define $\overline{z}^\varepsilon$ by

$$\begin{cases} -\Delta \overline{z}^\varepsilon = \overline{\mu}^\varepsilon & \text{in } \mathcal{D}'(\Omega), \\ \overline{z}^\varepsilon \in H^1_0(\Omega), \end{cases}$$

and z by

$$\begin{cases} -\Delta \overline{z} = \overline{\mu} & \text{in } \mathcal{D}'(\Omega), \\ \overline{z} \in H^1_0(\Omega). \end{cases}$$

We have

$$\int_{\Omega} |\text{grad}(\bar{z}^{\varepsilon} - \bar{z})|^2 dx = <\bar{\mu}^{\varepsilon} - \bar{\mu}, \bar{z}^{\varepsilon} - \bar{z}>$$

$$= <\bar{\mu}^{\varepsilon}, (\bar{z}^{\varepsilon} - \bar{z})^+> - <\bar{\mu}^{\varepsilon}, (\bar{z}^{\varepsilon} - \bar{z})^-> - <\bar{\mu}, \bar{z}^{\varepsilon} - \bar{z}>$$

$$\leq <\mu^{\varepsilon}, (\bar{z}^{\varepsilon} - \bar{z})^+> - <\bar{\mu}, \bar{z}^{\varepsilon} - \bar{z}>,$$

which tends to zero since μ^{ε} converges strongly in $H^{-1}(\Omega)$ while $(\bar{z}^{\varepsilon} - \bar{z})^+$ and $(\bar{z}^{\varepsilon} - \bar{z})$ converges weakly to zero in $H_0^1(\Omega)$. This proves that

$$\bar{\mu}^{\varepsilon} \to \bar{\mu} \quad \text{strongly in } H^{-1}(\Omega),$$

which is the desired result.   ■

Example 2.9. Spherical holes periodically distributed on a hyperplane of $\mathbb{R}^N$

We now perforate $\mathbb{R}^N$ by holes whose centers are periodically distributed on an $(N-1)$ manifold of $\mathbb{R}^N$. For simplicity, let us take this manifold to be the hyperplane $\{x_N = 0\}$ and consider only spherical holes. More precisely, we construct in the hyperplane $\{x_N = 0\} = \mathbb{R}^{N-1}$ a lattice with cell size 2ε. Each node of this lattice is the center of a ball (of $\mathbb{R}^N$) of radius b^{ε}. This defines the holes $\hat{T}_i^{\varepsilon}$. Another way to look at this construction is to fill the strip $|x_N| \leq \varepsilon$ by cubes $\hat{P}_i^{\varepsilon}$ of $\mathbb{R}^N$ with sides parallel to the axes and with size 2ε, and to place at the center of each such cube a small ball $\hat{T}_i^{\varepsilon}$ of radius b^{ε}. We then remove from Ω these holes $\hat{T}_i^{\varepsilon}$ to obtain the domain Ω_{ε} (see Figure 5).

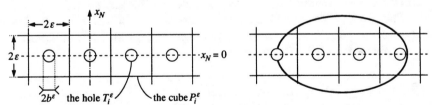

Figure 5. Small balls periodically distributed on a hyperplane of $\mathbb{R}^N$.

Now define the functions $\hat{w}^{\varepsilon}$ by first setting

$$\hat{w}^{\varepsilon} = 0 \quad \text{if } |x_N| \geq \varepsilon.$$

If $|x_N| \leq \varepsilon$, on each cube $\hat{P}_i^{\varepsilon}$ we set

$$\begin{cases} \hat{w}^{\varepsilon} = 0 & \text{in } \hat{T}_i^{\varepsilon}, \\ \Delta\hat{w}^{\varepsilon} = 0 & \text{in } \hat{B}_i^{\varepsilon} - \hat{T}_i^{\varepsilon}, \\ \hat{w}^{\varepsilon} = 1 & \text{in } \hat{P}_i^{\varepsilon} - \hat{B}_i^{\varepsilon}, \\ \hat{w}^{\varepsilon} & \text{continuous at the interfaces,} \end{cases} \tag{2.9}$$

where $\widehat{B}_i^\varepsilon \subset \widehat{P}_i^\varepsilon$ is the ball of radius ε concentric to $\widehat{T}_i^\varepsilon$.

Let $\delta_{\{x_N=0\}}$ be the Dirac mass supported by the hyperplane $\{x_N = 0\}$; that is,

$$< \delta_{\{x_N=0\}}, \varphi > = \int_{x_N=0} \varphi(x', 0) \, dx', \quad \forall \varphi \in \mathcal{D}(\mathbb{R}^N).$$

Let $\hat{\mu}$ be the measure defined by

$$\begin{cases} \hat{\mu} = \dfrac{\pi}{C_0} \delta_{\{x_N=0\}} & \text{if } N = 2, \\[3mm] \hat{\mu} = \dfrac{S_N(N-2)C_0^{N-2}}{2^{N-1}} \delta_{\{x_N=0\}} & \text{if } N \geq 3, \end{cases} \tag{2.10}$$

where C_0 is a constant, $C_0 > 0$.

Theorem 2.10. *If we choose*

$$\begin{cases} b^\varepsilon = \exp\left(-\dfrac{C_0}{\varepsilon}\right) & \text{if } N = 2, \\[3mm] b^\varepsilon = C_0 \, \varepsilon^{N-1/N-2} & \text{if } N \geq 3, \end{cases} \tag{2.11}$$

then $\widehat{w}^\varepsilon$ and $\hat{\mu}$ defined by (2.9) and (2.10) satisfy the hypotheses (H.1) *to* (H.5).

Note that beside hypothesis (H.1) to (H.5), hypothesis (H.0), (H.1)″, and (H.5)″ (see Remark 2.4) are also satisfied in this example.

The proof of Theorem 2.10 is very close to that of Theorem 2.2. The only delicate point is the proof of hypothesis (H.5), which is a straightforward application of the following result:

Lemma 2.11. *Let $N \geq 2$. Assume that the centers of the spheres $\partial\widehat{B}_i^\varepsilon$, $1 \leq i \leq n(\varepsilon)$, of radius ε are contained in the hyperplane $\{x_N = 0\}$ and are located at the nodes of a lattice in $\mathbb{R}^{N-1}$ with cell size 2ε. If $\widehat{\delta}_i^\varepsilon$ are the Dirac masses supported by $\partial\widehat{B}_i^\varepsilon$, then*

$$\sum_{i=1}^{n(\varepsilon)} \widehat{\delta}_i^\varepsilon \to \frac{S_N}{2^{N-1}} \delta_{\{x_N=0\}} \quad \text{strongly in } W_{\text{loc}}^{-1,p}(\mathbb{R}^N), \quad \forall p < \infty.$$

Proof. The proof follows along the lines of the proof of Lemma 2.3. One begins by introducing the auxiliary function $\widehat{q}^\varepsilon \in W^{1,\infty}(\mathbb{R}^N)$ vanishing outside the strip $|x_N| \leq \varepsilon$ and such that

$$\frac{\partial \widehat{q}^\varepsilon}{\partial r} = \frac{1}{\varepsilon} r \quad \text{in each ball } \widehat{B}_i^\varepsilon.$$

Clearly,

$$\widehat{q}^{\varepsilon} \to 0 \quad \text{strongly in } W^{1,\infty}(\mathbb{R}^N).$$

On the other hand, one has

$$-\Delta\widehat{q}^{\varepsilon} = -\frac{N}{\varepsilon}\chi_{\widehat{B}}^{\varepsilon} + \sum_{i=1}^{n(\varepsilon)}\delta_i^{\varepsilon},$$

where $\chi_{\widehat{B}}^{\varepsilon}$ is the characteristic function of the balls $\widehat{B}_i^{\varepsilon}$ of radius ε. It is easily proved that

$$\chi_{\widehat{B}}^{\varepsilon} \to \frac{S_N}{2^{N-1}}\delta_{\{x_N=0\}} \quad \text{weakly-}\star \text{ in the sense of measures.}$$

In fact this convergence takes place in $W_{\text{loc}}^{-1,p}(\mathbb{R}^N)$ strongly. In order to see that, one can, for instance, make use of the analogue in $W^{-1,p}$ of Lemma 2.8 and take into account the fact that

$$0 \le \frac{1}{\varepsilon}\chi_{\widehat{B}}^{\varepsilon} \le \frac{1}{\varepsilon}\chi_S^{\varepsilon}$$

where χ_S^{ε} is the characteristic function of the strip $|x_N| < \varepsilon$. It is then sufficient to note that

$$\frac{1}{2\varepsilon}\chi_S^{\varepsilon} = \frac{dg^{\varepsilon}}{dx_n},$$

where g^{ε} converges strongly in $L_{\text{loc}}^p(\mathbb{R}^N)$ to the Heaviside function. ∎

Many variants can be derived from the previous examples; we give some of them. To simplify their description (and especially the figures), we assume in Examples 2.12 and 2.13 below that $N = 3$.

Example 2.12. Cylinders distributed like a forest or like a fence

Let us consider the case where the holes are cylinders of infinite lengh with axes parallel to the x_3-axis and intersecting the hyperplane $\{x_3 = 0\}$ on spheres of $\mathbb{R}^2$ of radius $a^{\varepsilon} = \exp\left(-\frac{C_3}{\varepsilon^2}\right)$, which are periodically distributed in volume in $\mathbb{R}^2$. If these holes are seen as trunks of trees (of infinite length), we have a *forest of cylinders* (see Figure 6).

We use here functions w^{ε} which are independent of x_3, that is,

$$w_3^{\varepsilon}(x_1, x_2, x_3) = w^{\varepsilon}(x_1, x_2), \tag{2.12}$$

where $w^\varepsilon(x_1, x_2)$, defined over $\mathbb{R}^2$ by (2.1), is the function constructed for the holes of $\mathbb{R}^2$ distributed in volume in $\mathbb{R}^2$. The measure that appears in the limit problem is (see (2.3))

$$\mu_3 = \frac{\pi}{2}\frac{1}{C_3},\qquad (2.13)$$

and, clearly, μ_3 is independent of x_3 as well as x_1 and x_2. The fact that hypotheses (H.1) to (H.5) are fulfilled is obvious here; it was already verified in Theorem 2.2.

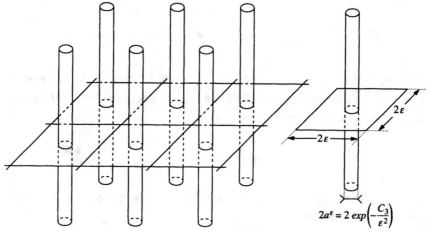

Figure 6. The Forest of cylinders.

One can also consider the case of cylindrical holes distributed like a fence. The cylinders, with axes parallel to the x_3-axis, lie along the plane $\{x_1 = 0\}$ (see Figure 7). Their radius b^ε is now equal to the radius of the holes distributed on a hyperplane of $\mathbb{R}^2$, which is here the line $x_2 = 0$ of $\mathbb{R}^2$. Hence $b^\varepsilon = \exp\left(-\dfrac{(C_3)}{\varepsilon}\right)$ and the test functions to be used are

$$\widehat{w}_{32}^\varepsilon(x_1, x_2, x_3) = \widehat{w}^\varepsilon(x_1, x_2),$$

with $\widehat{w}^\varepsilon$ defined in $\mathbb{R}^2$ by (2.9). The measure that appears is supported by the hyperplane $\{x_2 = 0\}$ (see (2.10)) and is equal to

$$\widehat{\mu}_{32} = \frac{\pi}{C_3}\delta_{\{x_2=0\}}.$$

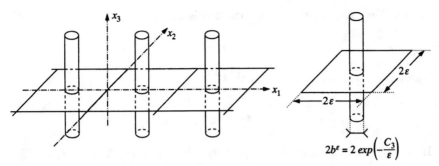

Figure 7. Cylinders as bars.

Example 2.13. *Trusses and grids*

From the example of cylinders distributed like a forest, one derives a three-dimensional truss. This is a connected hole in $\mathbb{R}^3$ formed by the union of cylinders of radius $\exp(-\dfrac{(C}{\varepsilon^2})$, and lying along all the edges of a lattice of $\mathbb{R}^3$ of cell size 2ε (see Figure 8).

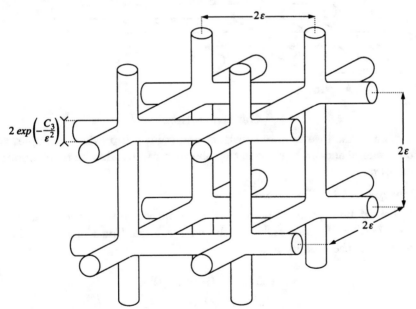

Figure 8. A truss.

In this example one uses functions w^ε which are products of the following form:

$$w^\varepsilon(x_1, x_2, x_3) = w_1^\varepsilon(x_2, x_3)w_2^\varepsilon(x_1, x_3)w_3^\varepsilon(x_1, x_2),$$

where the functions w_j^ε depend on two variables only and are defined as in

(2.12). The measure which appears in this case is

$$\mu = \mu_1 + \mu_2 + \mu_3,$$

where μ_j is defined as in (2.13), that is,

$$\mu = 3\frac{\pi}{2}\frac{1}{C}.$$

It is quite easy to verify that hypotheses (H.1) to (H.4) hold true, but the proof of (H.5) is more technical.

In a similar way one can construct, by crossing cylinders distributed like a fence, a hole in $\mathbb{R}^3$ with a grid form. Its ribs are cylinders of radius $\exp\left(-\dfrac{C}{\varepsilon}\right)$, at a distance 2ε from one another (see Figure 9).

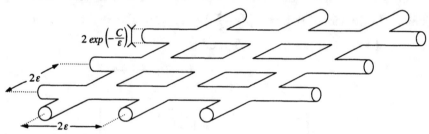

Figure 9. A grid.

Generalizations

The method we used to construct and study the previous examples is quite general and, as we show in the following, can be adapted to all elliptic operators.

Example 2.14. Bilaplacian

Let Ω^ε be a bounded domain of $\mathbb{R}^N$ from which we remove spherical holes of radius a^ε periodically distributed in volume. Consider the Dirichlet problem for the biLaplacian

$$\begin{cases} -\Delta^2 u^\varepsilon = f & \text{in } \mathcal{D}'(\Omega^\varepsilon), \\ u^\varepsilon \in H_0^2(\Omega^\varepsilon). \end{cases}$$

Introduce the functions w^ε defined on each cube P_i^ε of size 2ε as follows:

$$\begin{cases} w^\varepsilon = 0, \quad \dfrac{\partial w^\varepsilon}{\partial n} = 0 & \text{in the sphere of radius } c^\varepsilon, \\[2mm] w^\varepsilon = 1, \quad \dfrac{\partial w^\varepsilon}{\partial n} = 0 & \text{in the sphere of radius } \varepsilon, \\[2mm] \Delta^2 w^\varepsilon = 0 & \text{in the annulus } B^\varepsilon - B^{c^\varepsilon}. \end{cases}$$

Following along the lines of the proof of Theorem 2.2 (with more complicated computations), we prove that choosing

$$\begin{cases} c^\varepsilon = \exp\left(-\dfrac{C_0}{\varepsilon^4}\right) & \text{for } N = 4, \\ c^\varepsilon = C_0 \varepsilon^{N/(N-4)} & \text{for } N \geq 5, \end{cases}$$

the extensions $\widetilde{u^\varepsilon}$ of u^ε by zero in the holes converge weakly in $H_0^2(\Omega)$ to the solution u of

$$\begin{cases} -\Delta^2 u + \mu u = f & \text{in } \mathcal{D}'(\Omega), \\ u \in H_0^2(\Omega), \end{cases}$$

where μ is a positive constant.

It should be emphasized that the size of the holes has to be adapted to the order of the operator; this is connected to the capacity of the functional space in which we are working. As $H^2(\mathbb{R}^N) \subset C^{0,\lambda}(\mathbb{R}^N)$ for $N \leq 3$ with $\lambda > 0$, it is possible to construct examples with "strange" terms for the bilaplacian only for $N \geq 4$ (see also Remark 1.7).

Example 2.15. The operator $-\mathrm{div}(A\,\mathrm{grad})$ with constant coefficients

Consider the case of holes periodically distributed in volume and introduce the operator $-\mathrm{div}(A\,\mathrm{grad})$ where A is a coercive matrix with constant coefficients. It is then possible to replace A by its symmetric part $^sA = \dfrac{(A + {}^tA)}{2}$ because

$$-\mathrm{div}(A\,\mathrm{grad}) = -\sum_{i,j=1}^{N} a_{ij} \frac{\partial^2}{\partial x_i \partial x_j} = -\mathrm{div}(^sA\,\mathrm{grad}).$$

Let B denote the inverse matrix of ^sA and set

$$\sigma^2(x) = B\,x \cdot x.$$

Then $\sigma(x)$ is an elliptic distance equivalent to $r = |x|$ since for some $0 < \alpha \leq \beta < \infty$, one has

$$\frac{1}{\beta} r^2 \leq \sigma^2 \leq \frac{1}{\alpha} r^2.$$

The Green function of the operator $-\mathrm{div}(A\,\mathrm{grad})$ is

$$\begin{cases} \dfrac{1}{\sigma^{N-2}} & \text{if } N \geq 3, \\ \log \sigma & \text{if } N = 2, \end{cases}$$

which corresponds to the potentials $\dfrac{1}{r^{N-2}}$ and $\log r$ for the Laplacian.

Let us consider holes T_i^ε periodically distributed in volume and defined by (see Figure 10)

$$\begin{cases} T_i^\varepsilon = \left\{ x \in \mathbb{R}^2 \;:\; \sigma(x) \leq \exp\left(-\dfrac{C_0}{\varepsilon^2}\right) \right\} & \text{if } N = 2, \\[2mm] T_i^\varepsilon = \left\{ x \in \mathbb{R}^N \;:\; \sigma(x) \leq C_0 \varepsilon^{N/(N-2)} \right\} & \text{if } N \geq 3. \end{cases}$$

These holes are small ellipsoids identical in size to the balls considered in the case of the laplacian (see equation (2.4)). Define the functions w^ε as in Figure 10. The analogue of Theorem 2.2 is then obvious. For its proof it is sufficient to replace the balls by the ellipsoids, r by σ, and to perform the same computations as in the proof of Theorem 2.2, with - div(A grad) in place of $-\Delta$.

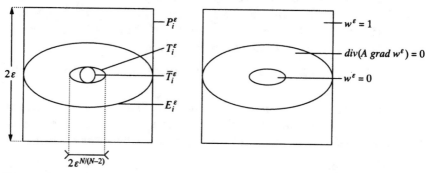

Figure 10. The case of operator $-\mathrm{div}(A$ grad$)$ with constant coefficients.

If the holes, instead of being small ellipsoids related to the operator, are the small balls $\overline{T}_i^\varepsilon$ of radius a^ε defined in (2.4), then, as in Example 2.6, one has to inscribe these balls into the ellipsoids and to compare the new test function $\overline{w}^\varepsilon$ with the test function w^ε defined in Figure 10.

Example 2.16. The operator $-\mathrm{div}(A$ grad$)$ with continuous coefficients

Suppose that we are in the case of small spherical holes of radius a^ε given by (2.4), distributed periodically in volume in $\mathbb{R}^N$, and consider the operator $-\mathrm{div}(A$ grad$)$ where the matrix A is coercive and has *continuous* coefficients in $\overline{\Omega}$.

We define, for every ε, a matrix A^ε by taking the averages of A on the cubes P_i^ε. By definition, the matrix A^ε is a step function with values in $\mathbb{R}^{N^2}$ and constant on each cube. One has

$$A^\varepsilon - A \to 0 \quad \text{strongly in } (L^\infty(\Omega))^{N^2}. \tag{2.14}$$

We can now build up, on each cube P_i^ε, functions w^ε related to the matrix ${}^sA^\varepsilon$ (whose coefficients are constant on this cube) in the same way as we did in the preceding example. The only difference is that the ellipsoid E_i^ε, which replaces the ball B_i^ε of radius ε, changes its principal axes and lengths of semi-axes on each cube. We still obtain

$$-\mathrm{div}({}^sA\ \mathrm{grad}\ w^\varepsilon) = \mu^\varepsilon - \gamma^\varepsilon, \qquad (2.15)$$

where μ^ε is a positive measure supported by the boundaries of the ellipsoids E_i^ε with

$$\mu^\varepsilon \to \mu \quad \text{strongly in } H^{-1}(\Omega),$$

whereas γ^ε is a measure supported by the holes T_i^ε and therefore is "invisible" in the Dirichlet problem on Ω^ε. Since w^ε is equal to 1 outside the ellipsoids E_i^ε and A^ε is constant on each cube P_i^ε, we have

$$\begin{aligned}
-\mathrm{div}\ ({}^sA^\varepsilon\ \mathrm{grad}\ w^\varepsilon) &= -\mathrm{div}(A^\varepsilon\ \mathrm{grad}\ w^\varepsilon) \\
&= -\mathrm{div}({}^tA^\varepsilon\ \mathrm{grad}\ w^\varepsilon) \quad \text{in } \mathcal{D}'(\Omega).
\end{aligned} \qquad (2.16)$$

Notice finally that

$$-\mathrm{div}({}^tA\ \mathrm{grad}\ w^\varepsilon) = -\mathrm{div}({}^tA^\varepsilon\ \mathrm{grad}\ w^\varepsilon) - \mathrm{div}(({}^tA - {}^tA^\varepsilon)\ \mathrm{grad}\ w^\varepsilon). \quad (2.17)$$

Then, using (2.14), (2.15) and (2.16) and taking into account that w^ε is bounded in $H^1(\Omega)$, we deduce that hypothesis (H.5)$_A$ is satisfied (see Generalizations at the end of Section 1). The theory of Section 1 can thus be applied to operators with continuous coefficients in the case of spherical holes of radius a^ε given by (2.4) which are periodically distributed in volume in $\mathbb{R}^N$.

Let us remark that the "size" of the holes, always the same, is the size related to the Laplacian, irrespective of the (continuous) coefficients of the operator.

Remark also that in the case of continuous coefficients, the following hypothesis (H.6) is satisfied in addition to hypotheses (H.0), (H.1) to (H.4) and (H.5)$_A$:

$$\left\{ \begin{aligned}
&\text{for every sequence } v^\varepsilon \text{ such that } v^\varepsilon = 0 \text{ on } T_i^\varepsilon, 1 \le i \le n(\varepsilon), \\
&\text{satisfying } v^\varepsilon \rightharpoonup v \text{ weakly in } H^1(\Omega) \text{ with } v \in H^1(\Omega), \text{ one has} \\
&< -\mathrm{div}((A - {}^tA)\ \mathrm{grad}\ w^\varepsilon), \varphi v^\varepsilon >_{H^{-1}(\Omega), H_0^1(\Omega)} \to 0, \\
&\text{for all } \varphi \in \mathcal{D}(\Omega).
\end{aligned} \right. \qquad \text{(H.6)}$$

This hypothesis, which means that the antisymmetric part of A does not play any role in the Dirichlet problem for the domain Ω^ε, is very useful in

Sections 3 and 4 when studying the correctors and variational inequalities with oscillating obstacles.

Remark 2.17. Other functions w^ε can be found in Hrouslov [12] and Marcenko and Hrouslov [15]. Our test functions are quite close to theirs but, when performing explicit computations, they take advantage of the fact that their laplacians are concentrated measures instead of being distributed functions.

3. Weak lower semicontinuity of the energy. Correctors

Let us turn back to the "abstract" framework of Section 1. We suppose that there exist w^ε and μ satisfying hypotheses (H.1) to (H.5).

Proposition 3.1. *Assume that hypotheses* (H.1) *to* (H.5) *are satisfied. For any sequence z^ε and z such that*

$$\begin{cases} z^\varepsilon \rightharpoonup z & \text{weakly in } H_0^1(\Omega), \\ z^\varepsilon = 0 & \text{on } T_i^\varepsilon, \ 1 \le i \le n(\varepsilon), \end{cases} \tag{3.1}$$

one has

$$\liminf_{\varepsilon \to 0} \int_\Omega |\text{grad } z^\varepsilon|^2 \, dx \ge \int_\Omega |\text{grad } z|^2 \, dx + < \mu, z^2 > . \tag{3.2}$$

The classical weak lower semicontinuity of the energy $\int_\Omega |\text{grad } z|^2 \, dx$ implies that for any sequence v^ε that converges weakly in $H_0^1(\Omega)$ to v, one has

$$\liminf_{\varepsilon \to 0} \int_\Omega |\text{grad } v^\varepsilon|^2 \, dx \ge \int_\Omega |\text{grad } v|^2 \, dx.$$

The fact that the elements z^ε of the sequence under consideration vanish on the holes T_i^ε improves the last information and implies that at the limit a new energy appears, namely,

$$\int_\Omega |\text{grad } z|^2 \, dx + < \mu, z^2 > .$$

Here (and, of course, in (3.2)), $< \mu, z^2 >$ is the duality pairing between μ which belongs to $W^{-1,\infty}(\Omega)$ by hypothesis (H.4), and z^2 which belongs to $W^{1,1}(\Omega)$ since $z \in H_0^1(\Omega)$.

Proposition 3.2. *Assume that hypotheses of Proposition 3.1 are satisfied. If, furthermore, z^ε satisfies*

$$\int_\Omega |\text{grad } z^\varepsilon|^2 \, dx \to \int_\Omega |\text{grad } z|^2 \, dx + < \mu, z^2 >, \tag{3.3}$$

then

$$z^\varepsilon - w^\varepsilon z \to 0 \quad \text{strongly in } W_0^{1,1}(\Omega). \tag{3.4}$$

Remark 3.3. The strong convergence in (3.4) takes place in $W_0^{1,1}(\Omega)$, but we would like to have it in $H_0^1(\Omega)$, which is the natural space for the problem.

As a matter of fact, we see at the end of the proof of Proposition 3.2 that one can improve (3.4) by using a Sobolev embedding theorem. This will guarantee a strong convergence in $W_0^{1,q}(\Omega)$ where $q = \frac{N}{(N-1)}$ if $N \geq 3$ and $q < 2$ if $N = 2$. This result can be improved again if one assumes that the limit z is more regular than $H_0^1(\Omega)$, say $z \in W_0^{1,p}(\Omega)$ with $p > 2$. In this context, the most pleasant situation is when $z \in C_0^1(\Omega)$ because in this case we have the convergence

$$z^\varepsilon - w^\varepsilon z \to 0 \quad \text{strongly in } H_0^1(\Omega).$$

Let us suppose for the moment that Propositions 3.1 and 3.2 hold true. We have the following corollary:

Theorem 3.4. *Let u^ε be the solution of the Dirichlet problem* (1.1). *Then, under hypotheses* (H.1) *to* (H.5), *there exists r^ε such that*

$$\begin{cases} \widetilde{u^\varepsilon} = w^\varepsilon \, u + r^\varepsilon, \\ r^\varepsilon \to 0 \quad \text{strongly in } W_0^{1,1}(\Omega), \end{cases} \tag{3.5}$$

where u is the solution of (1.5).

This result asserts that the solution u^ε of the Dirichlet problem in Ω^ε can be written (up to a remainder r^ε which converges strongly to zero) as the product of u, the solution of the limit problem (1.5), by the function w^ε. If we adopt the boundary layer language, we can say that $(1 - w^\varepsilon)u$ is a corrector since

$$\begin{cases} \widetilde{u^\varepsilon} - u \to 0 & \text{weakly in } H_0^1(\Omega), \\ \widetilde{u^\varepsilon} - u - (1 - w^\varepsilon)u \to 0 & \text{strongly in } W_0^{1,1}(\Omega). \end{cases}$$

In the case where the holes are periodically distributed as in Examples 2.1 or 2.9, $u + (1 - w^\varepsilon)u$ is the beginning of the asymptotic expansion of u^ε obtained by the multiple scale method (see Sanchez-Palencia [21]).

As in Remark 3.3, one can notice that in fact r^ε actually converges strongly to zero in a space $W_0^{1,q}(\Omega)$, $q > 1$, which is "better" than $W_0^{1,1}(\Omega)$. This result is improved again if u is more regular than $H_0^1(\Omega)$. For example, if $u \in C_0^1(\Omega)$, then

$$r^\varepsilon \to 0 \quad \text{strongly in } H_0^1(\Omega).$$

Proof of Theorem 3.4. From Theorem 1.2 one has

$$\widetilde{u^\varepsilon} \to u \quad \text{weakly in } H_0^1(\Omega),$$

where u is the solution of (1.5). Multiplying (1.1) by u^ε and (1.5) by u and integrating by parts, one obtains

$$\int_\Omega |\text{grad } \widetilde{u^\varepsilon}|^2 \, dx = \int_\Omega f\widetilde{u^\varepsilon} \, dx \to \int_\Omega fu \, dx = \int_\Omega |\text{grad } u|^2 \, dx + < \mu, u^2 > .$$

The convergence of the energy in Ω^ε to the new energy in Ω is thus established and Theorem 3.4 follows from Proposition 3.2. ∎

Proof of Proposition 3.1. Let $\varphi \in \mathcal{D}(\Omega)$. Consider the integral:

$$\int_\Omega |\text{grad } (z^\varepsilon - w^\varepsilon \varphi)|^2 \, dx = \int_\Omega |\text{grad } z^\varepsilon - w^\varepsilon \text{grad } \varphi - \varphi \text{ grad } w^\varepsilon|^2 \, dx,$$

and let us expand it. Taking into account that

$$< -\Delta w^\varepsilon, z^\varepsilon \varphi > = \int_\Omega \varphi \text{ grad } z^\varepsilon \text{ grad } w^\varepsilon \, dx + \int_\Omega z^\varepsilon \text{grad } \varphi \text{ grad } w^\varepsilon \, dx,$$

one has

$$\begin{cases} \int_\Omega |\text{grad } (z^\varepsilon - w^\varepsilon \varphi)|^2 \, dx = \int_\Omega |\text{grad } z^\varepsilon|^2 \, dx \\ \quad + \int_\Omega |w^\varepsilon|^2 |\text{grad } \varphi|^2 \, dx + \int_\Omega |\varphi|^2 |\text{grad } w^\varepsilon|^2 \, dx \\ \quad - 2 \int_\Omega w^\varepsilon \text{grad } z^\varepsilon \text{ grad } \varphi \, dx + 2 \int_\Omega w^\varepsilon \varphi \text{ grad } w^\varepsilon \text{ grad } \varphi \, dx \\ \quad + 2 \int_\Omega z^\varepsilon \text{grad } \varphi \text{ grad } w^\varepsilon \, dx - 2 < -\Delta w^\varepsilon, z^\varepsilon \varphi > . \end{cases} \quad (3.6)$$

Let us extract a subsequence ε' such that $\int_\Omega |\text{grad } z^{\varepsilon'}|^2 \, dx$ converges. We can pass to the limit in each term of the right-hand side of (3.6). Using (3.1), the Rellich-Kondrashov theorem, Proposition 1.1, and (H.5) we obtain

$$\begin{cases} \lim_{\varepsilon' \to 0} \int_\Omega |\text{grad } (z^{\varepsilon'} - w^{\varepsilon'} \varphi)|^2 \\ \quad = \lim_{\varepsilon' \to 0} \int_\Omega |\text{grad } z^{\varepsilon'}|^2 \, dx + \int_\Omega |\text{grad } \varphi|^2 \, dx + < \mu, \varphi^2 > \quad (3.7) \\ \quad - 2 \int_\Omega \text{grad } z \text{ grad } \varphi \, dx - 2 < \mu, \varphi z > . \end{cases}$$

Let us now choose the subsequence ε' such that

$$\lim_{\varepsilon' \to 0} \int_\Omega |\text{grad } z^{\varepsilon'}|^2 \, dx = \lim_{\varepsilon \to 0} \inf \int_\Omega |\text{grad } z^\varepsilon|^2 \, dx.$$

Since the left-hand side of (3.7) is nonnegative, we have

$$\lim_{\varepsilon \to 0} \inf \int_\Omega |\text{grad } z^{\varepsilon'}|^2 \, dx \geq 2 \int_\Omega \text{grad } z \, \text{grad } \varphi \, dx - \int_\Omega |\text{grad } \varphi|^2 \, dx$$
$$+ 2 < \mu, \varphi z > - < \mu, \varphi^2 > .$$

This holds true for all $\varphi \in \mathcal{D}(\Omega)$. Now choose φ such that φ converges strongly to z in $H_0^1(\Omega)$. The above inequality then yields (3.2) (note that there is no problem in the last two terms since $\mu \in W^{-1,\infty}(\Omega)$ in view of (H.4)). ∎

Remark 3.5. The ideal thing in this proof would be to take $\varphi = z$ from the very beginning. This is obviously possible if $z \in \mathcal{D}(\Omega)$ and this would of course simplify many things. If $z \in H_0^1(\Omega)$ only, then $w^\varepsilon z$ does not belong to $H^1(\Omega)$ under the sole hypothesis (H.1). Accordingly, (3.6) does not make sense with $\varphi = z$. This is the reason we had to approximate z by smooth functions φ.

One could be tempted to make hypothesis (H.1)', that is,

(H.1)' $\qquad\qquad w^\varepsilon \in W^{1,\infty}(\Omega),$

which allows one to write down (3.6) with $\varphi = z \in H_0^1(\Omega)$. As we already said in Remark 2.4, this individual regularity assumption on w^ε is satisfied in many examples. But even if other hypotheses are added to it such as, (H.0) or (H.5)″ (see again Remark 2.4), they would not be sufficient to guarantee the passage to the limit in (3.6) under the sole hypothesis that $\varphi = z \in H_0^1(\Omega)$. We are always obliged to assume more regularity on z.

Proof of Proposition 3.2. Let us go back to (3.7). Taking into account assumption (3.3), we have established that for all $\varphi \in \mathcal{D}(\Omega)$,

$$\lim_{\varepsilon \to 0} \int_\Omega |\text{grad } (z^\varepsilon - w^\varepsilon \varphi)|^2 = \int_\Omega |\text{grad } (z - \varphi)|^2 \, dx + < \mu, (z - \varphi)^2 > .$$

If $z \in \mathcal{D}(\Omega)$, one may take $\varphi = z$ and so we have proved that

$$z^\varepsilon - w^\varepsilon z \to 0 \quad \text{strongly in } H_0^1(\Omega).$$

If z is not smooth, let us fix φ such that

$$\|z - \varphi\|_{H_0^1(\Omega)} \leq \delta,$$

for some $\delta > 0$. Then, it follows that

$$\lim_{\varepsilon \to 0} \int_\Omega |\text{grad } (z^\varepsilon - w^\varepsilon \varphi)|^2 \leq (1 + 2\|\mu\|_{W^{-1,\infty}(\Omega)})\delta^2,$$

and there exists ε_0 such that, for any $\varepsilon < \varepsilon_0$, one has

$$\|z^\varepsilon - w^\varepsilon \varphi\|^2_{H^1_0(\Omega)} \leq C_1 \delta^2.$$

On the other hand,

$$\|z^\varepsilon - w^\varepsilon z\|_{W^{1,1}_0(\Omega)} \leq \|z^\varepsilon - w^\varepsilon \varphi\|_{W^{1,1}_0(\Omega)} + \|w^\varepsilon (z - \varphi)\|_{W^{1,1}_0(\Omega)}. \qquad (3.8)$$

Thus one has

$$\|z^\varepsilon - w^\varepsilon z\|_{W^{1,1}_0(\Omega)} \leq C'_1 \delta + C_2 \delta,$$

whenever $\varepsilon < \varepsilon_0$, which proves (3.4).

We used in (3.8) a rough estimate of $w^\varepsilon (z - \varphi)$ in $W^{1,1}_0(\Omega)$. Recalling the Sobolev embedding $H^1_0(\Omega) \subset L^{2\star}(\Omega)$ where $2\star = 2N/(N-2)$, and setting $\dfrac{1}{q} = \dfrac{1}{2} + \dfrac{1}{2\star}$, one can write

$$\|w^\varepsilon (z - \varphi)\|_{W^{1,q}_0(\Omega)} = \|\text{ grad } (w^\varepsilon (z - \varphi))\|_{L^q(\Omega)}$$
$$\leq \|\text{ grad } w^\varepsilon\|_{L^2(\Omega)}\|z - \varphi\|_{L^{2\star}(\Omega)} + \|\text{ grad } (z - \varphi)\|_{L^2(\Omega)}\|w^\varepsilon\|_{L^{2\star}(\Omega)}$$

which leads to

$$z^\varepsilon - w^\varepsilon z \to 0 \quad \text{strongly in } W^{1,q}_0(\Omega).$$

This improves (3.4). Suppose further that z belongs to a space of functions which are more regular than $H^1_0(\Omega)$. By estimating $z - \varphi$ in this space, one can improve again the space of convergence to finally get the result stated in Remark 3.3. ∎

Generalizations

The preceding results can easily be transposed to the case where, in place of the Laplacian, we deal with the operator $-\text{div}(A \text{ grad})$ where A is a coercive matrix. The method consists of expanding the expression

$$\int_\Omega A \text{ grad } (z^\varepsilon - w^\varepsilon \varphi) \text{ grad } (z^\varepsilon - w^\varepsilon \varphi) \, dx.$$

If A is a symmetric matrix, hypotheses (H.1) to (H.4) and (H.5)$_A$ (see Generalizations, end of Section 1) allow one to pass to the limit. If A is not symmetric, it is sufficient to add to these hypotheses the following one,

$$\begin{cases} \text{for every sequence } v^\varepsilon \text{ such that } v^\varepsilon = 0 \quad \text{on } T_i^\varepsilon, \ 1 \le i \le n(\varepsilon), \\ \text{satisfying } v^\varepsilon \rightharpoonup v \text{ weakly in } H^1(\Omega) \text{ with } v \in H^1(\Omega), \text{ one has} \\ < -\text{div}((A - {}^tA) \text{ grad } w^\varepsilon), \varphi v^\varepsilon >_{H^{-1}(\Omega), H_0^1(\Omega)} \to 0, \\ \text{for all } \varphi \in \mathcal{D}(\Omega). \end{cases} \quad \text{(H.6)}$$

This hypothesis means that the skew-symmetric part of A does not play any role when passing to the limit. Remark that (H.6) is satisfied, for instance, when A has continuous coefficients as seen in Example 2.16.

To conclude this section, note that we did not make use here of any property specific to the order 2 (suh as truncation or maximum principle). Consequently, all the results can easily be extended to operators of higher order.

4. Variational inequalities with highly oscillating obstacles

In this section we again use the abstract framework defined in Section 1. We suppose that there exist w^ε and μ satisfying hypotheses (H.1) to (H.5). Moreover, we assume that

(H.0) $\qquad\qquad 0 \le w^\varepsilon \le 1 \quad$ a.e. in Ω.

As shown in Remark 2.4, this hypothesis is satisfied in all the examples discussed in Section 2.

Let ψ be a given measurable function defined in Ω. Introduce the following unilateral convex sets,

$$\begin{cases} K(\psi) = \{v \in H_0^1(\Omega) \ : \ v \ge \psi \quad \text{a.e. in } \Omega\}, \\ K^\varepsilon(\psi) = K(\psi^\varepsilon) = \{v \in H_0^1(\Omega) \ : \ v \ge \psi^\varepsilon \quad \text{a.e. in } \Omega\}, \end{cases}$$

where ψ^ε is defined by (see Figure 11)

$$\psi^\varepsilon = \begin{cases} \psi & \text{in } \Omega^\varepsilon, \\ 0 & \text{in the holes } T_i^\varepsilon, \ 1 \le i \le n(\varepsilon). \end{cases}$$

These functions define a sequence of obstacles that oscillate rapidly between ψ and 0. One could also take into consideration the case where the obstacles oscillate between ψ and θ, but if $\theta \in H_0^1(\Omega)$, an elementary translation allows one to take $\theta = 0$. This is the situation we study here.

If we suppose that $K(\psi)$ is non empty, i.e. if we assume that

$$\exists V \in H_0^1(\Omega), \quad V \geq \psi \quad \text{a.e. in } \Omega, \tag{4.1}$$

the convex sets $K^\varepsilon(\psi)$ are non empty too, as one has

$$V^+ \in K^\varepsilon(\psi), \quad \forall \varepsilon. \tag{4.2}$$

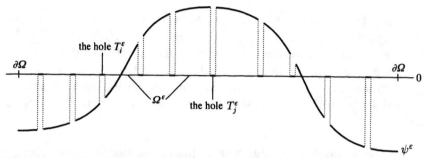

Figure 11. The function ψ^ε oscillates between 0 and ψ.

Let us consider the variational inequality: Find y_ε such that

$$\begin{cases} y^\varepsilon \in K^\varepsilon(\psi), \\ \displaystyle\int_\Omega \text{grad } y^\varepsilon \text{ grad } (v - y^\varepsilon) \, dx \geq < f, v - y^\varepsilon >_{H^{-1}(\Omega), H_0^1(\Omega)}, \\ \forall v \in K^\varepsilon(\psi), \end{cases} \tag{4.3}$$

where f is a given function in $H^{-1}(\Omega)$. As $K^\varepsilon(\psi)$ is a non empty convex closed set, this variational inequality has a unique solution $y^\varepsilon \in H_0^1(\Omega)$. In addition to assumption (4.1), which is the minimal one on ψ, we require that

$$\begin{cases} \mathcal{K}(\psi) \text{ is dense in } K(\psi), \\ \mathcal{K}(\psi) = \{\phi \in \mathcal{D}(\Omega) \, : \, \phi \geq \psi \text{ a.e. in } \Omega\}. \end{cases} \tag{4.4}$$

This density assumption is in fact a regularity assumption on ψ. Although this hypothesis is not absolutely necessary provided that hypothesis (H.5) is slightly modified (see Theorem 4.3), we shall keep it as it makes the presentation in the sequel easier.

Theorem 4.1. *Under hypotheses* (H.0) *to* (H.5) *and assumptions* (4.1) *and* (4.4), *the solutions y^ε of the variational inequalities* (4.3) *satisfy*

$$y^\varepsilon \rightharpoonup y \quad \text{weakly in } H_0^1(\Omega),$$

where y is the unique solution of

$$\begin{cases} y \in K(\psi) \\ \displaystyle\int_\Omega \text{grad } y \ \text{grad } (v-y) \ dx - \ < \mu y^-, v-y >_{H^{-1}(\Omega), H^1_0(\Omega)} \\ \qquad\qquad\qquad\qquad \geq < f, v-y >_{H^{-1}(\Omega), H^1_0(\Omega)}, \\ \forall v \in K(\psi). \end{cases} \qquad (4.5)$$

Moreover,

$$\begin{cases} (y^\varepsilon)^+ \to y^+ & \text{strongly in } H^1_0(\Omega), \\ (y^\varepsilon)^- - w^\varepsilon y^- \to 0 & \text{strongly in } W^{1,1}_0(\Omega). \end{cases} \qquad (4.6)$$

The solution y^ε of the variational inequality (4.3) is constrained to be greater than ψ in Ω^ε and to be nonnegative on the holes T^ε_i. When passing to the limit, this results in the fact that in the regions where y is nonnegative, it is only the obstacle ψ which plays a role. In contrast, in the regions where y is nonpositive, the obstacle 0 is "seen" by y^ε, and the price to be paid is the term $- < \mu y^-, v-y >_{H^{-1}(\Omega), H^1_0(\Omega)}$ in the variational inequality (4.5). This term makes sense since μy^- belongs to $H^{-1}(\Omega)$ as $\mu \in W^{-1,\infty}(\Omega)$ and $y^- \in H^1_0(\Omega)$.

These phenomena can also be observed in (4.6) where the behavior of y^ε is analyzed in a more precise way. The positive part $(y^\varepsilon)^+$ of y^ε does not see the oscillations of the obstacle ψ^ε and it converges strongly. The negative part $(y^\varepsilon)^-$ is zero on the holes, as $y^\varepsilon \geq 0$ on the holes, and it behaves like the solution of a variational inequality in the domain Ω^ε with Dirichlet conditions on $\partial\Omega^\varepsilon$. This explains the corrector result for $(y^\varepsilon)^-$, which is analogous to the result obtained for the solutions of equations (see Theorem 3.4).

Note that, as in Remark 3.3, $(y^\varepsilon)^- - w^\varepsilon y^-$ tends actually strongly to 0 in a better space $W^{1,q}_0(\Omega)$, $q > 1$, than $W^{1,1}_0(\Omega)$. This result can be improved if y^- is more regular than $H^1_0(\Omega)$.

Proof. The proof is done in several steps.
First step. Take $v = V^+$ in inequality (4.3) (this is possible thanks to (4.2)). We get

$$\int_\Omega |\text{grad } y^\varepsilon|^2 \ dx \leq \int_\Omega \text{grad } y^\varepsilon \ \text{grad } V^+ \ dx - \ < f, V^+ - y^\varepsilon >_{H^{-1}(\Omega), H^1_0(\Omega)} .$$

The sequence y^ε is therefore bounded in $H^1_0(\Omega)$ and by extracting a subsequence, one has

$$y^{\varepsilon'} \to y \quad \text{weakly in } H^1_0(\Omega).$$

Let us prove that

$$y^\varepsilon \geq w^\varepsilon \psi^+ - \psi^- \quad \text{a.e. in } \Omega. \tag{4.7}$$

In order to verify this assertion observe that, since $y^\varepsilon \geq 0$ and $w^\varepsilon = 0$ on the holes, one has

$$w^\varepsilon \psi^+ - \psi^- = -\psi^- \leq 0 \quad \text{in } T_i^\varepsilon, \ 1 \leq i \leq n(\varepsilon).$$

On the other hand, $y^\varepsilon \geq \psi$ and $w^\varepsilon \leq 1$ on Ω^ε imply that

$$w^\varepsilon \psi^+ - \psi^- \leq \psi^+ - \psi^- = \psi \quad \text{in } \Omega^\varepsilon,$$

and therefore (4.7) is proven.

Passing now to the limit a.e. in (4.7) yields

$$y \geq \psi^+ - \psi^- = \psi$$

which means that

$$y \in K(\psi). \tag{4.8}$$

Second step. Let us introduce the following functionals defined on $H_0^1(\Omega)$,

$$J(v) = \int_\Omega |\text{grad } v|^2 \, dx - 2 < f, v >,$$

$$J^0(v) = \int_\Omega |\text{grad } v|^2 \, dx + < \mu, (v^-)^2 > -2 < f, v >,$$

where the term $< \mu, (v^-)^2 >$ is the duality pairing between μ which belongs to $W^{-1,\infty}(\Omega)$ (see (H.4)), and $(v^-)^2$ which belongs to $W_0^{1,1}(\Omega)$ as v^- is in $H_0^1(\Omega)$.

It is well known that the solution y^ε of the variational inequality (4.3) is also the unique solution of the minimization problem

$$\begin{cases} y^\varepsilon \in K^\varepsilon(\psi), \\ J(y^\varepsilon) = \inf_{v \in K^\varepsilon(\psi)} J(v). \end{cases} \tag{4.9}$$

Third step. Let $\phi \in K(\psi)$, that is, $\phi \in \mathcal{D}(\Omega)$ with $\phi \geq \psi$ a.e. in Ω. Let us begin by proving that

$$v^\varepsilon = \phi^+ - w^\varepsilon \phi^- \in K^\varepsilon(\psi),$$

which will allow us to take v^ε as a test function in (4.9). Remark first that ϕ^+ and ϕ^- belong to $W^{1,\infty}(\Omega)$ and have compact support in Ω. Note next that w^ε belongs to $H^1(\Omega)$ by (H.1) and so v^ε belongs to $H_0^1(\Omega)$. On the other hand, $w^\varepsilon = 0$ on the holes T_i^ε so that $v^\varepsilon = \phi^+ \geq 0$ on $T_i^\varepsilon, 1 \leq i \leq n(\varepsilon)$. Finally, as $w^\varepsilon \leq 1$ on Ω^ε by (H.0), one has $v^\varepsilon \geq \phi^+ - \phi^- \geq \psi$ on Ω^ε.

Using v^ε in (4.9) yields

$$J(y^\varepsilon) \leq J(v^\varepsilon). \tag{4.10}$$

But

$$
\begin{aligned}
J(v^\varepsilon) &= \int_\Omega |\text{grad } \phi^+ - w^\varepsilon \text{grad } \phi^- - \phi^- \text{grad } w^\varepsilon|^2 \, dx \\
&\quad - 2 < f, \phi^+ - w^\varepsilon \phi^- >= \int_\Omega |\text{grad } \phi^+|^2 \, dx \\
&\quad + \int_\Omega |w^\varepsilon|^2 |\text{grad } \phi^-|^2 \, dx + \int_\Omega |\phi^-|^2 |\text{grad } w^\varepsilon|^2 \, dx \\
&\quad - 2 \int_\Omega w^\varepsilon \text{grad } \phi^+ \text{grad } \phi^- \, dx - 2 \int_\Omega \phi^- \text{grad } \phi^+ \text{grad } w^\varepsilon \, dx \\
&\quad + 2 \int_\Omega w^\varepsilon \phi^- \text{grad } \phi^- \text{grad } w^\varepsilon \, dx - 2 < f, \phi^+ - w^\varepsilon \phi^- > .
\end{aligned}
$$

It is easy to pass to the limit in almost all the terms of this equality thanks to hypothesis (H.3). For the term $\int_\Omega |\phi^-|^2 |\text{grad } w^\varepsilon|^2 \, dx$ we use the facts that $|\phi^-|^2$ is a continuous function with compact support and that $|\text{grad } w^\varepsilon|^2$ tends to μ weakly-$\star$ in the sense of measures. Hence

$$
\begin{aligned}
\lim_{\varepsilon \to 0} J(v^\varepsilon) &= \int_\Omega |\text{grad } \phi^+|^2 \, dx + \int_\Omega |\text{grad } \phi^-|^2 \, dx + < \mu, (\phi^-)^2 > \\
&\quad - 2 \int_\Omega \text{grad } \phi^+ \text{grad } \phi^- \, dx - 2 < f, \phi^+ - \phi^- > \\
&= \int_\Omega |\text{grad } \phi|^2 \, dx + < \mu, (\phi^-)^2 > -2 < f, \phi >= J^0(\phi).
\end{aligned}
$$

We then deduce from (4.10) that

$$\limsup_{\varepsilon \to 0} J(y^\varepsilon) \leq J^0(\phi), \quad \forall \phi \in \mathcal{K}(\psi),$$

which, thanks to the density assumption (4.4), implies

$$\limsup_{\varepsilon \to 0} J(y^\varepsilon) \leq \inf_{v \in \mathcal{K}(\psi)} J^0(v). \tag{4.11}$$

Fourth step. Decompose y^ε as $(y^\varepsilon)^+ - (y^\varepsilon)^-$. Then

$$J(y^\varepsilon) = \int_\Omega |\text{grad} \, ((y^\varepsilon)^+ - (y^\varepsilon)^-)|^2 \, dx - 2 < f, y^\varepsilon >$$

$$= \int_\Omega |\text{grad} \, (y^\varepsilon)^+|^2 \, dx + \int_\Omega |\text{grad} \, (y^\varepsilon)^-|^2 \, dx - 2 < f, y^\varepsilon > .$$

Let us pass to the lim inf (as ε tends to zero) in each of the preceding terms. In the first term, one uses the classical lower semicontinuity of the energy $\int_\Omega |\text{grad} \, v|^2 \, dx$, taking into account that $(y^{\varepsilon'})^+$ converges weakly in $H_0^1(\Omega)$ to y^+. Thus

$$\liminf_{\varepsilon' \to 0} \int_\Omega |\text{grad} \, (y^{\varepsilon'})^+|^2 \, dx \geq \int_\Omega |\text{grad} \, y^+|^2 \, dx. \tag{4.12}$$

For the second term, remark that $(y^{\varepsilon'})^-$, which converges weakly to y^- in $H_0^1(\Omega)$, vanishes on the holes $T_i^{\varepsilon'}$ since $y^\varepsilon \in K^\varepsilon(\psi)$ implies that y^ε is nonnegative on the holes. Using again the lower semicontinuity of the energy but now with the new energy, one has (Proposition 3.1)

$$\liminf_{\varepsilon' \to 0} \int_\Omega |\text{grad} \, (y^{\varepsilon'})^-|^2 \, dx \geq \int_\Omega |\text{grad} \, y^-|^2 \, dx + < \mu, (y^-) >^2 . \tag{4.13}$$

Finally, from (4.12) and (4.13) we have

$$\liminf_{\varepsilon' \to 0} J(y^{\varepsilon'}) \geq \int_\Omega |\text{grad} \, y^+|^2 \, dx + \int_\Omega |\text{grad} \, y^-|^2 \, dx$$
$$+ < \mu, (y^-)^2 > -2 < f, y >,$$

that is,

$$\liminf_{\varepsilon' \to 0} J(y^{\varepsilon'}) \geq J^0(y). \tag{4.14}$$

Fifth step. Recalling (4.8), (4.11), and (4.14), we have proved that

$$\begin{cases} y \in K(\psi), \\ J^0(y) = \inf_{v \in K(\psi)} J^0(v). \end{cases}$$

The function y is therefore the unique solution of this minimization problem. By using Minty's trick and Lemma 4.2 below, one can prove as usual, that y is also the unique solution of the variational inequality (4.5).

As y is uniquely determined, the whole sequence y^ε converges to y and the first part of Theorem 4.1 is proved.

In order to prove (4.6), notice that (4.11) implies that

$$\limsup_{\varepsilon \to 0} J(y^\varepsilon) \le \inf_{v \in K(\psi)} J^0(v) = J^0(y).$$

Then, from (4.12) through (4.14) one gets

$$\int_\Omega |\text{grad } (y^\varepsilon)^+|^2 \, dx \to \int_\Omega |\text{grad } y^+|^2 \, dx$$

$$\int_\Omega |\text{grad } (y^\varepsilon)^-|^2 \, dx \to \int_\Omega |\text{grad } y^-|^2 \, dx + < \mu, (y^-)^2 > .$$

From these two results, both assertions of (4.6) are easily obtained, the first one by classical arguments and the second one by using Proposition 3.2 (recall that $(y^\varepsilon)^-$ vanishes on the holes). This completes the proof of Theorem 4.1. ∎

Lemma 4.2. *Consider u and v in $H_0^1(\Omega)$. Then, if $t \to 0$,*

$$\frac{|(u + tv)^-|^2 - |u^-|^2}{t} \to -2u^- v \quad \text{strongly in } W_0^{1,1}(\Omega).$$

Proof. One starts by showing that if a and b are two real numbers and if r is defined by

$$|(a + b)^-|^2 = (a^-)^2 - 2a^- b + r,$$

then

$$0 \le r \le b^2.$$

For this, it it sufficient to consider all the possible cases: $a \ge 0$, $a \le 0$, $(a + b) \ge 0$, $(a + b) \le 0$. One deduces that

$$\left\| \frac{|(u + tv)^-|^2 - |u^-|^2}{t} + 2u^- v \right\|_{L^1(\Omega)} \le |t| \, \|v\|_{L^2(\Omega)}^2,$$

and hence

$$X(t) = \frac{|(u + tv)^-|^2 - |u^-|^2}{t} \to -2u^- v \quad \text{strongly in } L^1(\Omega).$$

The gradient of $X(t)$ is given by

$$\text{grad } X(t) = \frac{2}{t} \{-(u + tv)^- \text{grad } (u + tv) + u^- \text{grad } u\}$$

$$= -2 \frac{(u + tv)^- - u^-}{t} \text{grad } u - 2(u + tv)^- \text{grad } v.$$

It is easily seen that the first term converges a.e. by considering the sets where $u(x) < 0$, $u(x) > 0$ and $u(x) = 0$ and recalling that grad $u = 0$ a.e. on the set where $u(x) = 0$. On the other hand, the application $x \mapsto x^-$ is Lipschitz continuous from $\mathbb{R}$ to $\mathbb{R}$ and thus

$$\left| \frac{(u + tv)^- - u^-}{t} \text{ grad } u \right| \leq |v| \, |\text{grad } u| \quad \text{a.e. in } \Omega.$$

By virtue of Lebesgue theorem, the first term converges strongly in $L^1(\Omega)$ to $-2v$ grad u^-.

The second term converges strongly in $L^1(\Omega)$ to $-2u^-$ grad v because $(u + tv)^-$ converges strongly in $L^2(\Omega)$ to u^-. Therefore

$$\text{grad } X(t) \to \text{grad } (-2u^- v) \quad \text{strongly in } L^1(\Omega)$$

and the proof of Lemma 4.3 is complete.   ∎

In the statement of Theorem 4.1, we made the regularity assumption (4.4) on the obstacle ψ. We mentioned that this assumption is not necessary if (H.5) is (slightly) modified. With this modification ψ only needs to satisfy the "minimal" assumption (4.1).

Theorem 4.3. *Assume that hypotheses* (H.0) *to* (H.4) *and* (4.1) *are satisfied. Suppose further that*

$$\begin{cases} \text{for every sequence } v^\varepsilon \text{ such that } v^\varepsilon = 0 \quad \text{on } T_i^\varepsilon, \ 1 \leq i \leq n(\varepsilon), \\ \text{satisfying } v^\varepsilon \rightharpoonup v \text{ weakly in } H_0^1(\Omega) \text{ with } v \in H_0^1(\Omega), \text{ one has} \qquad \text{(H.5)}''' \\ < -\Delta w^\varepsilon, v^\varepsilon >_{H^{-1}(\Omega), H_0^1(\Omega)} \to < \mu, v >_{H^{-1}(\Omega), H_0^1(\Omega)} . \end{cases}$$

Then the results of Theorem 4.1 hold true.

Clearly (H.5)''' implies (and is almost equivalent to) (H.5). Hypothesis (H.5)''' is satisfied in all the examples we treated before since (see Remark 1.6)

$$(\text{H.5})' \Longrightarrow (\text{H.5})''' \Longrightarrow (\text{H.5}),$$

and (H.5)' was satisfied in the examples.

Proof. The proof is identical to the proof of Theorem 4.1, except for the third step which has to be modified as follows. Let $v \in K(\psi)$, fix $n > 0$, and set $v_n = \sup(v, -n)$. As $v_n \geq v$, v_n still belongs to $K(\psi)$. On the other hand, $(v_n)^-$ belongs to $L^\infty(\Omega)$. Furthermore, as $(v_n)^- \in H_0^1(\Omega) \cap L^\infty(\Omega)$ and $w^\varepsilon \in H^1(\Omega) \cap L^\infty(\Omega)$, the product $w^\varepsilon (v_n)^-$ belongs to $H_0^1(\Omega) \cap L^\infty(\Omega)$ so that

$$v^\varepsilon = (v_n)^+ - w^\varepsilon (v_n)^- \in K^\varepsilon(\psi).$$

Let us now consider $J(v^\varepsilon)$ which, as previously, is the sum of seven terms. Letting $\varepsilon \to 0$ with n fixed, almost all the terms pass to the limit. There are two terms where this passage to the limit is not completely obvious:

$$\int_\Omega w^\varepsilon (v_n)^- \text{ grad } (v_n)^- \text{ grad } w^\varepsilon \, dx \text{ and } \int_\Omega |(v_n)^-|^2 |\text{grad } w^\varepsilon|^2 \, dx.$$

For the first, one passes to the limit since $w^\varepsilon (v_n)^-$ grad $(v_n)^-$ converges strongly in $(L^2(\Omega))^N$ by virtue of (H.0) and of Lebesgue theorem. For the second one, we use hypothesis (H.5)''' which, together with (H.0), enables us to prove Proposition 1.1 for test functions $\varphi \in H_0^1(\Omega) \cap L^\infty(\Omega)$. Note that $|(v_n)^-|^2$ is such a function.

We finally obtain

$$\lim_{\varepsilon \to 0} J(v^\varepsilon) = \int_\Omega |\text{grad } v_n|^2 \, dx + < \mu, |(v_n)^-|^2 > -2 < f, v_n >,$$

where we just have to let n tend to infinity to obtain (4.11). ∎

Generalizations

The proof of Theorem 4.1 relies on the equivalence between the variational inequality (4.5) and the corresponding minimization problem (4.9). One could therefore think that the possible generalizations to operators $-\text{div}(A \text{ grad})$ are restricted to the case of symmetric matrices A where this equivalence holds true. Using the same method, in the following, we actually treat the case where A is non necessarily symmetric by assuming that the skew-symmetric part of A does not play any role in the passage to the limit.

To be precise, let A be a coercive matrix with coefficients in $L^\infty(\Omega)$ and consider the variational inequality

$$\begin{cases} \overline{y}^\varepsilon \in K^\varepsilon(\psi), \\ \displaystyle\int_\Omega A \text{grad } \overline{y}^\varepsilon \text{ grad } (v - \overline{y}^\varepsilon) \, dx \geq < f, v - \overline{y}^\varepsilon >_{H^{-1}(\Omega), H_0^1(\Omega)}, \quad (4.15) \\ \forall v \in K^\varepsilon(\psi), \end{cases}$$

where f belongs to $H^{-1}(\Omega)$ and ψ is a given measurable function satisfying (4.1). The convex set $K^\varepsilon(\psi)$ is defined as in the Laplacian case. As regarding the holes, we suppose that there exist w^ε and μ satisfying (H.0) to (H.4), (H.5)$_A$, and (H.6)$_A$, with

$$
(H.5)_A \begin{cases}
\text{for every sequence } v^\varepsilon \text{ such that } v^\varepsilon = 0 \quad \text{on } T_i^\varepsilon,\ 1 \le i \le n(\varepsilon), \\
\text{satisfying } v^\varepsilon \rightharpoonup v \text{ weakly in } H_0^1(\Omega) \text{ with } v \in H_0^1(\Omega), \text{ one has} \\
< -\mathrm{div}(^t A\,\mathrm{grad}\,w^\varepsilon), \varphi v^\varepsilon >_{H^{-1}(\Omega), H_0^1(\Omega)} \to\ < \mu, \varphi v >_{H^{-1}(\Omega), H_0^1(\Omega)} \\
\text{for all } \varphi \in \mathcal{D}(\Omega).
\end{cases}
$$

(H.6)$_A$ $-\mathrm{div}\,((A - {}^t A)\,\mathrm{grad}\,w^\varepsilon) \to 0$ strongly in $H^{-1}(\Omega)$.

It should be noted that (H.5)$_A$ is the same hypothesis that we assumed when passing to the limit in Dirichlet problems in Ω^ε for the operator $-\mathrm{div}(A\,\mathrm{grad})$ (see Generalizations at the end of Section 1). Hypothesis (H.6)$_A$ is stronger than (H.6) which was done in order to have the lower semicontinuity of the energy to the new energy (see Generalizations at the end of Section 3).

Let us finally remark that hypotheses (H.0) to (H.4), (H.5)$_A$ and (H.6)$_A$ are all satisfied if A is a matrix with continuous coefficients, see Example 2.16 and especially (2.14) to (2.17) for the verification of (H.6)$_A$.

Theorem 4.4.   *Under assumptions* (H.0) *to* (H.4), (H.5)$_A$, (H.6)$_A$, (4.1) *and* (4.4), *the solutions* $\overline{y}^\varepsilon$ *of the variational inequalities* (4.15) *satisfy*

$$
\overline{y}^\varepsilon \rightharpoonup \overline{y} \quad \text{weakly in } H_0^1(\Omega),
$$

where $\overline{y}$ *is the unique solution of*

$$
\begin{cases}
\overline{y} \in K(\psi), \\
\displaystyle\int_\Omega A\,\mathrm{grad}\,\overline{y}\,\mathrm{grad}\,(v - \overline{y})\,dx \ + < \mu\overline{y}, v - \overline{y} > \\
\qquad\qquad\qquad\qquad \ge < f, v - \overline{y} >_{H^{-1}(\Omega), H_0^1(\Omega)}, \\
\forall v \in K(\psi).
\end{cases}
\tag{4.16}
$$

Moreover,

$$
\begin{cases}
(\overline{y}^\varepsilon)^+ \to (\overline{y})^+ & \text{strongly in } H_0^1(\Omega), \\
(\overline{y}^\varepsilon)^- - w^\varepsilon (\overline{y})^- \to 0 & \text{strongly in } W_0^{1,1}(\Omega).
\end{cases}
\tag{4.17}
$$

Note that the regularity hypothesis (4.4) on ψ can be removed by slightly modifying (H.5)$_A$ as we did in Theorem 4.3.

Proof. The proof is a slight variant of the proof of Theorem 4.1.

First step. It is analogous to the first step of the proof of Theorem 4.1. It consists of proving that a subsequence $\bar{y}^{\varepsilon'}$ converges weakly in $H_0^1(\Omega)$ to an element $\bar{y}$ of $K(\psi)$.

Second step. Multiplying (4.15) by 2, adding to the left-hand side the term $\int_\Omega A \operatorname{grad}(v - \bar{y}^\varepsilon) \operatorname{grad}(v - \bar{y}^\varepsilon) \, dx$ which "reinforces" the inequality, one has

$$
\left\{
\begin{aligned}
& \int_\Omega A \operatorname{grad} v \operatorname{grad} v \, dx + \int_\Omega ({}^t A - A) \operatorname{grad} v \operatorname{grad} \bar{y}^\varepsilon \, dx \\
& \quad - 2 < f, v > \; \geq \int_\Omega A \operatorname{grad} \bar{y}^\varepsilon \operatorname{grad} \bar{y}^\varepsilon \, dx - 2 < f, \bar{y}^\varepsilon > .
\end{aligned}
\right.
\tag{4.18}
$$

Third step. Let $\phi \in \mathcal{K}(\psi)$ and take $v = \phi^+ - w^\varepsilon \phi^- \in K^\varepsilon(\psi)$ in the left-hand side of (4.18). When expanding the integral, we see that each term passes to the limit. In particular, hypothesis $(H.6)_A$ is used for the term $-\int_\Omega \varphi^- ({}^t A - A) \operatorname{grad} w^\varepsilon \operatorname{grad} \bar{y}^\varepsilon \, dx$. One obtains that the left-hand side term of (4.18) converges to

$$
\int_\Omega A \operatorname{grad} \phi \operatorname{grad} \phi \, dx + \int_\Omega ({}^t A - A) \operatorname{grad} \varphi \operatorname{grad} \bar{y} \, dx
$$
$$
+ < \mu, |\phi|^2 > -2 < f, \phi >
$$

for all $\phi \in \mathcal{K}(\psi)$ and by density, for all $\phi \in K(\psi)$.

Fourth step. Decompose $\bar{y}^\varepsilon$ as $\bar{y}^\varepsilon = (\bar{y}^\varepsilon)^+ - (\bar{y}^\varepsilon)^-$ and pass to the lim inf in the right-hand side of (4.18). Taking into account the lower semicontinuity of the energy to the new energy, since $(\bar{y}^\varepsilon)^-$ vanishes on the holes, one has

$$
\liminf_{\varepsilon' \to 0} \int_\Omega A \operatorname{grad} (\bar{y}^{\varepsilon'})^- \operatorname{grad} (\bar{y}^{\varepsilon'})^- \, dx
$$
$$
\geq \int_\Omega A \operatorname{grad} (\bar{y})^- \operatorname{grad} (\bar{y})^- \, dx + < \mu, |(\bar{y})^-|^2 > .
$$

Consequently, the lim inf of the right hand side of (4.18) is greater than or equal to

$$
\int_\Omega A \operatorname{grad} \bar{y} \operatorname{grad} \bar{y} \, dx + < \mu, |(\bar{y})^-|^2 > -2 < f, \bar{y} > .
$$

Fifth step. Collecting the results obtained in the first, third, and fourth steps, one obtains

$$
\left\{
\begin{aligned}
& \bar{y} \in K(\psi), \\
& \int_\Omega A \operatorname{grad} (\phi - \bar{y}) \operatorname{grad} (\phi - \bar{y}) \, dx + < \mu, |\phi^-|^2 - |(\bar{y})^-|^2 > \\
& \qquad + 2 \int_\Omega A \operatorname{grad} \bar{y} \operatorname{grad} (\phi - \bar{y}) \, dx \geq 2 < f, \phi - \bar{y} >, \\
& \forall \phi \in K(\psi).
\end{aligned}
\right.
$$

Take now $\phi = tv + (1-t)\overline{y}$ with $0 < t < 1$ and $v \in K(\psi)$. Dividing by $2t$ and letting $t \to 0$, we deduce thanks to Lemma 4.2 that $\overline{y}$ is the solution of (4.16).

Choose finally $\phi = \overline{y}$ in the limit of the left-hand side of (4.18) obtained in the third step. One deduces that, in fact, the lim inf taken in the fourth step are true limits and (4.17) follows directly.

Remark 4.5. The method used in the proofs of this section relies heavily on *truncation*, in opposition to what was done in Sections 1, 2, and 3. For this reason it seems impossible to use it for variational inequalities with oscillating obstacles for higher-order operators.

Remark 4.6. Variational inequalities with highly oscillating obstacles of the same type as those we studied here were first investigated by Carbone and Colombini [2]. Their examples concern spherical holes periodically distributed in volume in $\mathbb{R}^N$ (see Example 2.1 above) or on a hyperplane of $\mathbb{R}^N$ (see Example 2.9 above).

In the present work, we have included these examples in a more general presentation that links the variational inequalities with highly oscillating obstacles to the corresponding Dirichlet problems. This presentation enables us to give the corrector result (4.6) which seems to be new.

It should be noted that the proof of Theorem 4.1 actually consists of proving that under hypotheses (H.0) to (H.5), the functionals

$$v \mapsto \int_\Omega |\text{grad } v|^2 + I_{K^\varepsilon(\psi)}(v),$$

where the indicator function $I_{K^\varepsilon(\psi)}$ is defined by

$$\begin{cases} I_{K^\varepsilon(\psi)} = 0 & \text{if } v \in K^\varepsilon(\psi), \\ I_{K^\varepsilon(\psi)} = +\infty & \text{if } v \notin K^\varepsilon(\psi), \end{cases}$$

Γ-converge to the functional

$$v \mapsto \int_\Omega |\text{grad } v|^2 + < \mu, |v^-|^2 > + I_{K(\psi)}.$$

The essential steps of the proof of this assertion are the third and fourth steps of the proof of Theorem 4.1.

In recent years, De Giorgi, Dal Maso and Longo [11], Dal Maso and Longo [10], Dal Maso [8], [9] and Attouch and Picard [1] showed that for a given sequence of obstacles ψ^ε (not necessarily of the type previously described that oscillate between 0 and ψ), there exists a subsequence $\psi^{\varepsilon'}$ such that, roughly speaking, the functionals

$$v \mapsto \int_\Omega |\text{grad } v|^2 + I_{K^\varepsilon(\psi)}(v)$$

Γ-converge to the functional

$$v \mapsto \int_\Omega |\text{grad } v|^2 + \int_\Omega F(x, v) \, d\mu,$$

where F is a nonincreasing convex function in v and μ a nonnegative measure of $H^{-1}(\Omega)$. To correctly state this result, one has to introduce a "rich" family B of Borel sets on which one localizes the problem and to write the constraint as

$$v(x) \geq \psi^\varepsilon(x) \quad \text{quasi everywhere,}$$

where quasi everywhere refers to the H^1- capacity. One has also to introduce in the limit functional a term $\nu(B)$, where ν is a positive Borel measure. Our result may thus be considered as a very particular case of this general compactness result for the Γ-convergence.

To conclude let us quote John Lennon and Paul McCartney (Fixing a Hole, in *Sgt. Pepper's Lonely Hearts Club Band*, The Beatles, ed., Northern Songs Ltd., London (1967)):

> I'm fixing a hole where the rain gets in
> And stops my mind from wandering
> Where it will go. . .

References

[1] H. Attouch and C. Picard, Asymptotic analysis of variational problems with constraints of obstacle type, *J. Funct. Anal.*, **15**, (1983), 329–386.

[2] L. Carbone and F. Colombini, On convergence of functionals with unilateral constraints, *J. Math. pures et appl.*, **59**, (1980), 465–500.

[3] D. Cioranescu, Calcul des variations sur des sous-espaces variables, *C. R. Acad. Sci. Paris, Série A*, **291**, (1980), 19–22.

[4] D. Cioranescu, Calcul des variations sur des sous-espaces variables. Applications(*). *C. R. Acad. Sci. Paris, Série A*, **291**, (1980), 87–90.

(*) Some of the results stated in this Note are not valid. As a matter of fact, Theorem 1 does not apply to the torsion problem, but to the Dirichlet problem. In order for the results to be correctly stated, it is sufficient to replace everywhere "torsion élastique" by "Dirichlet," to replace "Cte" by "0" in the definition of $H^{(s)}$ and in (1), and to remove the last line of (1).

[5] D. Cioranescu and J. Saint Jean Paulin, Homogenization in open sets with holes, *J. Math. Anal. Appl.*, **71**, (1978), 590–607.

[6] D. Cioranescu and J. Saint Jean Paulin, Homogénéisation de problèmes d'évolution dans des ouverts à cavités, *C. R. Acad. Sci. Paris, Série A*, **286**, (1978), 899–902.

[7] R. Courant and D. Hilbert, *Methods of mathematical physics*, Volume I. Interscience, New York, (1953).

[8] G. Dal Maso, Limiti di soluzioni di problemi variazionali con ostacoli bilaterali, *Atti Accad. Naz. Lincei, Rend. Cl. Sci. Fis. Mat. Natur.*, **69**, (1980), 333–337.

[9] G. Dal Maso, Asymptotic behaviour of minimum problems with bilateral obstacles, *Ann. mat. pura ed appl.*, **129**, (1981), 327–366.

[10] G. Dal Maso and P. Longo, Γ-limits of obstacles, *Ann. mat. pura ed appl.*, **128**, (1981), 1–50.

[11] E. De Giorgi, G. Dal Maso and P. Longo, Γ-limiti di ostacoli, *Atti Accad. Naz. Lincei, Rend. Cl. Sci. Fis. Mat. Natur.*, **68**, (1980), 481–487.

[12] E. Ja. Hrouslov, The method of orthogonal projections and the Dirichlet problem in domains with fine-grained boundary, *Math. USSR Sb.*, **17**, (1972), 37–59.

[13] E. Ja. Hrouslov, The first boundary problem in domains with a complicated boundary for higher order equations, *Math. USSR Sb.*, **32**, (1977), 535–549.

[14] E. Ja. Hrouslov, The asymptotic behaviour of solutions of the second boundary value problem under fragmentation of the boundary of the domain, *Math. USSR Sb.*, **35**, (1979), 266–282.

[15] V. A. Marcenko and E. Ja. Hrouslov, *Boundary value problems in domains with fine-grained boundary* (in russian). Naukova Dumka, Kiev, (1974).

[16] J. Nečas, *Les méthodes directes en théorie des équations elliptiques*. Masson, Paris, (1967).

[17] G. C. Papanicolaou and S. R. S. Varadhan, Diffusion in regions with many small holes. In *Stochastic differential systems, filtering and control, Proceedings IFIP WG 7/1 Conference, Vilnius, 1978*, ed. B. Grigelionis, Lecture Notes in Control and Information Sciences, **25**, Springer, Berlin, (1980), 190–206.

[18] J. Rauch and M. Taylor, Potential and scattering theory on wildly perturbed domains, *J. Funct. Anal.*, **18**, (1975), 27–59.

[19] J. Rauch and M. Taylor, Electrostatic screening, *J. Math. Phys.*, **16**, (1975), 284–288.

[20] J. Saint Jean Paulin, Étude de quelques problèmes de mécanique et d'éléctrotechnique liés aux méthodes d'homogénéisation. Thèse d'Etat, Université Paris VI, (1981).

[21] E. Sanchez-Palencia, Boundary value problems in domains containing perforated walls. In *Nonlinear partial differential equations and their applications, Collège de France Seminar, Volume III*, ed. H. Brezis and J. L. Lions, Research Notes in Mathematics, **70**, Pitman, London, (1980), 309–325.

[22] L. Tartar, Cours Peccot, Collège de France (1977). Partialy written in F. Murat and L. Tartar, *H*-convergence, translated in the present volume.

[23] L. Tartar, Quelques remarques sur l'homogénéisation. In *Functional Analysis and Numerical Analysis, Proceedings of the Japan-France Seminar 1976*, ed. H. Fujita, Japan Society for the Promotion of Science, Tokyo, (1978), 469–482.

Afterword

This article is the translation of the French paper *Un terme étrange venu d'ailleurs I and II*, which has been published in *Nonlinear partial differential equations and their applications, Collège de France Seminar, Volumes II and III*, H. Brezis and J. L. Lions, eds., Research Notes in Mathematics, **60** and **70**, Pitman, London, (1982), 98–138 and 154–178.

We would like to quote here some (but not all the) papers which appeared since the publication of *Un terme étrange venu d'ailleurs* and are concerned with the same problems.

The Γ-convergence of integral functionals involving oscillating obstacles or Dirichlet problems was a subject for many papers. Extensive references on this topic can be found in the book Dal Maso [26] and in the paper Dal Maso and Garroni [30]. Let us also mention the paper of Defranceschi and Vitali [33] which deals with the vector valued case. On the other hand, the method introduced by Marcenko and Hrouslov in the book [15] was developed in the setting of non linear problems by Skrypnik [35] [36] [37].

As said in the introduction, the method we followed in the present paper is the method introduced by Tartar for studying homogenization problems. It consists of using special test functions (here w^ε). In the same framework, an error estimate was obtained in Kacimi and Murat [34]. The case of the wave equation was investigated in Cioranescu, Donato, Murat and Zuazua [28]. The homogenization of the Stokes problem was considered in Allaire [24] and the problem where the laplacian is perturbed by a term with quadratic growth was solved in Casado-Diaz [25].

Our contribution to the study of the Dirichlet problems was unfortunately limited by the fact that we *assume the existence* of w^ε satisfying hypotheses (H.1) to (H.5), which in the examples forced us to make certain geometrical assumptions. This limitation has recently been removed by Dal Maso and Garroni [30] who followed the same method but used different test function, constructed from the function z^ε defined by

$$\begin{cases} -\Delta z^\varepsilon = 1 & \text{in } \mathcal{D}'(\Omega^\varepsilon), \\ z^\varepsilon \in H_0^1(\Omega^\varepsilon). \end{cases}$$

Defining roughly speaking w^ε as $\dfrac{z^\varepsilon}{z}$ (this has to be made precise since z, limit of $\tilde{z}^\varepsilon$, can vanish) they were able to remove our assumptions and to solve the general Dirichlet problem, irrespective of any geometrical assumption on the holes. This break through made then possible to solve the general (without geometrical assumptions) case of Dirichlet problems for monotone operators, see Dal Maso and Murat [31] [32], Casado-Diaz [26] and Casado-Diaz and Garroni [27].

Additional references

[24] G. Allaire, Homogenization of the Navier-Stokes equations in open sets perforated with tiny holes, I and II, *Arch. Rat. Mech. Anal.*, **113**, (1991), 209–259 and 261–298.

[25] J. Casado-Diaz, Sobre la homogeneización de problemas no coercivos y problemas en dominios con ajujeros. Ph. D. Thesis, University of Seville, 1993.

[26] J. Casado-Diaz, Asymptotic behaviour and correctors for Dirichlet problems in perforated domains with general monotone operators, *Proc. Roy. Soc. Edinburgh*, to appear.

[27] J. Casado-Diaz and A. Garroni, The limit of Dirichlet problems in perforated domains with general pseudo-monotone operators, to appear.

[28] D. Cioranescu, P. Donato, F. Murat and E. Zuazua, Homogenization and corrector for the wave equation in domains with small holes, *Ann. Sc. Norm. Sup. Pisa*, **18**, (1991), 251–293.

[29] G. Dal Maso, *An Introduction to Γ-Convergence*. Birkhäuser, Boston, (1993).

[30] G. Dal Maso and A. Garroni, New results on the asymptotic behaviour of Dirichlet problems in perforated domains, *Math. Mod. Meth. Appl. Sci.*, **4** (1994), no. 3, 373–407.

[31] G. Dal Maso and F. Murat, Dirichlet problems in perforated domains for homogeneous monotone operators on H_0^1. In *Calculus of variations, homogenization and continuum mechanics, (Marseilles-Luminy, 1993)*, World Scientific.

[32] G. Dal Maso and F. Murat, Asymptotic behaviour and correctors for Dirichlet problems in perforated domains with homogeneous monotone operators. To appear in *Annali della Scuola Normale Superiore di Pisa*, 1997.

[34] H. Kacimi and F. Murat, Estimation de l'erreur dans des problèmes de Dirichlet où apparaît un terme étrange. In *Partial Differential Equations and the Calculus of Variations: Essays in Honor of Ennio De Giorgi, Volume II*, F. Colombini, A. Marino, L. Modica and S. Spagnolo, eds., Progress in Nonlinear Differential Equations and their Applications, **2**, Birkhäuser, Boston, (1989), 661–696.

[35] I.V. Skrypnik, *Nonlinear elliptic boundary value problems*. Teubner-Verlag, Leipzig, (1986).

[36] I.V. Skrypnik, *Methods of investigation of nonlinear elliptic boundary value problems*, *Nauka*, Moscow, (1990), (in russian).

[37] I.V. Skrypnik, Averaging nonlinear Dirichlet problems in domains with channels, *Soviet Math. Dokl.*, **42**, (1991), 853–857.

Design of Composite Plates
of Extremal Rigidity[*]

L.V. Gibiansky and A.V. Cherkaev

Abstract. We consider design problems for plates possessing an extremal rigidity. The plate is assumed to be assembled from two isotropic materials characterized by different values of their elastic moduli; the amount of each material is given. We look for a distribution of the materials which renders the plate's rigidity for either its maximal or minimal value. The rigidity is defined here as work produced by an extremal load on deflection of the points of the plate. The optimal distribution of the materials is characterized by some infinitely often alternating sequences of domains occupied by each of the materials (see [1], [2]). This leads to the appearance of anisotropic composites; their structures are to be determined at each point of the plate.

The exact bounds on the elastic energy density are obtained; the microstructures of the optimal composites are found. The effective properties of such composites (we call them matrix laminate composites) are explicitly calculated. These composites extend the set of available materials. The initial problems are formulated and solved for an extended set of design parameters.

We use the results to solve the problem of optimal design of the plate with variable thickness, given mass, and with additional restrictions on the range of the thickness values. A rule is found that allows us to distinguish the cases in which the optimal composite is assembled of elements having only maximal or minimal thickness.

The number of optimal design problems for the clamped square plate of maximal and minimal rigidity are solved numerically.

Part I. Optimal structures of composites.

§1. Statement of the problem.

Let the bending of a plate be described by the equation:

$$\int_\Omega \nabla\nabla\eta \cdot\cdot \mathbf{D} \cdot\cdot \nabla\nabla w d\mathbf{x} = \int_\Omega f\eta d\mathbf{x}, \ \forall \eta \in \overset{\circ}{W}{}_2^2, \tag{1.1}$$

---

[*]The article is the translation of an article originally written in Russian and published as the report of *Ioffe Physico-Technical Institute, Academy of Sciences of USSR, Publication 914, Leningrad, 1984.*

© Springer Nature Switzerland AG 2018

A. V. Cherkaev, R. Kohn (eds.), *Topics in the Mathematical Modelling of Composite Materials*, Modern Birkhäuser Classics, https://doi.org/10.1007/978-3-319-97184-1_5

95

or by an equivalent system:

$$\nabla\nabla \cdot \mathbf{M} = f, \quad \mathbf{M} = \mathbf{D} \cdot\cdot\boldsymbol{\epsilon}, \quad \boldsymbol{\epsilon} = \nabla\nabla w, \qquad (1.2)$$

where

Ω is a bounded two-dimensional domain with smooth boundary,

$\mathbf{x} \in \Omega$ is a vector of coordinates of a point in Ω,

$f \in L_2(\Omega)$ is a normal load density,

$w \in W_2^2(\Omega)$ is a normal displacement of a plate,

$\boldsymbol{\epsilon} = \{\varepsilon_{ij}\}$ ($\varepsilon_{ij} \in L_2(\Omega)$), $\mathbf{M} = \{M_{ij}\}$ ($M_{ij} \in L_2(\Omega)$) are 2×2 symmetric tensors of strain and bending moments correspondingly,

$\mathbf{D} = \{D_{ijkl}\}$ ($D_{ijkl} \in L_\infty(\Omega)$) is a fourth rank self-adjoint stiffness tensor of a plate,

symbol $\cdot\cdot$ denotes a double convolution.

We are also given homogeneous boundary conditions on $\partial\Omega$.

We assume that a plate is made of two isotropic materials with stiffness tensors $\mathbf{D}_1$, $\mathbf{D}_2$ that are given by dyadic decompositions [3,4]:

$$\begin{aligned} \mathbf{D}_1 &= \kappa_1\mathbf{a}_1\mathbf{a}_2 + \mu_1(\mathbf{a}_2\mathbf{a}_2 + \mathbf{a}_3\mathbf{a}_3), \\ \mathbf{D}_2 &= \kappa_2\mathbf{a}_1\mathbf{a}_1 + \mu_2(\mathbf{a}_2\mathbf{a}_2 + \mathbf{a}_3\mathbf{a}_3), \end{aligned} \qquad (1.3)$$

where κ_1, κ_2 and μ_1, μ_2 are bulk and shear moduli of the first and second materials, and

$$\mathbf{a}_1 = \frac{1}{\sqrt{2}}(\mathbf{ii} + \mathbf{jj}), \quad \mathbf{a}_2 = \frac{1}{\sqrt{2}}(\mathbf{ii} - \mathbf{jj}), \quad \mathbf{a}_3 = \frac{1}{\sqrt{2}}(\mathbf{ij} + \mathbf{ji}). \qquad (1.4)$$

Here $\mathbf{a}_1$ is a bulk tensor (i.e. tensor that is proportional to the identity matrix), $\mathbf{a}_2$, $\mathbf{a}_3$ are mutually orthogonal deviators such that

$$\mathbf{a}_2 \cdot\cdot\mathbf{a}_3 = 0,$$

$\mathbf{i}$, $\mathbf{j}$ are orthonormal vectors. The quantities κ_i, μ_i can be expressed in terms of cylindrical stiffness d_i and Poisson's ratio ν_i of materials by $\kappa_i = d_i(1 + \nu_i)$, $\mu_i = d_i(1 - \nu_i)$; we also assume that either

$$\kappa_1 < \kappa_2, \ \mu_1 < \mu_2 \qquad (1.5)$$

or

$$\kappa_1 > \kappa_2, \ \mu_1 < \mu_2. \qquad (1.6)$$

The material with stiffness $\mathbf{D}_1$ occupies a part Ω_1 of Ω and the material with stiffness $\mathbf{D}_2$ occupies the part $\Omega_2 = \Omega - \Omega_1$ of Ω. The stiffness tensor

$\mathbf{D}(\mathbf{x})$ of this composite plate is given by

$$\mathbf{D}(\mathbf{x}) = \mathbf{D}_1\chi(\mathbf{x}) + \mathbf{D}_2\left(1 - \chi(\mathbf{x})\right), \qquad (1.7)$$

where $\chi \in L_\infty(\Omega)$ is a characteristic function of the region Ω_1:

$$\chi(\mathbf{x}) = \begin{cases} 1, & \text{if } \mathbf{x} \in \Omega_1, \\ 0, & \text{if } \mathbf{x} \in \Omega_2. \end{cases}$$

In addition to the stiffness tensor $\mathbf{D}(\mathbf{x})$ it is convenient to introduce a compliance tensor:

$$\mathbf{C}(\mathbf{x}) = \mathbf{D}^{-1}(\mathbf{x}) = \mathbf{C}_1\chi(\mathbf{x}) + \mathbf{C}_2(1 - \chi(\mathbf{x})). \qquad (1.8)$$

Here $\mathbf{C}_1$, $\mathbf{C}_2$ are the compliance tensors of phases, with $\mathbf{C}_1 = \mathbf{D}_1^{-1}$, $\mathbf{C}_2 = \mathbf{D}_2^{-1}$.

Let us now consider a sequence of partitions Ω_1^s, Ω_2^s of Ω such that for $s \to \infty$ in a neighborhood of every point $\mathbf{x} \in \Omega$ there are regions that belong to Ω_1 and Ω_2 for large enough s. Such partitions model a composite material which is described by a G-limit system of equations [5]:

$$\nabla\nabla \cdot\cdot\mathbf{M}^0 = f, \ \mathbf{M}^0 = \mathbf{D}_0 \cdot\cdot\epsilon^0, \ \epsilon^0 = \nabla\nabla w^0, \qquad (1.9)$$

where w^0, $\mathbf{M}^0$, ϵ^0 are weak limits of the sequences $\{w^s\}$, $\{\mathbf{M}^s\}$, $\{\varepsilon^s\}$ (in the corresponding spaces $W_2^2(\Omega)$, $L_2(\Omega)$, $L_2(\Omega)$) of the solutions of (1.2), (1.7) generated by a sequence of partitions $\{\chi^s\}$. Equivalently, these weak limits are averages of the values of w, $\mathbf{M}$, ϵ over neighborhoods of each point of Ω. The tensor $\mathbf{D}_0$, the G-limit of the sequence $\{\mathbf{D}^s\}$, is the effective stiffness tensor of a composite material of a plate. Since the sequence $\{\chi^s\}$ is bounded in $L_\infty(\Omega)$, it weakly* converges (with respect to $L_\infty(\Omega)$) to some function $m(\mathbf{x})$:

$$\chi^s(\mathbf{x}) \xrightarrow[L_\infty(\Omega)]{} m(\mathbf{x}), \ 0 \le m(\mathbf{x}) \le 1. \qquad (1.10)$$

This limit is interpreted as a volume fraction of the material with a compliance tensor $\mathbf{D}_1$ in a neighborhood of a point $\mathbf{x}$. Note that the G-limiting tensor $\mathbf{D}_0$ depends on $m(\mathbf{x})$ as well as on the geometrical structure of the composite material of the plate, that is, on the properties of $\{\chi^s\}$ [5]. In this paper we are looking for the composites that are optimal in the sense that associated elastic energy density is extremal. This density is defined at every point of Ω by

$$\Pi = \epsilon^0 \cdot\cdot\mathbf{D}_0 \cdot\cdot\epsilon^0 = \mathbf{M}^0 \cdot\cdot\mathbf{C}_0 \cdot\cdot\mathbf{M}^0, \qquad (1.11)$$

where by $\mathbf{C}_0 = \mathbf{D}_0^{-1}$ we denote the effective compliance tensor of the material.

The most "rigid" composite possesses the minimal compliance, that is, if the quadratic form $\Pi(\mathbf{C}_0, \mathbf{M}^0) = \mathbf{M}^0 \cdot \cdot \mathbf{C}_0 \cdot \cdot \mathbf{M}^0$ associated with the compliance tensor $\mathbf{C}_0$ of this composite is minimal among the composites of all microstructures for a given tensor $\mathbf{M}^0$. The most "soft" composite possesses the minimal stiffness; that is, the form $\Pi(\mathbf{D}_0, \boldsymbol{\epsilon}^0) = \boldsymbol{\epsilon}^0 \cdot \cdot \mathbf{D}_0 \cdot \cdot \boldsymbol{\epsilon}^0$ associated with the stiffness tensor $\mathbf{D}_0$ is minimal among the composites of all microstructures for a given strain tensor $\boldsymbol{\epsilon}^0$. In this sense these composites "respond" to the stressed state in the best way.

To characterize the limiting properties of optimal composites we estimate the forms $\Pi(\mathbf{D}_0, \boldsymbol{\epsilon}^0)$, $\Pi(\mathbf{C}_0, \mathbf{M}^0)$ by the functions $\underline{\Pi}$ and $\overline{\Pi}$ that depend only on

(1) weak-limit values of the strain tensor $\boldsymbol{\epsilon}^0$ (moments $\mathbf{M}^0$),
(2) stiffness tensors $\mathbf{D}_1$, $\mathbf{D}_2$ (compliance tensors $\mathbf{C}_1$, $\mathbf{C}_2$),
(3) volume fraction $m(\mathbf{x})$;

that is, we obtain the inequalities

$$\Pi(\mathbf{D}_0, \boldsymbol{\epsilon}^0) \geq \overline{\Pi}(m, \mathbf{D}_1, \mathbf{D}_2, \boldsymbol{\epsilon}^0), \ \forall \chi^s \underset{L_\infty(\Omega)}{\longrightarrow} m, \qquad (1.12)$$

$$\Pi(\mathbf{C}_0, \mathbf{M}^0) \geq \underline{\Pi}(m, \mathbf{C}_1, \mathbf{C}_2, \mathbf{M}^0), \ \forall \chi^s \underset{L_\infty(\Omega)}{\longrightarrow} m. \qquad (1.13)$$

The bounds (1.12) and (1.13), valid for all sequences $\{\chi^s\}$, are exact if there exist the sequences $\{\underline{\chi}^s\}$ and $\{\overline{\chi}^s\}$ for which the inequalities (1.12), (1.13) become equalities. What follows is the method of obtaining the bounds for the energy density and the demonstration that they are exact.

Remark 1.1. The tensors $\mathbf{C}^0$ and $\mathbf{D}_0$ are determined by the values of the quadratic form Π up to the terms $\mathbf{C}_\perp$ ($\mathbf{C}_\perp \cdot \cdot \mathbf{M}^0 = 0$) and $\mathbf{D}_\perp$ ($\mathbf{D}_\perp \cdot \cdot \mathbf{C}^0 = 0$) orthogonal to $\mathbf{M}^0$ and $\mathbf{C}^0$, correspondingly. To minimize the quadratic forms $\Pi(\mathbf{C}_0, \mathbf{M}^0)$ and $\Pi(\mathbf{D}_0, \boldsymbol{\epsilon}^0)$, it is sufficient to find a minimum with respect to either of the tensors $\mathbf{C}_0$ or $\mathbf{D}_0$ from the corresponding equivalence classes. To completely describe the limiting tensors $\mathbf{C}_0$ and $\mathbf{D}_0$ it is necessary to bound the minimum of the linear combinations

$$\sum_{i=1}^{3} \gamma_i \mathbf{M}^0(i) \cdot \cdot \mathbf{C}_0 \cdot \cdot \mathbf{M}^0(i) + \delta_i \boldsymbol{\epsilon}^0(i) \cdot \cdot \mathbf{D}_0 \cdot \cdot \boldsymbol{\epsilon}^0(i)$$

of the quadratic forms (1.11) that correspond to the three linearly independent tensors of bending moments $\mathbf{M}^0(i)$ and strain tensors $\boldsymbol{\epsilon}^0(i)$), $i = 1, 2, 3$.

§2. Derivation of the bounds for the stiffness.

Let us be given a scalar function h^s defined by $h^s = h(\Delta(\chi^s), w^s)$. The first argument is a characteristic function χ^s of the partition of the given region Ω and the second is some tensor.

We assume that a sequence of characteristic functions $\{\chi^s\}$ converges (weakly* with respect to $L_\infty(\Omega)$) to m (cf. (1.10)) and that $\Delta(\chi^s)$ converges to a function Δ_0 in the same sense:

$$\Delta_0 = \lim_{\chi^s \to m} \Delta(\chi^s) = m\Delta(1) + (1-m)\Delta(0), \qquad (2.1)$$

which is determined only by a value of m. Let us also be given a sequence of tensors w^s that converges (weakly with respect to $L_p(\Omega)$) to a tensor w^0, $(1 \le p < \infty)$:

$$w^s \xrightarrow[L_p(\Omega)]{} w^0.$$

We would like to estimate the limit of the sequence

$$\{h^s\} = \{h(\Delta(\chi^s), w^s)\}$$

by using the functions Δ_0, w^0.

If $h(\Delta, w)$ is an affine (convex) function in both arguments, then the functional

$$I(h) = \int_{\Omega_\epsilon(x)} h(\Delta(\chi^s), w^s) d\Omega,$$

where $\Omega_\epsilon(x)$ is some neighborhood of a point $x \in \Omega$, is weakly continuous (weakly semicontinuous):

$$\lim_{weak} I(h^s) = I(h^0), \qquad \left(\lim_{weak} I(h^s) \ge I(h^0) \right)$$

from which the bound follows:

$$\lim_{s \to \infty} h^s \ge h^0 \quad \text{a.e. in } \Omega. \qquad (2.3)$$

Example 1. One can easily check that the function $\Pi(\mathbf{D}(x), \epsilon)$ is convex in arguments $\mathbf{D}^{-1}$ and ϵ. Taking into account (1.7), and (1.10) we obtain

$$\lim_{weak} \Pi(\mathbf{D}(\chi^s), \epsilon^s) \ge \epsilon_0 \cdot \cdot \left[\lim_{weak} \mathbf{D}^{-1}(\chi^s) \right]^{-1} \cdot \cdot \epsilon_0$$

$$= \epsilon_0 \cdot \cdot \left[m\mathbf{D}_1^{-1} + (1-m)\mathbf{D}_2^{-1} \right]^{-1} \cdot \cdot \epsilon_0; \qquad (2.4)$$

analogously, we obtain:

$$\lim_{weak} \Pi(\mathbf{C}(\chi^s), \mathbf{M}^s) \geq \mathbf{M}^0 \cdot \cdot \left[\lim_{weak} \mathbf{C}^{-1}(\chi^s) \right]^{-1} \cdot \cdot \mathbf{M}^0$$

$$= \mathbf{M}^0 \cdot \cdot \left[m\mathbf{C}_1^{-1} + (1-m)\mathbf{C}_2^{-1} \right] \cdot \cdot \mathbf{M}^0. \qquad (2.5)$$

On the other hand, the energy Π of the composite is a weak limit of a sequence of energies of the inhomogeneous medium [7]:

$$\lim_{weak} \Pi(\mathbf{D}(\chi^s), \epsilon^s) = \Pi(\mathbf{D}_0, \epsilon^0),$$

$$\lim_{weak} \Pi(\mathbf{C}(\chi^s), \mathbf{M}^s) = \Pi(\mathbf{C}_0, \mathbf{M}^0).$$

Therefore the right-hand sides of (2.4) and (2.5) bound the value of the energy of the composite. These bounds are the best of bounds that can be obtained for arbitrary weakly convergent sequences of tensors $\mathbf{M}^s$, ϵ^s; the right-hand side of these bounds is the largest function of weak limits m and ϵ^0 (m and $\mathbf{M}^0$) that does not exceed the weak limit of the sequence Π_s.

Remark 2.1. In problems with one independent variable, the bounds of the type (2.4) and (2.5) are exact [8]. They are also exact in a problem of bounds for the energy of the operator $\nabla \cdot \mathbf{D} \cdot \nabla$ in the space of any number of independent variables and are known as the Reuss-Voigt bounds.

If the tensors w^s are not arbitrary, that is, a certain linear combination of some of their elements is bounded in L_p ($1 \leq p \leq \infty$), the class of functions $h(\Delta(\chi^s), w^s)$ that generates the semi-continuous functionals becomes wider: in addition to convex (affine) functions it includes the so-called quasiconvex (quasiaffine) functions h [9–12]. In the following we give examples of the functions that are not convex but are quasiconvex (quasiaffine).

Example 2. If the tensor ϵ satisfies (1.2): $\epsilon = \nabla\nabla w$, (i.e., $curl(\epsilon) = 0$), then the function $\varphi(\epsilon) = \det \epsilon$ is quasiaffine [7,11]:

$$\det \epsilon^s \xrightarrow[L_1(\Omega)]{} \det \epsilon_0.$$

It is convenient in the sequel to write a quadratic form

$$\det \epsilon = \epsilon_{11}\epsilon_{22} - \epsilon_{12}^2 = \frac{\partial^2 w}{\partial x^2} \frac{\partial^2 w}{\partial y^2} - \left(\frac{\partial^2 w}{\partial x \partial y} \right)^2$$

as [7]:

$$\det \epsilon = \epsilon \cdot \cdot \mathbf{D}_{I_2} \cdot \cdot \epsilon,$$
$$\mathbf{D}_{I_2} = \mathbf{a}_1 \mathbf{a}_1 - \mathbf{a}_2 \mathbf{a}_2 - \mathbf{a}_3 \mathbf{a}_3, \tag{2.6}$$

where $\mathbf{a}_1$, $\mathbf{a}_2$, $\mathbf{a}_3$ is the tensor basis (1.4).

Example 3. If the tensor $\mathbf{M}$ satisfies (1.2), then the function

$$\psi(\mathbf{M}) = -\mathbf{M} \cdot \cdot \mathbf{D}_{I_2} \cdot \cdot \mathbf{M}$$

is quasiconvex and

$$\lim_{weak} (-\mathbf{M}^s \cdot \cdot \mathbf{D}_{I_2} \cdot \cdot \mathbf{M}^s) \geq -\mathbf{M}^0 \cdot \cdot \mathbf{D}_{I_2} \cdot \cdot \mathbf{M}^0. \tag{2.7}$$

Proof. The tensor $\mathbf{M}^s$ that satisfies the differential restrictions

$$\nabla\nabla \cdot \cdot \mathbf{M}^s = f, \ (f \in L_2(\Omega))$$

can be expressed in the form:

$$\mathbf{M}^s = \overline{\mathbf{M}} + \frac{1}{2}\mathbf{R}^T \cdot (\nabla\mathbf{u}^s + (\nabla\mathbf{u}_s)^T) \cdot \mathbf{R},$$

where $\overline{\mathbf{M}}$ is a special solution of an equation $\nabla\nabla \cdot \cdot \overline{\mathbf{M}} = f$, $\mathbf{R} = \mathbf{ij} - \mathbf{ji}$ is a tensor of rotation by $\pi/2$, $\mathbf{u}^s = (u_1^s, u_2^s)$ is some vector.

Let us consider a part of the expression for $\mathbf{M} \cdot \cdot \mathbf{D}_{I_2} \cdot \cdot \mathbf{M}$ that is non-linear in $\mathbf{u}^s$:

$$\det\left(\frac{1}{2}\mathbf{R}^T \cdot (\nabla\mathbf{u}^s + (\nabla\mathbf{u}^s)^T) \cdot \mathbf{R}\right) = \frac{\partial u_1^s}{\partial x}\frac{\partial u_2^s}{\partial y} - \frac{1}{4}\left(\frac{\partial u_1^s}{\partial y} + \frac{\partial u_2^s}{\partial x}\right)^2$$
$$= \det(\nabla u_1^s, \nabla u_2^s) - \frac{1}{4}\left(\frac{\partial u_1^s}{\partial y} - \frac{\partial u_2^s}{\partial x}\right)^2.$$

Taking into account that the first term of the right-hand side is quasiaffine [11]:

$$\det(\nabla u_1^s, \nabla u_2^s) \xrightarrow[L_1(\Omega)]{} \det(\nabla u_1^0, \nabla u_2^0),$$

and the second is concave, we arrive at (2.7).

Let us assume that we found some quasiconvex functions $\psi_i(w)$ and quasiaffine functions $\varphi_j(w)$. By using these functions we can sharpen the bound (2.2) for the function $h(\Delta(\chi), w)$ which is convex in both arguments. Indeed, let us form a linear combination:

$$h - \sum_i \alpha_i \varphi_i - \sum_j \beta_j \psi_j = g(\Delta'(\chi, \alpha_i, \beta_j), w),$$

where α_i, β_j are some real coefficients and g is a scalar function of w and Δ'. The argument Δ' depends on χ and α_i, β_j. If the parameters α_i, β_j are chosen in such a way that:

$$g = g(\Delta', w) \quad \text{is convex}, \tag{2.8}$$

then we have the bound:

$$\lim_{weak} g(\Delta'_s, w^s) \geq g(\Delta'_0, w^0),$$

where we defined

$$\Delta'_0(m, \alpha_i, \beta_j) = \lim_{weak} \Delta'_0(\chi^s, \alpha_i, \beta_j)$$
$$= m\Delta'(1, \alpha_i, \beta_j) + (1 - m)\Delta'(0, \alpha_i, \beta_j).$$

On the other hand, if $\beta_j \geq 0$, then

$$\lim_{weak} g(\Delta'_s, w^s) =$$

$$\lim_{weak} \left[h(\Delta(\chi^s), w^s) - \sum_i \alpha_i \varphi_i(w^s) - \sum_j \beta_j \psi_j(w^s) \right] \leq$$

$$\lim_{weak} h(\Delta(\chi^s), w^s) - \sum_i \alpha_i \varphi_i(w^0) - \sum_j \beta_j \psi_j(w^0).$$

where we took into account quasiconvexity of the functions $\varphi_i(w^s)$ and $\psi_j(w^s)$. The corresponding bound for $h(\Delta, w)$ is then:

$$\lim_{weak} h(\Delta(\chi^s), w^s) \geq g(\Delta'_0, w^0) + \sum_i \alpha_i \varphi_i(w^0) + \sum_j \beta_j \psi_j(w^0). \tag{2.9}$$

We note that the right-hand side of (2.9) is quasiconvex in w, although it is not convex. The bound (2.9) is sharper than the bound (2.2) because of this; it reduces to the latter if $\alpha_i = \beta_j = 0$. The inequality (2.9) is valid for all admissible values of the parameters α_i, β_j; the sharpest bound in these parameters is of the kind:

$$\lim_{weak} h(\Delta(\chi^s), w^s) \geq \tag{2.10}$$

$$\max_{\alpha_i, \beta_j \text{ as in } (2.8), \beta_j \geq 0} \left\{ \sum_j \beta_j \psi_j(w^0) + \sum_i \alpha_i \varphi_i(w^0) + g(\Delta'_0(m, \alpha_i, \beta_j), w^0) \right\}.$$

Example 4. A composite of maximal stiffness.
   Let us put

$$\Delta(\chi^s) = \mathbf{C}^{-1}(\chi^s), \ w = \mathbf{M}, \ h(\Delta, w) = \mathbf{M} \cdot\cdot(\Delta(\chi^s))^{-1} \cdot\cdot \mathbf{M} = \mathbf{M} \cdot\cdot \mathbf{C} \cdot\cdot \mathbf{M},$$

$$\psi(\mathbf{M}) = -\mathbf{M} \cdot\cdot \mathbf{D}_{I_2} \cdot\cdot \mathbf{M}$$

(see Example 3). Then

$$\Delta'(\chi^s, \beta) = (\mathbf{C} + \beta \mathbf{D}_{I_2})^{-1}, \ g(\Delta'_0, \mathbf{M}^0) = \mathbf{M}^0 \cdot\cdot (\Delta')^{-1} \cdot\cdot \mathbf{M}^0.$$

The bound (2.10) then becomes:

$$\mathbf{M}^0 \cdot\cdot \mathbf{C}^0 \cdot\cdot \mathbf{M}^0 \geq \max_{\beta \text{ as in } (2.8), \beta > 0} \mathbf{M}^0 \cdot\cdot \left[(\Delta'_0)^{-1} - \beta \mathbf{D}_{I_2} \right] \cdot\cdot \mathbf{M}^0 \tag{2.11.}$$

Here (cf.(1.3), (1.8), (2.1), and (2.6))

$$\Delta'_0 = m(\mathbf{C}_1 + \beta \mathbf{D}_{I_2})^{-1} + (1 - m)(\mathbf{C}_2 + \beta \mathbf{D}_{I_2})^{-1}, \tag{2.12}$$

$$\mathbf{C}_i + \beta \mathbf{D}_{I_2} = \left(\frac{1}{\kappa_i} + \beta \right) \mathbf{a}_1 \mathbf{a}_1 + \left(\frac{1}{\mu_i} - \beta \right) (\mathbf{a}_2 \mathbf{a}_2 + \mathbf{a}_3 \mathbf{a}_3), \ i = 1, 2.$$

The condition of quasiconvexity (2.8) is equivalent to the requirement that the eigenvalues of the tensors $\mathbf{C}_i + \beta \mathbf{D}_{I_2}$ be nonnegative. Together with the condition $\beta \geq 0$, it characterizes the domain of admissible values of the parameter β:

$$\beta \geq 0, \ \beta \text{ as in } (2.8) \Leftrightarrow \beta \in \left[0, \frac{1}{\mu_{\max}} \right] = \left[0, \frac{1}{\mu_2} \right]. \tag{2.13}$$

Example 5. A composite of minimal stiffness.

Let us put

$$\Delta(\chi) = \mathbf{D}^{-1}(\chi), \ w = \epsilon, \ h(\Delta, w) = \epsilon \cdot \cdot (\Delta)^{-1} \cdot \cdot \epsilon = \epsilon \cdot \cdot \mathbf{D} \cdot \cdot \epsilon,$$

$$\varphi = \epsilon \cdot \cdot \mathbf{D}_{I_2} \cdot \cdot \epsilon$$

(see Example 2). Then

$$\Delta'(\chi, \alpha) = (\mathbf{D}(\chi) + \alpha \mathbf{D}_{I_2})^{-1}, \ g = \epsilon \cdot \cdot (\Delta')^{-1} \cdot \cdot \epsilon.$$

The bound (2.10) then becomes:

$$\varepsilon^0 \cdot \cdot \Delta_0 \cdot \cdot \varepsilon^0 \geq \max_{\alpha \text{ as in (2.8)}} \epsilon^0 \cdot \cdot \left[(\Delta_0')^{-1} - \alpha \mathbf{D}_{I_2}\right] \cdot \cdot \epsilon^0, \qquad (2.14)$$

where Δ' is (cf. (1.3), (1.7), (2.1), and (2.6)):

$$\Delta'_0 = m\left(\mathbf{D}_1 + \alpha \mathbf{D}_{I_2}\right)^{-1} + (1-m)(\mathbf{D}_2 + \alpha \mathbf{D}_{I_2})^{-1},$$
$$\mathbf{D}_i + \alpha \mathbf{D}_{I_2} = (\kappa_i + \alpha)a_1 a_1 + (\mu_i - \alpha)(a_2 a_2 + a_3 a_3), \ i = 1, 2.$$

The domain of admissible values of the parameter α can then be determined from the condition (2.8). Namely, if the the inequalities (1.5) hold, then

$$-\kappa_1 \leq \alpha \leq \mu_1. \qquad (2.16)$$

If the inequalities (1.6) hold, then

$$-\kappa_2 \leq \alpha \leq \mu_1. \qquad (2.17)$$

Remark 2.2. The process of sharpening of the bounds (2.2) is related to a problem of finding suitable nonconvex but quasiconvex functions φ_i, ψ_i. The greater the number of such functions used in the procedure, the sharper are the bounds obtained. As we show in the following, the inequalities (2.11) and (2.14) reduce to equalities for certain sequences $\{\overline{\chi^s}\}$, $\{\underline{\chi^s}\}$, or for composites of a special structure (see §3). This means that in this case all the necessary corollaries to the differential properties of the tensors ϵ and $\mathbf{M}$ are applied.

The bounds (2.11) and (2.14) can also be obtained by constructing polyconvex envelopes [12, 13]. To do this one has to write the expression for $h(\Delta, w)$ as a function that is convex in the arguments Δ, w, φ_i, ψ_j.

Remark 2.3. This method was used [14] in the problem of describing the set $G_m U$ (i.e., the G_m-closure of the set of the original materials) for the operator $\nabla \cdot \mathbf{D} \cdot \nabla$. In that problem it has turned out to be possible to describe the effective properties of all the composites built from the two given phases, i.e. to obtain the exact bounds for the tensor $\mathbf{D}_0$ of the effective properties of every composite, which are independent of the stress state of the system. At the same time, for a number of optimization problems related to the same operator, it was shown in [15] that the optimal structures are simple laminates (which obviously do not exhaust the set $G_m U$). Analogously, in the problems that we discuss in Part II, one can only look for the composites that are extremal only in the sense of the bounds (2.11) and (2.14).

Finally, we note that in [15] it was sufficient to use the inequalities of the kind (2.4) and (2.5), whereas in the case of the operator $\nabla\nabla \cdot\cdot \mathbf{D} \cdot\cdot \nabla\nabla$ it is necessary to take into account the quasiconvexity of the functions (2.6) and (2.7).

§3. The bounds for the stiffness of a composite.

In this section we obtain the explicit expressions for the bounds on the elastic energy density of composites.

We begin by considering the bounds for the elastic energy of composites of maximal stiffness. Taking (1.3), (1.8) and (2.12) into account, we can write the inequality (2.11) as

$$\mathbf{M}^0 \cdot\cdot \mathbf{C}^0 \cdot\cdot \mathbf{M}^0 \geq \Pi = \max_{\beta \in [0,1/\mu_2]} \mathbf{M}^0 \cdot\cdot \underline{\mathbf{C}} \cdot\cdot \mathbf{M}^0. \tag{3.1}$$

Here

$$\underline{\mathbf{C}} = \{m \, (\mathbf{C}_1 + \beta \mathbf{D}_{I_2})^{-1} + (1 - m) \, (\mathbf{C}_2 + \beta \mathbf{D}_{I_2})^{-1}\}^{-1} - \beta \mathbf{D}_{I_2} \tag{3.2}$$

is an isotropic self-adjoint tensor which is defined by the dyadic expression:

$$\underline{\mathbf{C}}(\beta) = \underline{\kappa}^{-1}(\beta)\mathbf{a}_1\mathbf{a}_1 + \underline{\mu}^{-1}(\beta) \, (\mathbf{a}_2\mathbf{a}_2 + \mathbf{a}_3\mathbf{a}_3),$$

$$\underline{\kappa}^{-1}(\beta) = \left(\frac{m}{1/\kappa_1 + \beta} + \frac{1 - m}{1/\kappa_2 + \beta} \right)^{-1} - \beta,$$

$$\underline{\mu}^{-1}(\beta) = \left(\frac{m}{1/\mu_1 - \beta} + \frac{1 - m}{1/\mu_2 - \beta} \right)^{-1} + \beta. \tag{3.3}$$

The critical value β_0 of the parameter β that maximizes the right-hand side of (3.1) depends on the tensor $\mathbf{M}^0$ which can be presented in tensor basis $\mathbf{a}_1$, $\mathbf{a}_2$, $\mathbf{a}_3$ (cf. (1.4)) as:

$$\mathbf{M}^0 = M_1\mathbf{a}_1 + M_2\mathbf{a}_2. \tag{3.4}$$

Here the axes $\mathbf{i}, \mathbf{j}$ of these tensor bases are taken to coincide with the principal axes of the tensor $\mathbf{M}^0$. In addition, we require that the inequality

$$\mathbf{M}_1 \mathbf{M}_2 \geq 0$$

be satisfied, where $\mathbf{M}_1$, $\mathbf{M}_2$ are the components of the tensor $\mathbf{M}^0$ in the basis $\mathbf{a}_i$, $i = 1, 2, 3$. The values $\mathbf{M}_1$, $\mathbf{M}_2$ are given in terms of the eigenvalues λ_1, λ_2 ($\lambda_1 \geq \lambda_2$) of the tensor of bending moments by the formulae

$$\mathbf{M}_1 = \frac{1}{\sqrt{2}}(\lambda_1 + \lambda_2), \ \ \mathbf{M}_2 = \frac{1}{\sqrt{2}}(\lambda_1 - \lambda_2).$$

The optimal value β_0 of β depends on the parameter ξ given by

$$\xi = \frac{\mathbf{M}_2}{\mathbf{M}_1}, \ \ (0 \leq \xi \leq \infty).$$

which characterizes the stress state. Namely, $\xi = 0$ corresponds to a bulk tensor $\mathbf{M}^0$, $\xi = 1$ corresponds to the uniaxial stretching, and $\xi = \infty$ corresponds to a case of pure shear.

The right hand side of (3.1) is maximized either by the root β_0 of the equation $\dfrac{\partial \Pi}{\partial \beta} = 0$, subject to conditions

$$\left. \frac{\partial^2 \Pi}{\partial \beta^2} \right|_{\beta = \beta_0} \leq 0, \ 0 \leq \beta_0 \leq \frac{1}{\mu_2}, \tag{3.5}$$

or by the boundary value of β. The equation (3.1) has the roots β_{01}, and β_{02} (see Table 1). If the moduli of the phases satisfy the inequalities (1.5), then, as one easily checks, only the root β_{01} satisfies the first inequality of (3.5).

By the second inequality of (3.5), the values of ξ change in three intervals. In the first (interval 1.1 of Table 1) the root β_{01} is negative and the maximum of (3.5) is reached for $\beta = 0$. In this case the bound (3.1) takes the form:

$$\mathbf{M}^0 \cdot\cdot \mathbf{C}^0 \cdot\cdot \mathbf{M}^0 \geq \underline{\Pi}_{11}.$$

The values of $\underline{\Pi}$ are given in Table 1a. In the interval 1.2 (Table 1a) the root β_{01} satisfies both inequalities (3.5) and is the optimal value of the parameter β. Here the inequality (3.1) becomes:

$$\mathbf{M}^0 \cdot\cdot \mathbf{C}^0 \cdot\cdot \mathbf{M}^0 \geq \underline{\Pi}_{12}.$$

Finally, if the values of ξ belong to the interval 1.3 (Table 1a), then $\beta_{01} > 1/\mu_2$ and the maximum of the right-hand side of (3.1) is reached for $\beta_0 = 1/\mu_2$, and the bound (3.1) is given by:

$$\mathbf{M}^0 \cdot\cdot \mathbf{C}^0 \cdot\cdot \mathbf{M}^0 \geq \underline{\mathrm{II}}_{13}.$$

Table 1a. Energy bounds for the stiff plate. Case (1.5) $(\Delta\kappa\Delta\mu \geq 0)$

Interval 1.1	Interval 1.2	Interval 1.3
$\xi \geq \xi_1$	$\xi_2 \leq \xi \leq \xi_1$	$\xi \leq \xi_2$
$\beta_0 = 0$	$\beta_0 = \beta_{01}$	$\beta_0 = \mu_2^{-1}$
$\underline{\mathrm{II}}_{11}$	$\underline{\mathrm{II}}_{12}$	$\underline{\mathrm{II}}_{13}$

Notations:

$$\xi_1 = \frac{\widetilde{(\mu^{-1})}\Delta(\kappa^{-1})}{\widetilde{(\kappa^{-1})}\Delta(\mu^{-1})}, \xi_2 = \frac{(1-m)\Delta(\kappa^{-1})}{\widetilde{(\kappa^{-1})} + \mu_2^{-1}},$$

$$\Delta a = a_1 - a_2, \quad <a> = ma_1 + (1-m)a_2, \quad \widetilde{(a)} = ma_2 + (1-m)a_1,$$

$$\beta_{01} = \frac{\Delta(\kappa^{-1})\widetilde{(\mu^{-1})} - \Delta(\mu^{-1})\widetilde{(\kappa^{-1})}\xi}{\Delta(\kappa^{-1}) + \Delta(\mu^{-1})\xi},$$

$$\underline{\mathrm{II}}_{11} = <\kappa>^{-1} \mathbf{M}_1^2 + <\mu>^{-1} \mathbf{M}_2^2,$$

$$\underline{\mathrm{II}}_{12} = <\kappa^{-1}> \mathbf{M}_1^2 + <\mu^{-1}> \mathbf{M}_2^2 - \frac{m(1-m)[\Delta(\kappa^{-1})\mathbf{M}_1 + \Delta(\mu^{-1})\mathbf{M}_2]^2}{\widetilde{(\kappa^{-1})} + \widetilde{(\mu^{-1})}},$$

$$\underline{\mathrm{II}}_{13} = \mu_2^{-1}\mathbf{M}_2^2 + \left[\left(\frac{m}{\kappa_1^{-1} + \mu_2^{-1}} + \frac{1-m}{\kappa_2^{-1} + \mu_2^{-1}}\right) - \frac{1}{\mu_2}\right]\mathbf{M}_1^2$$

Table 1b. Energy bounds for the stiff plate. Case (1.6) $(\Delta\kappa\Delta\mu \leq 0)$

Interval 2.1	Interval 2.2	Interval 2.3
$\xi \geq \xi_3$	$\xi_4 \leq \xi \leq \xi_3$	$\xi \leq \xi_4$
$\beta_0 = 0$	$\beta_0 = \beta_{02}$	$\beta_0 = \mu_2^{-1}$
$\underline{\mathrm{II}}_{21} = \underline{\mathrm{II}}_{11}$	$\underline{\mathrm{II}}_{22}$	$\underline{\mathrm{II}}_{23} = \underline{\mathrm{II}}_{13}$

Notations:

$$\xi_3 = -\frac{\widetilde{(\mu^{-1})}\Delta(\kappa^{-1})}{\widetilde{(\kappa^{-1})}\Delta(\mu^{-1})}, \xi_4 = -\frac{(1-m)\Delta(\kappa^{-1})}{\widetilde{(\kappa^{-1})}+\mu_2^{-1}},$$

$$\beta_{02} = \frac{\Delta(\kappa^{-1})\widetilde{(\mu^{-1})}+\Delta(\mu^{-1})\widetilde{(\kappa^{-1})}\xi}{\Delta(\kappa^{-1})-\Delta(\mu^{-1})\xi},$$

$$\underline{\Pi}_{22} = <\kappa^{-1}>\mathbf{M}_1^2 + <\mu^{-1}>\mathbf{M}_2^2 - \frac{m(1-m)[\Delta(\kappa^{-1})\mathbf{M}_1-\Delta(\mu^{-1})\mathbf{M}_2]^2}{\widetilde{(\kappa^{-1})}+\widetilde{(\mu^{-1})}}$$

Let us comment on the bounds just obtained.

In the interval 1.1 (Table 1a) of values of the parameter ξ (which includes the value $\xi = \infty$, i.e., the stress state of pure shear) the energy of the composite is bounded below by the energy of a certain isotropic medium whose compliance tensor does not depend on the stress state and is determined by the Reuss bound (2.5).

If the stress state is such that the parameter ξ belongs to the interval 1.2 (Table 1); (including the value $\xi = 1$ which is the case of bending with respect to one axis), then the bound $\underline{\Pi}_{12}$ remains a quadratic form in $\mathbf{M}^0$, in spite of the fact that $\beta = \beta(\xi)$. This means that in this case as well, the energy of the composite is bounded below by the energy of some anisotropic medium whose elastic moduli are also independent of ξ. Finally, in the interval 1.3 (Table 1a), including the value $\xi = 0$ which corresponds to the bulk stress field, the energy is bounded below by the energy of an isotropic medium whose shear modulus is μ_2 (cf. (1.3), (1.11) and Table 1a), that is, is determined only by the shear modulus of the second phase.

In §4 we show that all the bounds just derived can be realized by the composites of a special structure. The form of expressions $\underline{\Pi}_{11}$, $\underline{\Pi}_{12}$, $\underline{\Pi}_{13}$ suggests that the corresponding composites will be of different structures.

The treatment of the case when the elastic moduli of the phases satisfy the inequalities (1.6) is analogous. Here the inequality (3.5) is satisfied by the root β_{02} (cf. Table 1b). The optimal values β_0 of the parameter β and the corresponding bounds for the energy $\underline{\Pi}$ are listed in Table 1b.

The bounds for the energy of the composites of minimal stiffness are obtained in the same way. The inequality (2.14) is written, taking into account (3.1) and (2.15)–(2.17), in the form:

$$\epsilon^0 \cdot \cdot \mathbf{D}_0 \cdot \cdot \epsilon^0 \geq \max_{\alpha \in [-min(\kappa_1,\kappa_2),\mu_1]} \epsilon^0 \cdot \cdot \underline{\mathbf{D}}(\alpha) \cdot \cdot \epsilon^0, \tag{3.6}$$

where

$$\underline{\mathbf{D}}(\alpha) = \{m\,(\mathbf{D}_1 + \alpha\mathbf{D}_{I_2})^{-1} + (1-m)\,(\mathbf{D}_2 + \alpha\mathbf{D}_{I_2})^{-1}\}^{-1} - \alpha\mathbf{D}_{I_2}.$$

The tensor $\underline{\mathbf{D}}(\alpha)$ is given by a dyadic expansion:

$$\underline{\mathbf{D}}(\alpha) = \underline{\kappa}(\alpha)\mathbf{a}_1\mathbf{a}_1 + \underline{\mu}(\alpha)\,(\mathbf{a}_2\mathbf{a}_2 + \mathbf{a}_3\mathbf{a}_3),$$

$$\underline{\kappa}(\alpha) = \left(\frac{m}{\kappa_1 + \alpha} + \frac{1-m}{\kappa_2 + \alpha}\right)^{-1} - \alpha,$$

$$\underline{\mu}(\alpha) = \left(\frac{m}{\mu_1 - \alpha} + \frac{1-m}{\mu_2 - \alpha}\right)^{-1} + \alpha.$$

The critical value α_0 of the parameter α in (3.6) depends on the parameter

$$\zeta = \frac{\varepsilon_2}{\varepsilon_1}, \ \ \varepsilon_1 = \boldsymbol{\epsilon}\cdot\cdot\mathbf{a}_1, \ \ \varepsilon_2 = \boldsymbol{\epsilon}\cdot\cdot\mathbf{a}_2,$$

where the principal axes $\mathbf{i}, \mathbf{j}$ of the tensor basis $\mathbf{a}_1, \mathbf{a}_2, \mathbf{a}_3$ coincide with the principal axes of the strain tensor $\boldsymbol{\epsilon}$; moreover, $\varepsilon_1\varepsilon_2 \geq 0$. (If λ_1, λ_2 are the eigenvalues of the tensor $\boldsymbol{\epsilon}$, then

$$\varepsilon_1 = \frac{1}{\sqrt{2}}(\lambda_1 + \lambda_2), \ \ \varepsilon_2 = \frac{1}{\sqrt{2}}(\lambda_1 - \lambda_2).$$

The parameter ζ $(0 \leq \zeta \leq \infty)$ characterizes the deformation of a composite. Namely, the case $\zeta = 0$ corresponds to bulk-type strain, the case $\zeta = 1$ is that of axial strain, and the case $\zeta = \infty$ corresponds to the shear strain. The values $\alpha(\zeta)$ and $\overline{\Pi}(\zeta)$ are given in Table 2.

The bounds of the present case are different from the bounds for the composites of maximal stiffness because negative values of α are also allowed. When ζ is in the intervals 1.1 and 1.2, the energy is bounded below by the value $\overline{\Pi}_{11}$ $(\overline{\Pi}_{12})$ of the energy of the corresponding medium whose bulk modulus depends either only on the value of κ_1 (in the case of (1.5)) or only on the value of κ_2 (in the case of (1.6)).

§4. Structures that saturate the bounds.

Let us show that the bounds $\underline{\Pi}$ $(\overline{\Pi})$ are exact in the sense that for any tensor $\mathbf{M}^0$ $(\boldsymbol{\epsilon}^0)$ there can be found a composite built from the amounts m and $(1-m)$ of the materials with stiffness tensors $\mathbf{D}_1$ and $\mathbf{D}_2$ possessing the energy:

$$\Pi(\mathbf{C}_0(m), \mathbf{M}^0) = \underline{\Pi}, \ \ \Pi(\mathbf{D}_0(m), \boldsymbol{\epsilon}^0) = \overline{\Pi}.$$

To prove this for a case of a plate of maximal stiffness, we consider a laminate composite built from orthotropic materials with compliance tensors

C_1 and C_2. If the directions of the principal axes of tensors C_1, C_2 coincide with this laminate with the normal to layers n and the tangent to layers t, then the effective compliance tensor is given by (compare to [16]):

$$C_I(m) = C_1 + (1-m)\left[\Delta C - \frac{m\Delta C \cdot\cdot tttt \cdot\cdot \Delta C}{tt \cdot\cdot (C_1 + m\Delta C) \cdot\cdot tt} \right.$$
$$\left. - \frac{m\Delta C \cdot\cdot (nt + tn)(nt + tn) \cdot\cdot \Delta C}{(nt + tn) \cdot\cdot (C_1 + m\Delta C) \cdot\cdot (nt + tn)} \right]. \tag{4.1}$$

Here $\Delta C = C_2 - C_1$ and m is the volume fraction of the material with compliance C_1, the tensor C_I is a compliance tensor of the orthotropic medium, and n, t are the normal and tangent to the laminates. In the basis

$$a_1' = \frac{1}{\sqrt{2}}(nn + tt), a_2' = \frac{1}{\sqrt{2}}(nn - tt), a_3' = \frac{1}{\sqrt{2}}(nt + tn) \tag{4.2}$$

Table 2a. Energy bounds for the soft plate. Case (1.5) $(\Delta\kappa\Delta\mu \geq 0)$

Interval 1.1	Interval 1.2	Interval 1.3
$\zeta \geq \zeta_1$	$\zeta_2 \leq \zeta \leq \zeta_1$	$\xi \leq \xi_2$
$\alpha_0 = -\kappa_1$	$\alpha_0 = \alpha_{01}$	$\alpha_0 = \mu_1$
$\overline{\Pi}_{11}$	$\overline{\Pi}_{12}$	$\overline{\Pi}_{13}$

Notations:

$$\zeta_1 = -\frac{\tilde{\mu} + \kappa_1}{m\Delta\mu}, \zeta_2 = -\frac{m\Delta\kappa}{\tilde{\kappa} + \mu_1}, \alpha_{01} = \frac{\Delta\kappa\tilde{\mu} - \Delta\mu\tilde{\kappa}\zeta}{\Delta\kappa + \Delta\mu\zeta},$$

$$\overline{\Pi}_{11} = \kappa_1\epsilon_1^2 + \left[\left(\frac{m}{\kappa_1 + \mu_1} + \frac{1-m}{\kappa_1 + \mu_2} \right) - \kappa_1 \right] \epsilon_2^2,$$

$$\overline{\Pi}_{12} = <\kappa> \epsilon_1^2 + <\mu> \epsilon_2^2 - \frac{m(1-m)[\Delta\kappa\epsilon_1 + \Delta\mu\epsilon_2]^2}{\tilde{\kappa} + \tilde{\mu}},$$

$$\overline{\Pi}_{13} = \mu_1\epsilon_2^2 + \left[\left(\frac{m}{\kappa_1 + \mu_1} + \frac{1-m}{\kappa_2 + \mu_1} \right) - \mu_1 \right] \epsilon_1^2.$$

Table 2b. Energy bounds for the soft plate. Case (1.6) ($\Delta\kappa\Delta\mu \leq 0$)

Interval 2.1	Interval 2.2	Interval 2.3
$\zeta \geq \zeta_3$	$\zeta_4 \leq \zeta \leq \zeta_3$	$\zeta \leq \zeta_4$
$\alpha_0 = 0 - \kappa_2$	$\alpha_0 = \alpha_{02}$	$\alpha_0 = \mu_1$
$\overline{\overline{\Pi}}_{21}$	$\overline{\overline{\Pi}}_{22}$	$\overline{\overline{\Pi}}_{23} = \overline{\overline{\Pi}}_{13}$

Notations:

$$\zeta_3 = -\frac{\widetilde{\mu} + \kappa_2}{m\Delta\mu}, \zeta_4 = \frac{m\Delta\kappa}{\widetilde{\kappa} + \mu_1}, \alpha_{02} = \frac{\Delta\kappa\widetilde{\mu} + \Delta\mu\widetilde{\kappa}\zeta}{\Delta\kappa - \Delta\mu\zeta},$$

$$\overline{\overline{\Pi}}_{21} = \kappa_2\epsilon_1^2 + \left[\left(\frac{m}{\kappa_2 + \mu_1} + \frac{1 - m}{\kappa_2 + \mu_2}\right) - \kappa_2\right]\epsilon_2^2,$$

$$\overline{\overline{\Pi}}_{22} = <\kappa> \epsilon_1^2 + <\mu> \epsilon_2^2 - \frac{m(1 - m)[\Delta\kappa\epsilon_1 - \Delta\mu\epsilon_2]^2}{\widetilde{\kappa} + \widetilde{\mu}}.$$

this tensor is:

$$\mathbf{C}_I(m) = \mathbf{C}_{11}^I \mathbf{a}_1' \mathbf{a}_1' + \mathbf{C}_{22}^I \mathbf{a}_2' \mathbf{a}_2' + \mathbf{C}_{12}^I (\mathbf{a}_1' \mathbf{a}_2' + \mathbf{a}_2' \mathbf{a}_1') + \mathbf{C}_{33}^I \mathbf{a}_3' \mathbf{a}_3'. \quad (4.3)$$

The coefficients $\mathbf{C}_{ij}^I$ can be found easily from (4.1). If the two initial phases are isotropic with elastic moduli κ_1, μ_1 and κ_2, μ_2, the corresponding expressions are given by:[1]

$$\mathbf{C}_{11}^I(m) = \left\langle\frac{1}{\kappa}\right\rangle - \frac{m(1 - m)(\kappa_1^{-1} - \kappa_2^{-1})^2}{\widetilde{(\kappa^{-1})} + \widetilde{(\mu^{-1})}},$$

$$\mathbf{C}_{22}^I(m) = \left\langle\frac{1}{\mu}\right\rangle - \frac{m(1 - m)(\mu_1^{-1} - \mu_2^{-1})^2}{\widetilde{(\kappa^{-1})} + \widetilde{(\mu^{-1})}},$$

$$\mathbf{C}_{12}^I(m) = \frac{m(1 - m)(\kappa_1^{-1} - \kappa_2^{-1})(\mu_1^{-1} - \mu_2^{-1})}{\widetilde{(\kappa^{-1})} + \widetilde{(\mu^{-1})}}, \quad (4.4)$$

$$\mathbf{C}_{33}^I(m) = \left\langle\frac{1}{\mu}\right\rangle - \frac{m(1 - m)(\mu_1^{-1} - \mu_2^{-1})^2}{\widetilde{(\mu^{-1})}},$$

where

$$\langle\cdot\rangle = m(\cdot)_1 + (1 - m)(\cdot)_2,$$

$$\widetilde{(\cdot)} = m(\cdot)_2 + (1 - m)(\cdot)_1.$$

---

[1]A more elegant way to express the effective properties of laminates (see Chapter 8 of this book) was found after this paper had been published. (*Author's comment to the translation.*)

The value of the quadratic form $\Pi(\mathbf{C}, \mathbf{M})$ depends both on the invariants (4.4) of the compliance tensor of an orthotropic composite and on the angle ψ between the principal axis $\mathbf{n}$ of the tensor $\mathbf{C}$ and the principal axis $\mathbf{i}$ of the tensor of moments $\mathbf{M}$. We are looking for a minimum of the stored energy over this angle ψ. The minimum over ψ is reached for a critical value of the angle ψ. Proceeding according to [17] we can show that the equation $\dfrac{\partial \Pi}{\partial \psi} = 0$ has the solutions:

$$\psi = 0, \ \psi = \frac{\pi}{2},$$

$$\cos 2\psi = \frac{\mathbf{C}_{12}^I}{\mathbf{C}_{33}^I - \mathbf{C}_{22}^I} \frac{\mathbf{M}_1}{\mathbf{M}_2} = -\frac{(\kappa_1^{-1} - \kappa_2^{-1})(\widetilde{\mu^{-1}})\mathbf{M}_1}{(\mu_1^{-1} - \mu_2^{-1})(\widetilde{\kappa^{-1}})\mathbf{M}_2}. \tag{4.5}$$

where $\mathbf{M}_1 = (\lambda_1 + \lambda_2)/\sqrt{2}$, $\mathbf{M}_2 = (\lambda_1 - \lambda_1)/\sqrt{2}$, and λ_1 and λ_2 are the eigenvalues of the tensor $\mathbf{M}$.

Let us compute the value $\Pi(\mathbf{C}_I, \mathbf{M})$ corresponding to the critical values of ψ in (4.5). It can be easily checked that when $\psi = 0$, the bound $\underline{\mathrm{II}}_{22}$ is reached; the bound $\underline{\mathrm{II}}_{12}$ is reached when $\psi = \pi/2$. Finally, when ψ satisfies the third regime of (4.5), either the bound $\underline{\mathrm{II}}_{11}$ is reached (for the case (1.5)), or the bound $\underline{\mathrm{II}}_{21}$ (for the case (1.6)). The condition $|\cos 2\psi| \leq 1$, for which (4.5) makes sense, is satisfied if the parameter ξ either belongs to the interval 1.1 (in the case (1.5)) or the interval 2.1 of Table 1 (in the case (1.6)).

As for the bounds $\underline{\mathrm{II}}_{13}$ and $\underline{\mathrm{II}}_{23}$, they cannot be reached by laminate composites.

To determine the structures that correspond to the listed bounds, we consider a "matrix" composite (see Figure 1) which is built from two kinds of materials taken in quantities p_2 and $1 - p_2$. The first material (it has a compliance tensor $\mathbf{C}_I(p_1)$) is an orthotropic laminate composite built from the two materials with compliance tensors $\mathbf{C}_1$, $\mathbf{C}_2$. The parameter p_1 is the volume fraction of the material with compliance $\mathbf{C}_1$ in this composite; it has the same meaning as the parameter m in the formulae (4.4). The second material is the initial phase with compliance $\mathbf{C}_2$. The total amount m of the material with compliance $\mathbf{C}_1$ in the medium just described equals to (see Figure 1):

$$m = p_1 p_2. \tag{4.6}$$

We denote the compliance tensor of the matrix composite by $\mathbf{C}_{II}(p_2)$. It can be determined from the formula (4.1) where m should be changed to p_2, n to t, and $\mathbf{C}_1$ to $\mathbf{C}_I(p_1)$. Its dyadic expression is similar to the one in (4.3) and the coefficients $\mathbf{C}_{ij}^I$ should be changed to the coefficients $\mathbf{C}_{ij}^{II}$.

Here we list some values of $\mathbf{C}_{ij}^{II}$:

$$
\begin{aligned}
\mathbf{C}_{11}^{II} &= \langle \mathbf{C}_{11} \rangle_2 - \frac{p_2(1-p_2)(\Delta \mathbf{C}_{11} + \Delta \mathbf{C}_{12})^2}{(\widetilde{\mathbf{C}_{11}})_2 + (\widetilde{\mathbf{C}_{22}})_2 + 2(\widetilde{\mathbf{C}_{12}})_2}, \\
\mathbf{C}_{22}^{II} &= \langle \mathbf{C}_{22} \rangle_2 - \frac{p_2(1-p_2)(\Delta \mathbf{C}_{22} + \Delta \mathbf{C}_{12})^2}{(\widetilde{\mathbf{C}_{11}})_2 + (\widetilde{\mathbf{C}_{22}})_2 + 2(\widetilde{\mathbf{C}_{12}})_2}, \\
\mathbf{C}_{12}^{II} &= \langle \mathbf{C}_{12} \rangle_2 + \frac{p_2(1-p_2)(\Delta \mathbf{C}_{11} + \Delta \mathbf{C}_{12})(\Delta \mathbf{C}_{22} + \Delta \mathbf{C}_{12})}{(\widetilde{\mathbf{C}_{11}})_2 + (\widetilde{\mathbf{C}_{22}})_2 + 2(\widetilde{\mathbf{C}_{12}})_2}.
\end{aligned}
\tag{4.7}
$$

Here

$$
\begin{aligned}
\langle \mathbf{C}_{ij} \rangle_2 &= p_2 \mathbf{C}_{ij}^I + (1-p_2)\mathbf{C}_{ij}^2, \\
\langle \widetilde{\mathbf{C}_{ij}} \rangle_2 &= p_2 \mathbf{C}_{ij}^2 + (1-p_2)\mathbf{C}_{ij}^I, \\
\Delta \mathbf{C}_{ij} &= \mathbf{C}_{ij}^2 - \mathbf{C}_{ij}^I,
\end{aligned}
$$

and $\mathbf{C}_{ij}^I$, $\mathbf{C}_{ij}^2$ are the coefficients of the dyadic expansion of the tensors $\mathbf{C}_I$ and $\mathbf{C}_2$ in the basis (4.2).

The value of the energy $\Pi_{II} = \Pi(\mathbf{C}_{II}(p_1, p_2), \mathbf{M})$ of the composite with compliance $\mathbf{C}_{II}$ depends, in addition to angle ψ, on the two parameters p_1 and p_2 which are related by (4.6). Let us show that when the value of Π_{II} is minimal in p_1 and p_2, then the bounds $\underline{\Pi}_{13}$ and $\underline{\Pi}_{23}$ of Table 1 are reached.

Let $\psi = 0$. Taking into account (4.6), we see that the equation

$$
\frac{\partial \Pi}{\partial p_1} = 0
$$

has two roots p_a and p_b given by

$$
\begin{aligned}
p_a(m, \mathbf{C}_1, \mathbf{C}_2) &= \frac{(1+m)}{2} - \frac{\mathbf{M}_1}{2\mathbf{M}_2}\left(\frac{\kappa_2^{-1} + \mu_2^{-1}}{\mu_2^{-1} - \mu_1^{-1}} - (1-m)\right), \\
p_b(m, \mathbf{C}_1, \mathbf{C}_2) &= \frac{(1+m)}{2} - \frac{\mathbf{M}_2}{2\mathbf{M}_1}\left(\frac{\kappa_2^{-1} + \mu_2^{-1}}{\mu_2^{-1} - \mu_1^{-1}} - (1-m)\right).
\end{aligned}
\tag{4.8}
$$

The second of these roots satisfies a condition $\partial^2 \Pi / \partial p_1^2 \geq 0$ of the minimum of energy and the corresponding value of the form Π_{II} coincides the with the bounds

$$
\Pi_{II} = \underline{\Pi}_{13} = \underline{\Pi}_{23}.
$$

Moreover, the conditions that ξ belongs to the intervals (1.3) or (2.3) of

Table 1, lead to the inequalities

$$m \leq p_2 \leq 1,$$

which guarantee that equation (4.6) can be solved.

The optimal values of the structural parameters for all stress states are given in Table 3. The most rigid with respect to bulk loading microstructures are those of matrix composites such that the modulus μ of the matrix material is large. The laminates are optimal when the stress state is that of a pure shear ($\xi \to \infty$); when

$$\frac{\kappa_1}{\mu_1} = \frac{\kappa_2}{\mu_2},$$

the normal to the layers points in the direction of the zero bending moment. In the case of uniaxial bending ($\xi = 1$), the optimal structures are laminates with the layers directed along one of the principal axes of the tensor $\mathbf{M}$. The type of the optimal structure depends continuously on the values of the parameter ξ (see Table 3.)

Table 3a. Parameters of optimal stiff structures. Case (1.5) ($\Delta\kappa\Delta\mu \geq 0$)

Interval 1.1	Interval 1.2	Interval 1.3
$p_1 = 1$	$p_1 = 1$	$p_1 = p_b(m, \mathbf{C}_1, \mathbf{C}_2)$
$\cos 2\phi = -\dfrac{\overbrace{(\mu^{-1})}\Delta(\kappa^{-1})}{(\kappa^{-1})\Delta(\mu^{-1})\xi}$	$\phi = \pi/2$	$\phi = 0$

Notations: [□] - material $\mathbf{C}_1$, [▨] - material $\mathbf{C}_2$.

Table 3b. Parameters of optimal stiff structures. Case (1.5) ($\Delta\kappa\Delta\mu \leq 0$)

Interval 2.1	Interval 2.2	Interval 2.3
$p_1 = 1$	$p_1 = 1$	$p_1 = p_b(m, \mathbf{C}_1, \mathbf{C}_2)$
$\cos 2\phi = -\dfrac{\widetilde{(\mu^{-1})}\Delta(\kappa^{-1})}{(\kappa^{-1})\Delta(\mu^{-1})\xi}$	$\phi = 0$	$\phi = 0$

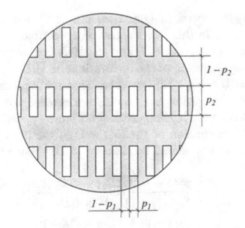

Figure 1.

The optimality of the bounds for the energy of composites of minimal stiffness can be checked an analogous way. The effective stiffness tensor of a laminate composite is given [16] by:

$$\mathbf{D}_I(m) = \mathbf{D}_1 + (1 - m)\left[\Delta\mathbf{D} - \frac{m\Delta\mathbf{D}\cdot\cdot\mathbf{nnnn}\cdot\cdot\Delta\mathbf{D}}{\mathbf{nn}\cdot\cdot(\mathbf{D}_1 + m\Delta\mathbf{D})\cdot\cdot\mathbf{nn}}\right], \qquad (4.9)$$

where $\Delta\mathbf{D} = \mathbf{D}_2 - \mathbf{D}_1$, $\mathbf{n}$ is a normal to the layers, and m is the volume fraction of the material with compliance $\mathbf{D}_1$. In the basis (4.3) this tensor is expressed by:

$$\mathbf{D}_I = \mathbf{D}_{11}^I a_1' a_1' + \mathbf{D}_{22}^I a_2' a_2' + \mathbf{D}_{12}^I(a_1' a_2' + a_2' a_1') + \mathbf{D}_{33}^I a_3' a_3'. \qquad (4.10)$$

The coefficients $\mathbf{D}^I_{ij}$ can be easily obtained from (4.9). If the initial materials are isotropic, then

$$
\begin{aligned}
\mathbf{D}^I_{11} &= \langle \kappa \rangle - \frac{m(1-m)(\kappa_1 - \kappa_2)^2}{\widetilde{\kappa} + \widetilde{\mu}}, \\
\mathbf{D}^I_{22} &= \langle \mu \rangle - \frac{m(1-m)(\mu_1 - \mu_2)^2}{\widetilde{\kappa} + \widetilde{\mu}}, \\
\mathbf{D}^I_{12} &= -\frac{m(1-m)(\kappa_1 - \kappa_2)(\mu_1 - \mu_2)}{\widetilde{\kappa} + \widetilde{\mu}}, \\
\mathbf{D}^I_{33} &= \langle \mu \rangle .
\end{aligned}
\tag{4.11}
$$

The equation $\dfrac{\partial \Pi}{\partial \psi} = 0$ has the solutions:

$$
\psi = 0, \ \psi = \frac{\pi}{2}, \ \cos 2\psi = \frac{\mathbf{D}^I_{12}}{\mathbf{D}^I_{33} - \mathbf{D}^I_{22}} \frac{\varepsilon_1}{\varepsilon_2},
\tag{4.12}
$$

where ψ is an angle between the principal axes of the strain tensor and the normal to laminates. In this case the last of the regimes in (4.12) does not satisfy the Legendre condition [17] and therefore is not optimal.

By computing the value of the quadratic form $\Pi(\mathbf{D}_I, \epsilon)$ corresponding to the angle $\psi = 0$ ($\psi = \pi/2$), we check that the bounds $\overline{\Pi}_{12}$ ($\overline{\Pi}_{22}$) are reached for laminates. All other bounds are reached, as in the previous case, for "rank two" matrix composites (cf. Figure 1).

In the dyadic expression of the tensor $\mathbf{D}_{II}$, the inverse of the tensor $\mathbf{C}_{II}$, the coefficients $\mathbf{D}^I_{ij}$ should be changed to $\mathbf{D}^{II}_{ij}$ that are given by

$$
\begin{aligned}
\mathbf{D}^{II}_{11} &= \langle D_{11} \rangle_2 - \frac{p_2(1 - p_2)(\Delta D_{11} + \Delta D_{12})^2}{(\widetilde{D_{11}})_2 + (\widetilde{D_{22}})_2 + 2(\widetilde{D_{12}})_2}, \\
\mathbf{D}^{II}_{22} &= \langle D_{22} \rangle_2 - \frac{p_2(1 - p_2)(\Delta D_{22} + \Delta D_{12})^2}{(\widetilde{D_{11}})_2 + (\widetilde{D_{22}})_2 + 2(\widetilde{D_{12}})_2}, \\
\mathbf{D}^{II}_{12} &= \langle D_{12} \rangle_2 - \frac{p_2(1 - p_2)(\Delta D_{11} + \Delta D_{12} \Delta D_{22} + \Delta D_{12})}{(\widetilde{D_{11}})_2 + (\widetilde{D_{22}})_2 + 2(\widetilde{D_{12}})_2} \\
\mathbf{D}^{II}_{33} &= \langle D_{33} \rangle_2 .
\end{aligned}
$$

Here

$$
\begin{aligned}
\langle D_{ij} \rangle_2 &= p_2 D^I_{ij}(p_1) + (1 - p_2) D^2_{ij}, \\
(\widetilde{D_{ij}})_2 &= p_2 D^2_{ij} + (1 - p_2) D^I_{ij}(p_1),
\end{aligned}
$$

and $D^I_{ij}(p_1)$, D^2_{ij} are the coefficients of the dyadic expansions of the tensors $\mathbf{D}_I$ and $\mathbf{D}_2$ in the basis (4.3).

The value of the energy

$$\Pi_{II} = \Pi(\mathbf{D}_{II}(p_1, p_2), \boldsymbol{\epsilon})$$

of the composite with the compliance $\mathbf{D}_{II}$ depends on two parameters, namely, the volume fractions p_1 and p_2 that are related by (4.6). The equation $\dfrac{\partial \Pi_{II}}{\partial p_1} = 0$ also has two roots :

$$p_a(m, \mathbf{D}_1, \mathbf{D}_2) = \frac{(1+m)}{2} - \frac{\varepsilon_1}{2\varepsilon_2}\left(\frac{\kappa_2 + \mu_2}{\mu_2 - \mu_1} - (1 - m)\right),$$

$$p_b(m, \mathbf{D}_1, \mathbf{D}_2) = \frac{(1+m)}{2} - \frac{\varepsilon_2}{2\varepsilon_1}\left(\frac{\kappa_2 + \mu_2}{\kappa_2 - \kappa_1} - (1 - m)\right).$$

The bound $\overline{\Pi}_{21}$ is reached when $p_1 = p_a$.

To demonstrate that the remaining bounds can also be reached, we consider rank two composites different from the rank two composites of the previous case by the fact that the material of the matrix is now the material of the inclusions and the material of inclusions is now the matrix material. This is equivalent to changing $m \Leftrightarrow (1 - m)$, $\mathbf{D}_1 \Leftrightarrow \mathbf{D}_2$ in the corresponding expressions. Following the steps of the previous case we verify that the bounds $\overline{\Pi}_{11}$, $\overline{\Pi}_{13}$, $\overline{\Pi}_{23}$ can be reached for the mentioned composites. The parameters of the optimal structures for all values of the parameter ζ are given in Table 4.

Thus we have shown that the laminates with compliance $\mathbf{D}_I$ and matrix composites with compliance $\mathbf{D}_{II}$ form a class of composites for which the bounds are reached, or the class of optimal structures. To choose an optimal element from the class we find the extremum of the energy with respect to the parameter p_1 and the angle ψ. We note that the value of the energy associated with the optimal matrix composite equals the energy of some isotropic material whose elastic moduli do not depend on the stress state and are determined only by κ_1, κ_2 and μ_1, μ_2 as well as the concentration m. In other words, all the equivalence classes for matrix composites for different stress states (see Remark 1.1) include the same material. The same can be said about the structures that are optimal for the values of the parameter ξ (ζ) that belong to any of the intervals of Tables 1 and 2.

Table 4a. Parameters of optimal soft structures. Case (1.5) ($\Delta\kappa\Delta\mu \geq 0$)

Interval 1.1	Interval 1.2	Interval 1.3
$p_1 = p_a(1 - m, \mathbf{D}_2, \mathbf{D}_1)$	$p_1 = 1$	$p_1 = p_b(1 - m, \mathbf{D}_2, \mathbf{D}_1)$
$\phi = 0$	$\phi = 0$	$\phi = 0$

Notations: [▢] - material $\mathbf{D}_1$, [▨] - material $\mathbf{D}_2$.

Table 4b. Parameters of optimal soft structures. Case (1.6) ($\Delta\kappa\Delta\mu \leq 0$)

Interval 2.1	Interval 2.2	Interval 2.3
$p_1 = p_a(m, \mathbf{D}_1, \mathbf{D}_2)$	$p_1 = 1$	$p_1 = p_b(1 - m, \mathbf{D}_2, \mathbf{D}_1)$
$\phi = 0$	$\phi = \pi/2$	$\phi = 0$

§5. The analysis of obtained results.

1. Let the initial materials have the same shear moduli μ and different bulk moduli κ_i.

Then, as shown in [18], the tensor of elastic moduli $\mathbf{D}_0$ is an isotropic tensor with the moduli

$$\kappa_0 = \frac{(\kappa_1 + \mu)(\kappa_2 + \mu)}{m\kappa_2 + (1 - m)\kappa_1 + \mu} - \mu, \quad \mu_0 = \mu,$$

that do not depend on the geometry of the composite.

Such a tensor $\mathbf{D}_0$ satisfies the bounds. The optimal values of the parameters $\alpha_0 = \mu$ and $\beta_0 = 1/\mu$ are independent of the stress state (see Tables 1 and 2).

2. Let the bulk moduli κ_i of the initial materials be equal:

$$\kappa_1 = \kappa_2 = \kappa.$$

Then, it can be seen from Table 1, the optimal value of the parameter β is $\beta_0 = 0$ regardless of the stress state. The optimal stiff composite has a laminate structure whose normal to layers forms an angle of $\pi/4$ with the principal axes of the tensor $\mathbf{M}$ (see Table 3).

The optimal value of the parameter α does not depend on the stress state and equals $-\kappa$; the optimal structure in this case is a laminate (see Table 4). These results agree with [16].

3. The results obtained in the present section are valid as well for the plane elasticity problem which also can be described by equation (1.1). For the plane elasticity w is the Airy function, $\mathbf{D}$ is the compliance tensor of the medium, and f is the Laplacian of the potential of the bulk forces. If the boundary conditions are inhomogeneous, contour integrals will appear on the right-hand side of (1.1). This does not change the homogenization procedure. The bounds and the structures that are optimal in the problem of a soft plate become the bounds and structures of a problem of a plane body of maximal stiffness because $\mathbf{D}$ has the meaning of compliance and not stiffness.

4. The bounds of stiffness just obtained improve the well known Reuss-Voigt inequalities (2.4) and (2.5). A distinctive feature of the improved bounds is their dependence on the invariant characteristic of the stress state which is the ratio $\xi(\zeta)$ of the bulk and deviator parts of the tensor $\mathbf{M}(\epsilon)$. We note, that the bounds (2.4) and (2.5) are the strongest possible, if the aforementioned dependence is ignored. Indeed, the exact bounds of Table 2 coincide with (2.4) and (2.5) for some types of the stress state. For the optimal soft plate this stress state is determined from the condition $\alpha = 0$; that is,

$$\zeta = \frac{\kappa_2 - \kappa_1}{\mu_2 - \mu_1} \frac{\widetilde{\mu}}{\widetilde{\kappa}} \, .$$

For the optimal stiff plate it is given by the interval 1.1 or the interval 2.1 of Table 1.

Note also that the Reuss-Voigt inequalities can be reached only in the "energy sense," that is, the value of the energy of the isotropic composite that can be computed from them coincides with the bounds (3.3) and (3.6). This, of course, does not mean that there exists an isotropic material with the moduli for which the Reuss-Voigt bounds are reached. The energy associated with these bounds is the energy of a composite which is essentially anisotropic with the optimal orientation of the layer normals.

5. A similar question was studied in [13], where there was solved a problem of optimal distribution of an elastic material in a two-dimensional region. It was assumed that a certain part of the region was occupied by holes and the stress state of the medium was described by the equations of the two-dimensional elasticity theory, that is, by equation (1.1) with respect to the Airy function w. The problem considered in [3] is, essentially, an asymptotic case of the problem of minimizing the stiffness previously considered ($\kappa_1 = \mu_1 = 1$, $\kappa_2 = \mu_2 = \infty$).

PART II OPTIMAL DESIGN OF SQUARE
CLAMPED PLATES. NUMERICAL RESULTS.

§6. Optimal design of a microstructure of composite plates with fixed volume fractions of phases.

The energy bounds obtained for the composites of maximal and minimal stiffness allow us to solve a number of optimal design problems. The simplest one is the optimal design of the structure for the composite plate of maximal (minimal) stiffness subject to the restriction that the volume fraction of the first phase is constant $m(\mathbf{x}) = m_0$ throughout the plate. We use the iterative procedure to find the numerical solution for this problem. Each iteration consists of the following steps.

(i) By using the known (from the previous iteration) field of the moments $\mathbf{M}$ (or the field of the deformations $\boldsymbol{\epsilon}$), we calculate the optimal field of the stiffness tensor according to Tables 3 or 4. This means that in every point $\mathbf{x}$ of the plate we choose the microstructure that is optimal for the given strain-stress state.

(ii) For the plate of given stiffness $\mathbf{D}(\mathbf{x})$ (found in Step (i)) we solve the direct problem of the plate theory, that is, we define the displacement field $w(\mathbf{x})$, tensor fields of moments $\mathbf{M}(\mathbf{x})$, and deformations $\boldsymbol{\epsilon}(\mathbf{x})$. We use the variational statement of the problem (1.2) that is solved by the method of local variations [19]. Convergence of the process is measured by the value of the maximal (over all points of the plates) change in the displacement field w.

We optimize the layout of the materials for a square clamped plate bent by the homogeneous ($f = const(\mathbf{x})$) load. The Poisson ratios of both materials are taken to be equal: $\kappa_1/\mu_1 = \kappa_2/\mu_2 = (1 + \nu)/(1 - \nu)$. The control parameters are the internal structural parameter p_1 (see §4) and the orientation of the axes of orthotropy of the stiffness tensor.

Figures 2 and 3 represent the optimal design of the rigid plate for the values of the parameters

$$m_0 = 0.5, \quad \frac{\kappa_1}{\mu_1} = \frac{\kappa_2}{\mu_2} = \frac{1 + \nu}{1 - \nu}, \quad \nu = 0.3, \quad \frac{\kappa_1}{\kappa_2} = \frac{\mu_1}{\mu_2} = 0.1. \qquad (6.1)$$

The areas of the different regimes (that correspond to Table 3) are denoted

by numbers. The bold lines show the directions of orthotropy that correspond to the maximal eigenvalue (i.e., the directions of maximal stiffness of
the plate). The dashed lines in the center show the directions of the second
axis of orthotropy in the points where the matrix composites are optimal.
As can be seen from the figures, the matrix composites are optimal in the
center of the plate (regime 3). Main stiffnesses are equal to each other in
the center of the plate; further to the sides the larger stiffness corresponds
to the direction tangential to the circles around the center of the plate. It
means that for such a loading the deformation $\epsilon_{\theta\theta}$ (where θ is the polar
angle) in the central regions is more important than the radial one. Further
from the center, the regime 3 is replaced by regime 2. It corresponds to the
laminate composite with the maximal rigidity direction pointed out tan-
gentially to the circles around the center of the plate. Closer to the sides,
regime 2 is replaced by the regime 1 that corresponds to the laminates
oriented by the angle

$$\phi = \pm \frac{1}{2} \arccos \left[-\frac{\kappa_1^{-1} - \kappa_2^{-1}}{\mu_1^{-1} - \mu_2^{-1}} \frac{m_1 \mu_2^{-1} + m_2 \mu_1^{-1}}{m_1 \kappa_2^{-1} + m_2 \kappa_1^{-1}} \frac{\mathbf{M}_1}{\mathbf{M}_2} \right]$$

to the eigendirection of the tensor $\mathbf{M}$. Independently of the sign of the angle
ϕ, the energy Π has the same value. Therefore two directions of laminates
are optimal in this regime. One of these directions is shown by the solid
lines in Figure 2; the other one is represented by the dashed lines. One
can see that regime 1 transforms into regime 2 so that the directions of the
largest principal stiffness twist like a spiral onto the boundary of these two
zones. Finally, near the sides of the plate, regime 2 of laminate composite
is also optimal. The directions of the laminate are almost orthogonal to
the clamped sides of the plate. In the neighborhood of the corners of the
plate regimes 2 and 1 coexist; it is due to the rapid change of the field $\mathbf{M}$
there.

Figure 4 represents the optimal design of the plate composed from the
materials with zero value of the Poisson ratios, namely,

$$m_0 = 0.5, \quad \frac{\kappa_1}{\mu_1} = \frac{\kappa_2}{\mu_2} = \frac{1+\nu}{1-\nu}, \quad \nu = 0, \quad \frac{\kappa_1}{\kappa_2} = \frac{\mu_1}{\mu_2} = 0.1. \qquad (6.2)$$

The main difference from the previous case is that the zone of regime
2 near the sides of the plate is replaced by the zone of regime 1.

L.V. Gibiansky and A.V. Cherkaev

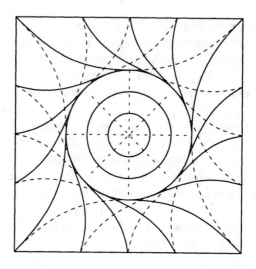

Figure 2.

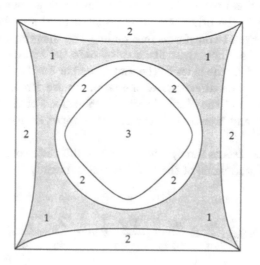

Figure 3.

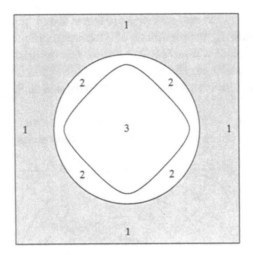

Figure 4.

Numerical experiments show that the optimal design almost does not depend on the volume fractions of the materials and on the ratio of their moduli.

If the stiffness of the first material is very small (both κ_1 and μ_1 are almost equal to zero), then Figure 2 can be interpreted as a project of the optimal beam construction with a fixed density of the beams.

The obtained results agree with physical intuition. Indeed, the center of the plate is subject to the bulk-type stress state close to "hydrostatic"; the most isotropic matrix composites are optimal in this region. In the other regions the laminate composites are optimal; they reinforce the most stressed direction (and act like stiffness ribs or stringers). It is interesting to mention that the circular laminates but not the radial ones are optimal near the center of the plate; that is in agreement with [19].

Figures 5 and 6 present the numerical results for the plate of minimal stiffness; the values of the parameters are given by (6.1). Zones of different regimes are numbered according to Table 4. The bold lines show the direction of maximal stiffness of the composite; the thin lines show the other eigendirection in the regions where the matrix composites are optimal. These lines can be treated as the directions of the "trenches" that weaken the plate most efficiently.

Matrix composites (i.e., regime 3) are optimal in the center of the plate; further to the sides they are replaced by the laminates pointed out to the center of the plate (regime 2). Then follows the zone of regime 1 (also matrix composites). The laminate composites that are directed along the sides of the plate are optimal near the boundary.

Note that the change of the lamination direction performs not by rotation of the angle of lamination, as for the optimal stiff plate, but through zone 1 of the matrix composite. Comparing the optimal designs of the soft plate (Figures 5 and 6) and stiff plate (Figures 2 and 3) we can see the following differences: (i) the directions of lamination differ by the right angle; (ii) for the soft plate the matrix composites are optimal in zone 1 instead of laminate composite for the same region of the rigid plate. It corresponds to the Tables 3 and 4; (iii) materials of matrix and inclusions interchange their places.

Optimal design of the soft plate turns out to be almost independent of the Poisson ratio of the materials; we present the results only for the value $\nu = 0.3$.

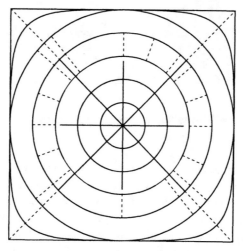

Figure 5.

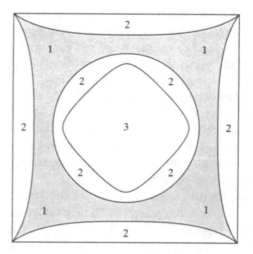

Figure 6.

§7. Optimal distribution of materials throughout the plate.

Here we drop the restriction $m(\mathbf{x}) = m_0$ and use volume fraction $m(\mathbf{x})$ as an additional control parameter. Let us fix the overall amount of the first material in the plate, that is, subject the control $m(\mathbf{x})$ to the integral restriction

$$\int_{\Omega} m(\mathbf{x})dx = m_0 \; area \; (\Omega), \; m_0 \in [0,1]. \tag{7.1}$$

We use the same numerical procedure as described in §6. but Step (i) of the iteration process now includes the solution of the system of equation (7.1) in conjunction with the equation

$$\frac{\partial \Pi}{\partial m} + \gamma = 0. \tag{7.2}$$

Here $\gamma = const(\mathbf{x})$ is a Lagrange multiplier for the restriction (7.1). Solution $m(\mathbf{x})$ of the system (7.1) and (7.2) gives the optimal distribution of the materials throughout the plate on each iteration. Note that $\Pi(m)$ is a convex function; therefore the iteration procedure converges quickly.

Figure 7 shows the optimal designs of the plates for different values of the parameter m_0 (m_0=0.95; 0.9; 0.8; and 0.5, respectively). Elastic moduli are chosen as in (6.1). When the value m_0 is sufficiently large (small amount of the stiff phase) the stiff material is concentrated in the most "dangerous" zones. The shaded regions in Figure 7 denote the zones occupied either by the pure stiff phase or by the composite material. White regions contain only soft material ($m = 1$). One can see that the most crucial regions are the middles of the sides of the plate; if one has more stiff material available, it is also placed in the center of the plate.

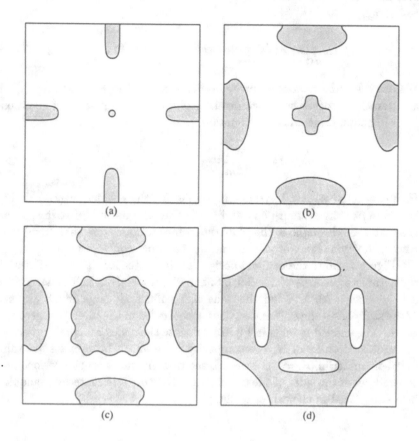

Figure 7.

Figure 8 shows zones of different regimes of control in the optimal plate (elastic constants are taken as in (6.1), $m_0 = 0.5$). Zones 1, 2, and 3 correspond to the regimes described in §6; regime 4 corresponds to the value $m = 0$ (pure stiff phase); regime 5 corresponds to $m = 1$ (pure soft phase). In all the examples that we studied, it turned out that the regions of small volume fraction of the first material ($m(\mathbf{x}) \leq 0.15$) are occupied by the matrix composites; composite zones (regimes 1, 2, 3, $m \neq 0$, $m \neq 1$ occupy 30 to 40 % of the plate.

Distribution of the stiff material (i.e., the function $1 - m(\mathbf{x})$ is shown in Figure 9 (it corresponds to Figure 7(c)).

Figure 10 shows the optimal design of the plate composed materials with a large ratio of the elastic moduli

$$\kappa_1/\kappa_2 = 10^{-10}, \ m_0 = 0.5.$$

We have compared the obtained results with the optimal design of the Kirchhoff plate of variable thickness [20–24] (for the same values of the parameters of the phases, the same value m_0, and the same loading). The regions where $m = 0$ and $m = 1$ almost coincide with the zones where the thickness h is equal to its maximal or minimal value $h = h_{max}$ or $h = h_{min}$. The difference is that the rest of the plate is occupied not by the isotropic material with intermediate value of the thickness but by the anisotropic composite of optimal microstructure (zones 1, 2, and 3 on Figure 8).

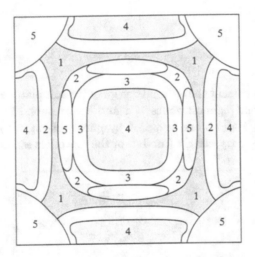

Figure 8.

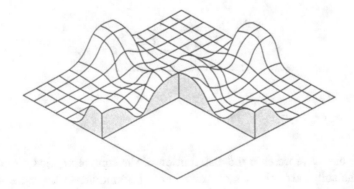

Figure 9.

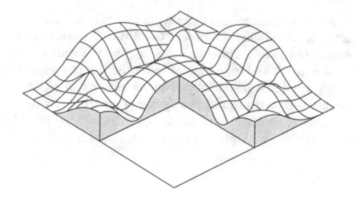

Figure 10.

Let us now present the optimal design of the soft plate. Figure 11 shows zones of different regimes: zones 1, 2 and 3 correspond to the regimes of Table 4; zones 4 and 5 are occupied by pure stiff ($m = 0$) and pure soft ($m = 1$) phases, respectively (moduli of the materials are defined by (6.1), $m_0 = 0.5$).

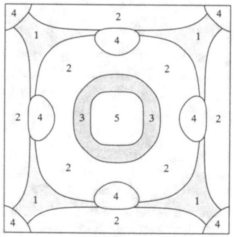

Figure 11.

Figure 12 represents the distribution of the control $m(\mathbf{x})$. One can see that the soft material is concentrated near the middles of the sides and in the center of the plate. The picture is qualitatively close to the distribution of the stiff phase in the optimal design of stiff plate.

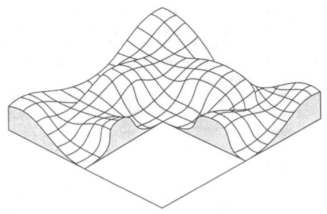

Figure 12.

§8. Effect of optimization.

Let us discuss the effect of optimization (i.e., improvement of the cost function) for the examples that we studied. We compare the value of the cost function (i.e., elastic energy stored in the plate) for the optimal design with the energy stored in the homogeneous plate. The last one is prepared from the material with the moduli

$$\kappa = (m\kappa_1^{1/3} + (1-m)\kappa_2^{1/3})^3, \quad \mu = (m\mu_1^{1/3} + (1-m)\mu_2^{1/3})^3. \qquad (8.1)$$

Such a choice of the comparison material is natural for the problem under study. Indeed, if the Poisson ratios of the materials are equal, we can treat them as the regions of different thickness for the Kirchhoff plate. Homogeneous plates of the same volume would possess the moduli that are given by (8.1). We discuss it in more detail later in Part 3. The values of the cost function are given in Tables 5 and 6 where we denote as I the value of the cost function for the optimal design, I_0 is the value of the function for the homogeneous plate with moduli defined by (8.1), I_1 and I_2 are the values of the cost function for the homogeneous plates with the moduli κ_1, μ_1, and κ_2, μ_2, respectively.

1. Note that the effect from structural optimization for the fixed value of the volume fraction $m(\mathbf{x}) = m_0$ is about half of the effect from structural optimization together with the optimization of the distribution $m(\mathbf{x})$ of the materials. In other words, we can get half of the possible effect by choosing the optimal microstructure and not changing the distribution of the materials.

2. Numerical calculations have shown that the effect of optimization slightly decreases with the increasing of the Poisson ratio.

3. Tables 5 and 6 also show the effect of optimization per unit volume of used material. It is characterized by the value $(1 - I/I_1)/(1 - m_0)$ for the stiff plate, and by the value $(-1 + I/I_2)/m_0$ for the soft plate. One

can see that for the stiff plate this effect is bigger when m_0 is close to zero. In other words, the first portions of the stiff material that are added to the homogeneous plate are used in the most efficient way. Most of the stiff material is concentrated in the stiffness ribs, and these ribs reinforce the most stressed parts of the plate.

The effect increases together with the ratio of the moduli of the initial materials and it is maximal for the perforated plate (i.e., when the ratio κ_1/κ_2 is near zero).

Table 5. Stiff plates

κ_2/κ_1	ν	m_0	$m(\mathbf{x})=m_0$		$\int_\Omega m(\mathbf{x})d\mathbf{x}=m_0$	
			I/I_0	$\frac{I_1-I}{I_1(1-m_0)}$	I/I_0	$\frac{I_1-I}{I_1(1-m_0)}$
10	0.3	0.95	0.864	5.40	0.664	8.77
10	0.3	0.90	0.798	4.25	0.606	5.62
10	0.3	0.80	0.756	2.97	0.550	3.53
10	0.3	0.50	0.815	1.58	0.535	1.72
10	0.3	0.20	0.940	1.08	0.758	1.12
3.375	0.3	0.50	0.947	1.03	0.712	1.27
10	0.3	0.50	0.815	1.58	0.535	1.72
125	0.3	0.50	0.526	1.96	0.335	1.98
10^{10}	0.3	0.50	0.322	2.00	0.210	2.00
10	0.5	0.50	0.916	1.53	0.563	1.71
10	0.3	0.50	0.815	1.58	0.535	1.72
10	0.0	0.50	0.745	1.61	0.535	1.72

Table 6. Soft plates

κ_2/κ_1	ν	m_0	$m(\mathbf{x})=m_0$		$\int_\Omega m(\mathbf{x})d\mathbf{x}=m_0$	
			I/I_0	$\frac{I-I_2}{I_2 m_0}$	I/I_0	$\frac{I-I_2}{I_2 m_0}$
10	0.3	0.05	1.28	7.83	1.54	13.4
10	0.3	0.10	1.50	7.75	1.85	12.0
10	0.3	0.20	1.80	7.74	2.26	10.9
10	0.3	0.50	1.98	8.01	2.50	10.7
10	0.3	0.80	1.46	8.53	1.73	10.3

PART III. OPTIMAL PLATE OF VARYING THICKNESS.

§9. Sufficient conditions of absence of intermediate values of the thickness.

The problem of optimal design of the two-phase plate was solved in the previous sections. We have found the optimal microstructures and optimal distribution of the materials throughout the plate for the case when only two materials are used in the design. Here we address the same problem of optimal design of the plate, but we assume that the set of available materials includes more (may be even infinitely many) phases; some integral characteristic of this set is fixed. As an example we discuss the problem of optimal design for the bending Kirchhoff plate. It is assumed to be prepared of one isotropic material but with different thicknesses in each point of the plate.

Differential equations of this problem coincide with (1.2) where the bulk and shear moduli depend on the plate's thickness as follows

$$\kappa(\mathbf{x}) = (1 + \nu)h^3(\mathbf{x}), \; \mu(\mathbf{x}) = (1 - \nu)h^3(\mathbf{x}). \qquad (9.1)$$

The scaling multiplier $E/12(1 - \nu^2)$ (where E is the Young modulus) is assumed to be equal to one. The volume of the plate is fixed:

$$\int_\Omega h(\mathbf{x})d\mathbf{x} = V, \qquad (9.2)$$

and the function $h(\mathbf{x})$ is assumed to be bounded in $L_\infty(\Omega)$:

$$0 < h_1 \le h(\mathbf{x}) \le h_2 < \infty. \qquad (9.3)$$

The problem of optimal distribution of isotropic material throughout the Kirchhoff plate in order to minimize the stored energy attracted a lot of attention. It has been studied numerically in [20,21,22 23,24]. In [20,24] there were found that the numerical solution is strongly nonhomogeneous and is not stable to the mesh refining. It can be explained by the fact that such a problem is ill-posed; the necessary conditions of optimality are contradictory (see [1]). One has to enlarge the set of admissible controls in order to make it G-closed, that is, to include in it all composites that can be prepared from the initial materials. Similar one-dimensional problems of optimal design for an axisymmetric plate were regularized by the mentioned procedure by introducing of axisymmetric stiffness ribs; see [22]. Here we study this problem in two dimensions that leads to the composites of more complicated structure.

Remark 9.1. Equations (1.2) originally were obtained by using the Kirchhoff assumptions that are valid only for the plate of constant thickness. Here we use the composites (see §1) with a fine alternation of the regions occupied by materials of different thickness. One can ask how well the G-limit equations (1.9) describe the real system. Rigorous asymptotic analyses of the three-dimensional elasticity equations show (see [25]) that the system (1.9) describes the equilibrium state of the plate if the wavelength of the thickness variations T is much larger than the thickness h itself, that is, if the ratio h/T tends to zero when h tends to zero.

Here we do not solve the whole problem of G-closure of the set of admissible controls. Instead we find particular cases when the optimal function $h(\mathbf{x})$ is equal either to h_1 or to h_2 everywhere. In other words we looking for the sufficient condition for the parameters of the problem that guarantee that the optimal mixture is composed only from the phases with minimal thickness h_1 and maximal thickness h_2. Subject to these sufficient conditions the problem reduces to the problem of optimal design of the two-phase plate that was solved earlier.

We use the procedure described in the §2 and §3. The energy bounds for the plate composed from N phases (with the moduli κ_i, μ_i, i=1,2,...,N) taken in the proportions m_1 are given by equations (3.1)–(3.3) where

$$
\underline{\kappa}^{-1} = \left(\sum_{i=1}^{N} \frac{m_i}{\kappa_i^{-1} + \beta} \right)^{-1} - \beta,
$$

$$
\underline{\mu}^{-1} = \left(\sum_{i=1}^{N} \frac{m_i}{\mu_i^{-1} - \beta} \right)^{-1} + \beta.
$$

$$(9.4)$$

When the number of materials increases to infinity, the summation is replaced by integration over some measure n; that is,

$$
\underline{\kappa}^{-1} = \left(\int_0^1 \frac{d\,n}{\kappa^{-1}(n) + \beta} \right)^{-1} - \beta,
$$

$$
\underline{\mu}^{-1} = \left(\int_0^1 \frac{d\,n}{\mu^{-1}(n) - \beta} \right)^{-1} + \beta.
$$

$$(9.5)$$

By substituting (9.1) into (9.5) we get

$$
\underline{\kappa}^{-1} = \underline{\kappa}^{-1}(h) = \left(\int_0^1 \frac{d\,n}{(h^3(n)(1+\nu))^{-1} + \beta} \right)^{-1} - \beta,
$$

$$
\underline{\mu}^{-1} = \underline{\mu}^{-1}(h) = \left(\int_0^1 \frac{d\,n}{(h^3(n)(1-\nu))^{-1} - \beta} \right)^{-1} + \beta.
$$

$$(9.5')$$

In fact here we introduce the small neighborhood $\Omega_{\epsilon(\mathbf{x})}$ of the point $\mathbf{x} \in \Omega$, and the function $h(n)$, that is, the value of the thickness emerging from (9.4) when m_i tends to zero and i tends to infinity. The function $n(h)$ (that is inverse to the function $h(n)$) is the area fraction of the subset of the set $\Omega_{\epsilon(x)}$ that is occupied by the materials with the thickness that is less than $h(n)$. In such a way we introduce the dependence of h on "fast" variable n for fixed value of the "slow" variable $\mathbf{x}$.

The function $h(n)$ is an additional control in the problem of optimal design of a composite microstructure. Similarly to (3.1), we get

$$\min_{h(n)} M^0 \cdot\cdot C_0 \cdot\cdot M^0 \geq \min_{h(n)} \max_{\beta \geq 0, \beta \ as \ in \ (2.8)} M^0 \cdot\cdot \underline{C} \cdot\cdot M^0. \qquad (9.6)$$

The moduli of the stiffness tensor $\underline{C}$ are given by (9.5); the function $h(n)$ is subject to the restrictions (9.3). We assume also that the average (over the set $\Omega_{\epsilon(x)}$) value of the function $h(n)$ is given

$$\int_0^1 h(n)dn = h(\mathbf{x}) \qquad (9.7)$$

The set of admissible values of the parameter β depends on the maximum value of the function $h(n)$:

$$\beta \in \left[0, \frac{1}{h_{\max}^3(1-\nu)}\right], \quad h_{\max} = \max_{\Omega_{\epsilon(x)}} h,$$

see (2.3) and (2.13).

Our plan is to show that the bound (9.6) can be presented in a form (3.1) subject to some restrictions on the parameter h_1, h_2, and ν. The bound (3.1) is exact and is realizable by the two-phase matrix composites of the special microstructure (see §4). It would prove that the two-phase matrix composites are optimal for the problem under study as well.

First note that if

$$\beta \in \left[0, \frac{1}{h_2^3(1-\nu)}\right], \qquad (9.8)$$

then $\beta \geq 0$ and β satisfy (9.8). It means that we only make the inequality (9.6) stronger by changing in it the admissible interval for the parameter β from (2.8) to (9.8) when taking the maximum in (9.6).

Then we require the function $h(n)$ to provide the minimum to each of the moduli (9.5) of the tensor of the bound $\underline{C}$. It is suffice to require the function $h(n)$ to provide maximum to the functional

$$\int_0^1 \omega_1 dn, \quad \omega_1 = \frac{1}{1/(h^3(1+\nu)) + \beta}$$

and to the functional

$$\int_0^1 w_2 dn, \; w_2 = \frac{1}{1/(h^3(1-\nu)) - \beta}$$

for all admissible values (9.8) of the parameter β.

Now we want to find the conditions that guarantee that both functionals are maximized by the function of the form

$$h(n) = \begin{cases} h_1, \; if \; 0 \le n \le n(h(\mathbf{x})) \\ h_2, \; if \; n(h(\mathbf{x})) \le n \le 1 \end{cases}$$

In this case the optimal structure is composed of the elements with thickness h_1 and h_2. It is true if for every β that satisfies (9.8) and for every $h \in [h_1, h_2]$ the values of the functions $w_1(h)$ and $w_2(h)$ lie below the lines that connect the points $(h_1, w_1(h_1))$ and $(h_2, w_1(h_2))$ and the points $(h_1, w_2(h_1))$ and $(h_2, w_2(h_2))$ respectively; see Figure 13.

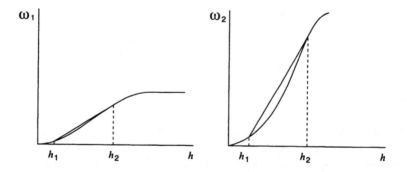

Figure 13.

Calculations that we omit show that the preceding conditions are equivalent to the following restrictions for the parameters h_1, h_2, and ν,

$$\frac{h_1}{h_2} \le \frac{\sqrt{(1+3\nu)^2 + 12(1+\nu)(1-3\nu)} - (1+3\nu)}{6(1+\nu)}, \; \nu \le \frac{1}{3}. \tag{9.9}$$

Note that this sufficient condition of optimality of two-phase composites does not depend on the stress state that enters the expressions (3.1) and (9.6).

If $\nu \le -1/3$, then the condition (9.9) is valid for any values of the parameters h_1 and h_2 (because $h_1/h_2 \le 1$ by definition). The value of the right-hand side of expression (9.9) for some other values of the Poisson ratio ν is given in Table 7.

Table 7.

ν	0	0.1	0.2	0.3	1/3
$(\frac{h_1}{h_2})_{max}$	0.438	0.303	0.180	0.053	0

If the parameters of the problem satisfy (9.9), then the energy bounds for the optimal composite coincide with the expressions given in Table 3; they are realizable by the laminate or matrix composites. In this case the results obtained in the previous sections are valid for the problem of optimization of the thickness for the Kirchhoff plate. In particular, it is true when $h_1 = 0$ and $\nu \leq 1/3$; that is, the optimal perforated plate should be made of the material of maximal thickness.

By using the same procedure one can find the sufficient conditions for the two-phase composites to be optimal in the problem of optimal design of the soft plate. These composites are optimal for any ration h_1/h_2 if absolute value $|\nu|$ of the Poisson ratio is less than 1/3, that is,

$$\forall \ h_1/h_2, \ if \ |\nu| \leq 1/3 \,.$$

For the other values of the Poisson ratio the sufficient condition of optimality of two-phase composites has a form:

$$\frac{h_1}{h_2} \leq \frac{1 - 3|\nu| + \sqrt{(1 - 3|\nu|)^2 + 12(1 - |\nu|)(1 + 3|\nu|)}}{2(1 + 3|\nu|)}, \ if \ |\nu| \geq 1/3.$$

$$(9.10)$$

One can also solve the similar problem for the three-layer plate when the moduli κ and μ depend linearly on the thickness of the plate; that is,

$$\kappa = h(1 + \nu), \ \mu = h(1 - \nu)$$

up to the scale multiplier that we omit. The results show that optimal two-phase composites are always optimal for the optimal design problem of a soft three-layer plate. For the optimal design problem of a stiff three-layer plate, the sufficient condition similar to (9.9) never holds for any choice of the parameters h_1, h_2, and ν. It is an additional confirmation of the suggested procedure. Indeed, it is known (see [26]) that for the latter problem there exists the smooth solution $h(\mathbf{x})$.

Acknowledgments. The authors are grateful to K.A. Lurie for helpful discussions.

References

[1] Lurie, K.A., Cherkaev, A.V., About the application of the Prager's theorem to the optimal design Problem of thin plates, *Izvestia AN SSSR, Mechanics of Solids* **6** (1976), (in Russian).

[2] Lurie, K.A., Cherkaev, A.V., Fedorov, A.V., Regularization of optimal design problems for bars and plates, I, II, *J. Opt. Th. Appl.* **37** (1982), 499–543.

[3] Lurie, K.A., Some problems of optimal bending and stretching of elastic plates, *Izvestia AN SSSR, Mech. Solids* **6** (1979) , 86–93, (in Russian).

[4] Rikhlevsky, J.K., About the Hook's law, *Applied Mathematics and Mechanics* **3** (1984), 420–436, (in Russian).

[5] Zhikov, V.V., Kozlov, S.M., Oleinik, O.A., Kha Tien Ngoan, Averaging and *G*-convergence of differential operators, *Uspehi Mat. Nauk* **34**(5) (1979), 209–219, (in Russian).

[6] K.A. Lurie and A.V. Cherkaev, Exact estimates of conductivity of composites formed by two isotropically conducting media taken in prescribed proportion, *Proc. Roy. Soc. Edinburgh* **99 A** (1984), 71–87 (first version: preprint 783, Ioffe Physico-technical Institute, Leningrad, 1982)

[7] Lurie, K.A., Cherkaev, A.V., Fedorov, A.V., On the existence of solutions to some problems of optimal design for bars and plates, *J. Opt. Th. Appl.* **43** (1984), 247–281.

[8] Filippov, A.F., Some questions of the theory of optimal control, *Vestnik MGU, Mathematics* **2** (1959), 25–32, (in Russian).

[9] Morrey, C.B., Integrals in the calculus of variations, Springer-Verlag, Berlin, 1966.

[10] Dacorogna, B., Weak continuity and weak lower semicontinuity of nonlinear functionals, *Lecture Notes in Math.* **922**, Springer-Verlag, New York, 1982.

[11] Ball, J.M., Currie, J.C., Olver, P.J. Null, Lagrangians, weak continuity, and variational problems of arbitrary order, *Funct. Anal.* **41**(2) (1981), 135–174.

[12] Ball, J.M., Convexity conditions and existence theorems in nonlinear elasticity, *Arch. Rat. Mech. Anal.* **63**(4) (1977), 337–403.

[13] Kohn, R.V., Strang, G., Optimal design and relaxation of variational problems, I, II, III, *Comm. Pure & Appl. Math.* **39** (1986), 113–137, 139–182, 353–377.

[14] K.A. Lurie and A.V. Cherkaev, Exact estimates of the conductivity of a binary mixture of isotropic materials," *Proc. Roy. Soc. Edinburgh* **104 A** (1986), 21–38. (first version: preprint 894, Ioffe Physico-Technical Institute, Leningrad, 1984).

[15] Raitum, U.E., Extension of the extremal problems for linear elliptic operators, *Dokladi AN SSSR* **243**(2), 281–283, (in Russian).

[16] Lurie, K.A., Fedorov, A.V., Cherkaev, A.V., Regularization of optimal design problems of bars and plates and correction of the contradictions in the system of necessary conditions of optimality, preprint, *Phys. Tech. Inst. AN SSSR* **667** (1980), (in Russian).

[17] Fedorov, A.V., Cherkaev, A.V., Optimal choice of the orientation of the axis for elastic symmetric orthotropic plate, *Izvestia AN SSSR, Mech. Solids* **3** (1983), 135–142, (in Russian).

[18] Lurie, K.A., Cherkaev, A.V., *G*-Closure of some particular sets of admissible material characteristic for the problem of bending of thin plates, *J. Opt. Th. Appl.* **42** (1984), 305–315.

[19] Kartvellishvilli, V.M., Mironov, A.A., Samsonov, A.M., Numerical method of solution for the problems of optimal design of reinforced constructions, *Izvestia AN SSSR, Mech. Solids* **2** (1981), 93–103, (in Russian).

[20] Olhoff, N., Optimal Design of Constructions, Mir, Moscow, 198, (in Russian).

[21] Banichuk, N.V., Shape Optimization of Elastic Bodies, Nauka, Moscow, 1980, (in Russian).

[22] Olhoff, N., Lurie, K.A., Cherkaev, A.V., Fedorov, A.V., Sliding regimes and anisotropy in optimal design of vibrating axisymmetric plates, *Denish Center Appl. Math. Mech.*, Report 192, 1980.

[23] Olhoff, N., Cheng, K.-T., An investigation concerning optimal design of solid elastic plates, *Int. J. Solid Struc.* **17** (1981), 305–321.

[24] Arman, J.-L. P., An Application of Optimal Control Theory for the System with Distributed Parameters to the Problems of Optimal Design of Constructions, Mir, Moscow, 1977, (in Russian).

[25] Kohn, R.V., Vogelius, M., A new model for thin plates with rapidly varying thickness I, *Proc. Un. of Maryland*, BN-988 (1982).

[26] Cherkaev, A.V., Some Problems of Optimal Design of for Elastic Elements of Constructions, PhD Thesis, Polytecnicheskij Institute, 1979, (in Russian).

Calculus of Variations
and Homogenization

François Murat and Luc Tartar

ABSTRACT. In this paper we study a class of problems which includes the following example. Let Ω be an open subset of $\mathbb{R}^N$ and let ω be a measurable subset of Ω with given measure γ'. Let $a(x) = \alpha$ on ω and β on $\Omega \setminus \omega$, and define u by $-div\,(a\,grad\,u) = 1$ in Ω and $u = 0$ on the boundary. We want to find an ω which maximizes $\int_\Omega u(x)\,dx$ among all the measurable sets ω with given measure γ'.

This optimal design problem has in general no solution. Indeed if there exists a smooth solution ω and if Ω is simply connected, then Ω is a ball. In general a maximizing sequence does not converge to a set ω, but to a "generalized domain", i.e. a more and more intricate mixture of materials α and β (homogenization phenomenon). Since we know the characterization of all the materials obtained in this way, i.e. the optimal bounds for homogenized materials, we can write a relaxed problem of the initial one. This relaxed problem has a solution, but the parameter is now a generalized (homogenized) domain.

We then study the necessary conditions for a generalized domain to be an optimal solution. This leads to some characterizations and to several geometrical remarks about the solutions of the relaxed problem and those of the initial one, if they exist.

Introduction

There are difficulties in several optimization problems when one tries to follow the classical method of the calculus of variations, which consists first of establishing the existence of an optimal solution by using a minimizing sequence (for minimization problems), and then computing the derivative of the cost function in order to obtain necessary conditions of optimality (i.e., Euler's equation if optimizing without constraints), see, for example, Lions (1968) where this method is described for several examples.

The first difficulty appears when one tries to prove the existence of a solution: for several problems an optimal solution does not exist in the class that was considered *a priori* and the optimizing (minimizing) sequences converge to something that we may call a "generalized solution" and which has to be precisely defined.

The second difficulty appears when one searches to differentiate the cost function in order to obtain necessary conditions of optimality (this is also needed in order to use numerical methods such as the steepest descent

© Springer Nature Switzerland AG 2018

A. V. Cherkaev, R. Kohn (eds.), *Topics in the Mathematical Modelling of Composite Materials*, Modern Birkhäuser Classics, https://doi.org/10.1007/978-3-319-97184-1_6

methods). In some cases the set on which one optimizes is not convex (or is not even a manifold), so that one cannot make variations or increments without enlarging the set that was considered *a priori*. This happens in domain optimization when one searches for a characteristic function: there is no natural way to go from one characteristic function to another one while remaining in the same set of characteristic functions ($t\chi_1 + (1-t)\chi_2$ is not a characteristic function for $0 < t < 1$ if $\chi_1 \neq \chi_2$). Of course one may build in a simple way some particular paths following, for example, Hadamard's idea (1907) which consists of a variation of the boundary of the domain in the normal direction. This permits us to obtain certain optimality conditions but not all of them since in this way it is not possible for a hole or a set to appear in a region where it did not exist.

In the 1940s, L.C. Young found the way to solve the first difficulty. The way to solve the second one is due to Pontryaguin: it consists of introducing Young's generalized functions, and then formulating the problem on a convex set and calculating its derivative. We call the procedure that consists of constructing generalized solutions, *relaxation.*For a better understanding we present these ideas in the first part this paper.

The results exposed here come from joint work on optimization problems which we started at the beginning of the 1970s. We were led to introduce the "right" class of generalized solutions, the homogenized domains, which are analogous to composite materials. This allowed us to solve the difficulties relative to the nonexistence problems and to write in this class (which has an underlying linear space structure: that of the $N \times N$ matrices with coefficients in L^∞) the necessary conditions of optimality. It is important to distinguish between the homogenization that we are using here and the techniques of asymptotic expantions studied in particular by Sanchez-Palencia (1980) and Bensoussan et al. (1978): there is no periodic structure in the problems that we consider here. Along the way we followed, we left some written marks (Tartar (1974), (1977b), Murat (1977), (1978a), Tartar (1983)) but many results that have been presented in various lectures, seminars, and conferences have not been written down. We hope that this defect will be corrected in the future. The reader should be warned that some theorems used here never appeared before in this form. A summary of this paper was published in Murat and Tartar (1985).

PART I. PRELIMINARIES

I.a An Abstract Formulation of Relaxation

When dealing with minimization problems of the form:

$$\inf_{x_0 \in X_0} F_0(x_0), \tag{1}$$

it is natural to try to use the following result in order to prove the existence of a solution (this is the direct method of the calculus of variations).

Proposition 1. *Let L_0 be a compact topological space and let F_0 be a real valued, lower semicontinuous function. Then problem (1) admits at least one solution.*

If it is not possible to apply this result, one can try to relax the problem by finding a compact topological space X and a lower semicontinuous function F on X adapted to problem (1) in the following sense:

$$\left\{ \begin{array}{l} X_0 \text{ is dense in X,} \\ \text{the restriction of F to } X_0 \text{ coincides with } F_0, \end{array} \right. \tag{2a}$$

$$\left\{ \begin{array}{l} \text{if } x_0^{(n)} \text{ is a sequence in } X_0 \text{ that converges to } x, \text{ then} \\ F(x) \leq \liminf F_0(x_0^{(n)}), \end{array} \right. \tag{2b}$$

$$\left\{ \begin{array}{l} \text{for every } x \in X \text{ there exists a sequence } x_0^{(n)} \text{ in } X_0 \\ \text{that converges to } x \text{ and satisfies } F(x) = \liminf F_0(x_0^{(n)}). \end{array} \right. \tag{2c}$$

Then one has the following.

Proposition 2. *Any limit point in X of a minimizing sequence of problem (1) is a solution of the following problem:*

$$\inf_{x \in X} F(x). \tag{3}$$

Problem (3) has at least one solution. Moreover,

$$\inf_{x \in X} F(x) = \inf_{x_0 \in X_0} F_0(x_0).$$

Finally, any solution of problem (1) is a solution of problem (3).

Remark 1.
As in any abstract setting, one can write simple variants: for example, it is possible to replace (2a) by the assumption that $X = X_0$ and that F is the lower semicontinuous envelope of F_0.

Remark 2.
We use the following important property of problem (3): X can be convex and F-differentiable in situations where there is no path in X_0 joining two given points. In this case we are able to write the optimality conditions for problem (3) and this are useful even when the optimal solution is in X_0 (i.e., when there exists a solution of (1)).

Remark 3.

It is possible to produce relaxed problems by compactification but the only interest of this notion consists of the cases where one can find X and calculate F in a simple way. In fact we take X as little as possible. This supposes that we know the behavior of the minimizing sequences.

Remark 4.

It is natural to introduce weak topologies in order to gain the compactness of certain sets or to produce convex sets X. The convexity is not an objective by itself but it will naturally appear from time to time.

I.b.   L. C. Young's Generalized Functions

Let Ω be an open bounded set of $\mathbb{R}^N$ and let K be a bounded set of $\mathbb{R}^p$. We consider functions $u : \Omega \longrightarrow \mathbb{R}^p$ and set:

$$X_0 = \{u \in (L^\infty(\Omega))^p \; : \; u(x) \in K \quad \text{a.e.}\}. \tag{4}$$

We are interested in the function $F_0 : X_0 \longrightarrow \mathbb{R}$ defined by

$$F_0(u) = \int_\Omega f_0\left(x, u\left(x\right)\right) dx, \tag{5}$$

where for the sake of simplicity f_0 is assumed to be continuous and bounded.

A *generalized function* is a nonnegative measure on $\Omega \times \mathbb{R}^p$ whose projection on Ω is dx; more precisely, let us define:

$$\left\{ \begin{array}{l} X = \{\mu \, \text{nonnegative measure on } \Omega \times \mathbb{R}^p \; : \\ \quad \text{support } \mu \subset \overline{\Omega} \times \overline{K}, \\ \quad \langle \mu, g \rangle = \int_\Omega g\left(x\right) dx \quad \forall g \in C_c^0(\Omega)\}, \end{array} \right. \tag{6}$$

where $C_c^0(\mathcal{O})$ denotes the set of continuous functions with compact support in the open set $\mathcal{O}$.

Remark 5.

A measure μ of X is described by a measurable family ν_x of probability measures on $\mathbb{R}^p$ with support $\nu_x \subset \overline{K}$:

$$\langle \mu, g\left(\cdot, \cdot\right)\rangle = \int_\Omega \langle \nu_x, g\left(x, \cdot\right)\rangle \, dx \quad \forall g \in C_c^0(\Omega \times \mathbb{R}^p).$$

Let us define the function $F : X \longrightarrow \mathbb{R}$ as follows,

$$F(\mu) = \langle \mu, f_0\left(\cdot, \cdot\right)\rangle, \tag{7}$$

which has a meaning since f_0 is continuous and bounded and since Ω is bounded.

The set X_0 is embedded in X by associating with every $u \in X$ the measure μ_u which is defined by:

$$\langle \mu_u, g \rangle = \int_\Omega g(x, u(x)) \, dx \quad \forall g \in C_c^0(\Omega \times \mathbb{R}^p). \tag{8}$$

When X is equipped with the weak $*$ topology of the space of measures, the problem of establishing that X_0 is dense in X and that X is compact is solved by the following.

Proposition 3. *For any sequence u_n in X_0 there exists a subsequence u_m and a measure $\mu \in X$ such that for all continuous g with compact support in $\Omega \times \mathbb{R}^p$ one has $\langle \mu_{u_m}, g \rangle \longrightarrow \langle \mu, g \rangle$. Conversely, for any $\mu \in X$ one can find such a sequence u_n in X_0.*

The space X is defined here independently of the choice of the function f_0. With this choice, the hypotheses of Proposition 2 turn out to be satisfied: indeed (2a) is deduced from Proposition 3, whereas (2b) and (2c) are deduced from the same Proposition and from the continuity of f_0. In certain cases the following result is sufficient.

Proposition 4. *Let u_n be a sequence in X_0 that converges weak $*$ in $(L^\infty(\Omega))^p$ to v. Then*

$$v(x) \in \overline{\text{conv} K} \quad a.e. \ x \in \Omega. \tag{9}$$

Conversely, every function v satisfying (9) can be obtained as the weak $*$ limit in $(L^\infty(\Omega))^p$ of a sequence of X_0.

For the proofs of Propositions 3 and 4 see, for example, Tartar (1979).

Remark 6.
In certain cases we can take X as the space of those v that satisfy (9) and use the previous proposition. As in the case of generalized functions, this space X is convex.

Remark 7.
If $K = \{0,1\} \subset \mathbb{R}$, u_n is a characteristic function and Proposition 4 implies that the weak $*$ limits of characteristic functions are all the functions θ satisfying $0 \le \theta(x) \le 1$ a.e. If u_n is moreover assumed to satisfy the condition $\int_\Omega u_n(x) \, dx = \gamma$ for some $\gamma > 0$, then we obtain the set $\{\theta : 0 \le \theta(x) \le 1 \text{ a.e. and } \int_\Omega \theta(x) \, dx = \gamma\}$. If X_0 is

the set of characteristic functions and $F_0(u) = \int_\Omega f_0(x, u(x)) \, dx$, we can choose $X = \{\theta \in L^\infty(\Omega) : 0 \le \theta(x) \le 1 \text{ a.e.}\}$ and define F by $F(\theta) = \int_\Omega (\theta(x) f_0(x, 1) + (1 - \theta(x)) f_0(x, 0)) \, dx$. Therefore the choice of X and F is not unique in a relaxation problem. It is useful to construct X and F as simple as possible.

I.c. Pontryagin's Principle

Let us consider as a basic example the following control problem. The state $y(t) \in \mathbb{R}^m$ is given by the state equation:

$$\begin{cases} \dfrac{dy}{dt} = A(t, y(t), u(t)), & t \in (0, T), \\ y(0) = y_0, \end{cases} \tag{10}$$

where the control $u(t)$ satisfies

$$u(t) = 0 \text{ or } 1 \text{ a.e. } t \in (0, T). \tag{11}$$

We want to minimize the cost function:

$$F_0(u) = \int_0^T B(t, y(t), u(t)) \, dt. \tag{12}$$

Pontryaguin's principle gives a necessary condition of optimality, that is, a necessary condition for a function u_0 to realize the infimum of F_0.

Let us take for X_0 the set of characteristic functions. Using Remark 7 and passing to the limit in (10) and (12) we get the following relaxed problem: the space X is

$$X = \{\theta \in L^\infty(0, T); \ 0 \le \theta \le 1 \quad \text{a.e. } t \in (0, T)\}, \tag{13}$$

the state $z(t) \in \mathbb{R}^m$ is given by the state equation:

$$\begin{cases} \dfrac{dz}{dt} = \theta(t) A(t, z(t), 1) + (1 - \theta(t)) A(t, z(t), 0), \ t \in (0, T), \\ z(0) = y_0, \end{cases} \tag{14}$$

and the problem is to minimize on X the cost function:

$$F(\theta) = \int_0^T \{\theta(t) B(t, z(t), 1) + (1 - \theta(t)) B(t, z(t), 0)\} \, dt. \tag{15}$$

Assume that the functions A and B are smooth with respect to y. Then the applications $\theta \longrightarrow z$ and $\theta \longrightarrow F$ are differentiable and we have the following relations between the increments $\delta\theta$, δz and δF:

$$
\begin{cases}
\dfrac{d\delta z}{dt} = [A(t,z,1) - A(t,z,0)]\,\delta\theta \\[2mm]
\qquad + [\theta\dfrac{\partial A}{\partial y}(t,z,1) + (1-\theta)\dfrac{\partial A}{\partial y}(t,z,0)]\,\delta z\,, \ t \in (0,T)\,, \\[2mm]
\delta z(0) = 0\,,
\end{cases}
\tag{14a}
$$

$$
\begin{cases}
\delta F = \displaystyle\int_0^T \{[B(t,z,1) - B(t,z,0)]\,\delta\theta \\[2mm]
\qquad + [\theta\dfrac{\partial B}{\partial y}(t,z,1) + (1-\theta)\dfrac{\partial B}{\partial y}(t,z,0)]\,\delta z\}\,dt\,.
\end{cases}
\tag{15a}
$$

In order to eliminate δz in the formula for δF we introduce the adjoint state $p : [0,T] \longrightarrow (\mathbb{R}^m)^* = \mathcal{L}(\mathbb{R}^m, \mathbb{R})$, as the solution of:

$$
\begin{cases}
-\dfrac{dp}{dt} = [\theta\dfrac{\partial A}{\partial y}(t,z,1) + (1-\theta)\dfrac{\partial A}{\partial y}(t,z,0)]^*\,p \\[2mm]
\qquad + [\theta\dfrac{\partial B}{\partial y}(t,z,1) + (1-\theta)\dfrac{\partial B}{\partial y}(t,z,0)]\,, \ t \in (0,T)\,, \\[2mm]
p(T) = 0\,.
\end{cases}
\tag{16}
$$

By integration by parts, δF becomes:

$$
\begin{cases}
\delta F = \displaystyle\int_0^T \{[B(t,z,1) - B(t,z,0)] \\[2mm]
\qquad + p(t)[A(t,z,1) - A(t,z,0)]\}\,\delta\theta\,dt\,.
\end{cases}
\tag{15b}
$$

Let us denote by $\varphi(t)$ the coefficient of $\delta\theta$ in (15b); that is,

$$
\varphi(t) = [B(t,z(t),1) - B(t,z(t),0)] + p(t)[A(t,z(t),1) - A(t,z(t),0)]\,.
$$

It is easy to see that at a point $\theta \in X$ where F achieves its minimum, one has either $\varphi(t) \geq 0$, $\varphi(t) \leq 0$ or $\varphi(t) = 0$ if $\theta(t) = 0$, $\theta(t) = 1$ or $0 < \theta(t) < 1$. In particular, if the minimum of F is achieved by a characteristic function u, then one has $\varphi(t) \geq 0$ if $u(t) = 0$, and $\varphi(t) \leq 0$ if $u(t) = 1$; that is,

$$
B(t,z(t),u(t)) + p(t)\,A(t,z(t),u(t)) \leq B(t,z(t),v) + p(t)\,A(t,z(t),v)
$$

$$
\forall\, v = 0 \text{ or } 1\,.
$$

This result is summarized in the following:

Proposition 5. *If u is an optimal solution of the minimization problem (12) where y and u satisfy (10) and (11), then*

$$\begin{cases} B(t, y(t), v) + p(t) A(t, y(t), v) \text{ is minimum} \\ \text{for } v \text{ satisfying (11) at the point } v = u(t), \end{cases} \tag{17}$$

where p is defined by

$$\begin{cases} -\dfrac{dp}{dt} = \dfrac{\partial A}{\partial y}(t, y, u)^* p + \dfrac{\partial B}{\partial y}(t, y, u), \; t \in (0, T), \\ p(T) = 0. \end{cases} \tag{18}$$

Remark 8.

Let us emphasize that there is not always a characteristic function that minimizes F_0; the control problem (10), (11), (12) may have no optimal solution. In contrast the relaxed problem always has at least one solution.

Remark 9.

If the control $u(t)$ is assumed to take its values in some set $K(t)$, the problem can be reduced to the previous one by introducing controls that oscillate between $\underline{u}$ and $\overline{u}$ and approximate the generalized controls $\theta(t)\underline{u}(t) + (1 - \theta(t))\overline{u}(t)$. The same analysis can then be used. If the function F_0 is changed by the addition of the term $C(y(T))$, the definition of p also has to be changed and $p(T) = 0$ has to be replaced by $p(T) = \dfrac{\partial C}{\partial y}(y(T))$.

PART II. HOMOGENIZATION

The concept of generalized control introduced in the previous section is inadequate when dealing with partial differential equations in more than one space variable. Topologies that are not the weak ones will appear. In fact we have to play with the weak convergence of inverse operators, which is difficult to recognize directly on the coefficients of the operators. The first works in this direction are due to Spagnolo (1968) and to Marino and Spagnolo (1969) (see also Spagnolo (1976)). The framework that we present here traces back to a joint work that we started in the beginning of the 1970s. This framework is general and can easily be adapted to all kinds of variational situations.

The problem that we consider here is the following.

$$\begin{cases} -\sum_{i,j} \dfrac{\partial}{\partial x_i} \left(a_{ij}(x) \dfrac{\partial u}{\partial x_j} \right) = f \text{ in } \Omega, \\ u = 0 \text{ on } \partial\Omega, \end{cases} \tag{19}$$

where Ω is a bounded open set in $\mathbb{R}^N$. Dirichlet conditions are considered for simplicity; the homogenization has a local character and the results presented here do not depend on the boundary conditions considered on $\partial\Omega$. The coefficients a_{ij} are bounded measurable functions that satisfy:

$$
\begin{cases}
a_{ij} = a_{ji} \quad \forall i, j = 1, \ldots, N, \\
\alpha |\xi|^2 \leq \displaystyle\sum_{i,j} a_{ij}(x)\xi_i \xi_j \leq \beta |\xi|^2, \ \forall \xi \in \mathbb{R}^N, \text{a.e. } x \in \Omega.
\end{cases}
\tag{20}
$$

The symmetry condition could be abandoned without difficulties; in the nonsymmetrical case one loses some inequalities.

The natural space for u, which corresponds to a finite energy condition, is the Sobolev space $H_0^1(\Omega)$; that is,

$$
\begin{cases}
u \in V = H_0^1(\Omega) \\
= \{u \in L^2(\Omega) \ : \ \dfrac{\partial u}{\partial x_i} \in L^2(\Omega), \ i = 1, \ldots, N, \ u \mid_{\partial\Omega} = 0\}.
\end{cases}
\tag{21}
$$

For mathematical reasons, the data f will vary in the dual space V':

$$
\begin{cases}
f \in V' = H^{-1}(\Omega) \\
= \{f \in \mathcal{D}'(\Omega) \ : \ f = -\displaystyle\sum_i \dfrac{\partial g_i}{\partial x_i}, g_i \in L^2(\Omega), \ i = 1, \ldots, N\}.
\end{cases}
\tag{22}
$$

We denote by $A(x)$ the matrix with entries $a_{ij}(x)$ and we write (19) in the form:

$$
-\operatorname{div}(A(x)\operatorname{grad}u) = f \text{ in } \Omega, \ u \in V,
\tag{19a}
$$

which is equivalent to:

$$
\begin{cases}
\displaystyle\int_\Omega (A(x)\operatorname{grad}u, \operatorname{grad}v)\,dx = \langle f, v\rangle, \qquad \forall v \in V, \\
u \in V.
\end{cases}
\tag{19b}
$$

A standard application of Lax-Milgram's lemma gives:

Proposition 6. *For any $f \in V'$ there exists a unique u in V which is the solution of (19). In other terms the operator $\mathcal{A} = \operatorname{div}(A\operatorname{grad}\cdot)$ is an isomorphism from V onto V'.*

The operator $\mathcal{A}$ depends linearly upon the coefficients a_{ij}, hence the operator $\mathcal{A}^{-1}$ is differentiable (and even analytic) in a_{ij}.

Proposition 7. *The map $(A, f) \longrightarrow u$ is Fréchet differentiable. To any increment $(\delta A, \delta f)$ there corresponds an increment δu which is the solution of*

$$
\begin{cases}
-\operatorname{div}(A\operatorname{grad}\delta u) = \delta f + \operatorname{div}(\delta A\operatorname{grad}u) \text{ in } \Omega, \\
\delta u \in V.
\end{cases}
\tag{23}
$$

Remark 10.

Note that $div \, (\delta A \, grad \, u)$ belongs to V'. One can see here the importance of a good choice for the functional spaces.

The set of matrices A satisfying (20) is compact when equipped with the weak $*$ topology of L^∞ but the map $A \longrightarrow u$ (for a fixed f) is not continuous for this topology; in other terms, if a sequence A_ε converges weakly $*$ to A_+, then the sequence u_ε of the solutions of (19) converges weakly in $H_0^1(\Omega)$ to some u_0 which in general is not the solution of the equation corresponding to A_+. The crucial problem in homogenization theory is to describe the equation satisfied by u_0 and to understand the corresponding convergence of the coefficients of A_ε. The one-dimensional example is instructive (but misleading): if $N = 1$ and if $\dfrac{1}{A_\varepsilon}$ converges to $\dfrac{1}{A_-}$ weakly $*$ in $L^\infty(\Omega)$, then u_0 is the solution of the equation corresponding to A_- (which in general differs from A_+).

Remark 11.

Equation (19) has several interpretations in physics. In electrostatics, u is the electrostatic potential and A represents the dielectric permittivity matrix. The material under investigation is nonhomogenous and made, for example, of a mixture of various simple materials. The parameter ε describes the size of the heterogeneities that are supposed to be small compared to the macroscopic size of the sample. Therefore there are two scales: a macroscopic one (the size of the sample Ω) and a microscopic one (the size of the heterogeneities, denoted by ε). A situation which is easy to describe is the case where the heterogeneities are periodically distributed, that is, the case where $a_{ij}^\varepsilon \, (x) = \overline{a}_{ij} \, (\dfrac{x}{\varepsilon})$, where the functions $\overline{a}_{ij}$ are periodic. One variant is the case where $a_{ij}^\varepsilon \, (x) = \overline{a}_{ij} \, (x, \dfrac{x}{\varepsilon})$, where the $\overline{a}_{ij}$ are periodic in the second variable. As already mentioned, let us emphasize that the results presented in this paper are valid without any assumption of periodicity. Only hypothesis (20) is assumed.

Here is the basic result in homogenization theory.

Proposition 8. *Let A_ε be a sequence satisfying (20). Then there exists a subsequence A_η and some A_* satisfying (20) such that for any $f \in V'$ the solution u_η of*

$$-div \, (A_\eta \, grad \, u_\eta) = f \ in \ \Omega, \, u_\eta \in V,$$

satisfies

$$\begin{cases} u_\eta \longrightarrow u_0 \ weakly \ in \ V \\ A_\eta \, grad \, u_\eta \longrightarrow A_* \, grad \, u_0 \ weakly \ in \ (L^2(\Omega))^N. \end{cases} \quad (24)$$

This immediately implies that u_0 is solution of

$$- div\,(A_* \, grad\,u_0) = f \text{ in } \Omega\,,\ u_0 \in V,$$

which defines u_0 in a unique way.

Remark 12.
In the nonsymmetric case, (20) has to be replaced by other inequalities. A good choice is:

$$\begin{cases} \alpha|\xi|^2 \leq (A\,(x)\,\xi,\,\xi) \quad \text{and} \quad (A^1\,(x)\,\xi,\,\xi) \geq \frac{1}{\beta'}|\xi|^2 \\ \forall \xi \in \mathbb{R}^N \text{ and a.e. } x \in \Omega, \end{cases}$$

where $A^{-1}\,(x)$ denotes the inverse matrix of $A\,(x)$. These inequalities are indeed maintained when passing to the limit as done in Proposition 8.

Remark 13.
In the symmetric case the coefficients a_{ij} are uniquely determined by the operator $\mathcal{A} = -div\,(A\,grad\cdot)$. In the nonsymmetric one this is no longer true and formulation (24) contains more information than the fact that u_0 is the solution of $div\,(A_*\,grad\,u_0) = f$ in $\Omega\,,\ u_0 \in V$.

Remark 14.
Here is the scheme of the proof which was presented by Tartar (1977a) and written in Murat (1978a). An English translation of this paper is given in this volume in our joint paper entitled "H-convergence," and we refer to it for details. For f belonging to a countable dense subset of V' we can extract a subsequence such that u_η converges weakly to some $u_0 = Bf$ and $A_\eta\,grad\,u_\eta$ converges weakly to some Cf. We show first that the application $f \longrightarrow Bf$ is invertible. The main difficulty then consists of proving that Cf can be written as a local function of $grad\,u_0$.

With this in view we note that when v_η converges weakly in $H^1_{loc}\,(\Omega)$ to v with $div\,(A_\eta\,grad\,v_\eta)$ remaining in a compact subset of $H^1_{loc}\,(\Omega)$ and when $A_\eta\,grad\,v_\eta$ converges weakly in $[L^2(\Omega)]^N$ to g, one has the equality:

$$\begin{aligned} (Cf,\,grad\,v) &= \lim_\eta (A_\eta\,grad\,u_\eta,\,grad\,v_\eta) \\ &= \lim_\eta (A_\eta\,grad\,v_\eta,\,grad\,u_\eta) = (g,\,grad\,u_0)\,. \end{aligned}$$

This equality takes place in $L^1(\Omega)$, whereas the limits are taken in the sense of distributions. This is a simple consequence of the div-curl lemma of compensated compactness, which in this case is easily proved by integration by parts (cf. Murat (1978b), Tartar (1979)).

It is then sufficient to produce enough test functions v_η. But it is easy to construct a sequence v_η that converges to an arbitrary function $v_0 \in H^1_0(\Omega)$, by following the abstract construction sketched previously. The end of the proof is easy.

Remark 15.

In the periodic case, that is, when $A_\varepsilon(x) = \bar{A}(x/\varepsilon)$ with $\bar{A}$ periodic, A_* is a constant matrix and is defined by an "explicit" formula which is easy to obtain in the following way.

For $\lambda \in \mathbb{R}^N$ we solve the problem

$$
\begin{cases}
- div\,(\bar{A}(y)\,(grad\,w_\lambda + \lambda)) = 0 \text{ in } \mathbb{R}^N, \\
w_\lambda \text{ periodic.}
\end{cases}
\tag{25}
$$

This defines w_λ up to an additive constant. Setting

$$
v_\varepsilon(x) = \lambda x + \varepsilon\, w_\lambda(x/\varepsilon)
$$

we have

$$
\begin{cases}
- div\,(A_\varepsilon(x)\,grad\,v_\varepsilon) = 0 \text{ in } \mathbb{R}^N \\
v_\varepsilon \longrightarrow \lambda x \text{ in } H^1_{loc}(\mathbb{R}^N)
\end{cases}
$$

and thus

$$
A_*\lambda = \fint \bar{A}(y)\,(grad\,w_\lambda + \lambda)\,dy,
\tag{26}
$$

where $\fint$ denotes the mean value over the period.

Many other generalizations concerning the periodical case may be found in Bensoussan et al., (1978) and in Sanchez-Palencia (1980).

Remark 16.

A simple consequence of the hypotheses of Proposition 8 and of the div-curl lemma (see Remark 15) is that:

$$
(A_\eta\,grad\,u_\eta\,,\,grad\,u_\eta) \longrightarrow (A_*\,grad\,u_0\,,\,grad\,u_0) \text{ in } \mathcal{D}'(\Omega)
\tag{27}
$$

which means that the energies converge to the energy of the limit problem in the sense of distributions.

This convergence can be proved to take place in the weak topology of $L^1_{loc}(\Omega)$ (and even of $L^1(\Omega)$ if the boundary is smooth) by applying Meyers' result, which is valid for a large class of equations (see Meyers (1963) and De Giorgi and Spagnolo (1973)).

Remark 17.

The local character of this convergence is fundamental. If the property of Proposition 8 is satisfied, we say that the sequence A_η *H−converges to* A_* *on* Ω and write $A_\eta \xrightarrow{H} A_*$. If $A_\eta \xrightarrow{H} A_*$ and $B_\eta \xrightarrow{H} B_*$ and if $A_\eta = B_\eta$ on an open set ω, one has $A_* = B_*$ on ω. This result also holds if ω is only measurable as can be seen by using Meyers' result mentioned in Remark 16. Therefore $A_\eta \xrightarrow{H} A_*$ on $\Omega = \omega_1 \cup \omega_2 \cup z$, with ω_1 and ω_2 open sets and $meas\, z = 0$, is equivalent to $A_\eta \mid_{\omega_j} \xrightarrow{H} A_* \mid_{\omega_j}$ on ω_j, $j = 1, 2$.

The next fundamental step is then to characterize the matrices $A_*(x)$ that can be obtained when A_ε belongs to a set that is smaller than the one defined by (20). An important problem is to know what materials are obtained by mixing in an arbitrary way some given materials, i.e. to find all possible H−limits of matrices $A(x)$ that take only some given values. A material means here that the eigenvalues of the matrix $A(x)$ even, with any arbitrary orientation for the eigenvectors. From a more practical point of view, we want to determine the properties of any composite material made of some basic materials. We fix the volume fractions of the materials, that is, the measures of the sets where $A(x)$ takes the given values. Let us notice that an isotropic material corresponds to a matrix $A(x) = a(x)I$ with a scalar a.

Let us begin with a simple result which is valid in the general case; we use the notation of the H−convergence introduced in the preceding remark.

Proposition 9. *Let us assume that* $A_\varepsilon = {}^t\!A_\varepsilon$ *and that* $A_\varepsilon \xrightarrow{H} A_*$. *If we have:*

$$\begin{cases} A_\varepsilon \longrightarrow A_+ \text{ weakly } * \text{ in } (L^\infty(\Omega))^{N^2}, \\ A_\varepsilon^{-1} \longrightarrow (A_-)^{-1} \text{ weakly } * \text{ in } (L^\infty(\Omega))^{N^2} \end{cases} \qquad (28)$$

then

$$A_-(x) \le A_*(x) \le A_+(x) \quad a.e. \ x \in \Omega. \qquad (29)$$

Remark 18.

This result is more precise than (20), which only implies that $\alpha I \le A_* \le \beta I$. The proof consists of using (24) and (27) in the inequalities

$$(A_\varepsilon(\lambda - grad\, u_\varepsilon), \lambda - grad\, u_\varepsilon) \ge 0 \text{ a.e.}$$

$$(A_\varepsilon^{-1}(\lambda - A_\varepsilon\, grad\, u_\varepsilon), \lambda - A_\varepsilon\, grad\, u_\varepsilon) \ge 0 \text{ a.e.}$$

in order to get, respectively, $A_* \le A_+$ and $A_- \le A_*$.

There are cases where both $A_* - A_-$ and $A_+ - A_*$ have 0 as eigenvalue but (29) is not optimal in the sense that for a class which is smaller than the class defined by (20), any symmetric matrix between A_- and A_+ is not an A_*.

The optimal results in this direction are very scarce. They are much more difficult to obtain than (29). The proof of such results consists of two steps: one obtains a necessary condition analogous to (29) (but more precise); or for any given matrix satisfying this necessary condition one constructs explicitly a sequence A_ε that H-converges to this matrix. This last construction generally uses computations that are much more explicit than those of formula (25).

Here is one of the few cases where an optimal characterization is known.

Proposition 10. *Assume that* $A_\varepsilon \xrightarrow{H} A_*$ *and that* $A_\varepsilon(x) = a_\varepsilon(x)I$ *where* $a_\varepsilon(x)$ *takes two values* α *and* β $(0 < \alpha < \beta < \infty)$. *Assume also that*

$$a_\varepsilon \longrightarrow \theta(x)\alpha + (1 - \theta(x))\beta \qquad \text{weakly} * \text{ in } L^\infty(\Omega) \qquad (30)$$

($\theta(x)$ thus denotes the local volume fraction at the point x of the material characterized by the scalar coefficient α). Then we have

$$A_*(x) \in M_{\theta(x)} \quad a.e. \ x \in \Omega \qquad (31)$$

where for $0 \leq \theta \leq 1$, the set M_θ is defined in the following way. Consider at every point x the N eigenvalues $(\mu_1(x), \ldots, \mu_N(x))$ of $A_*(x)$ and define two numbers:

$$\begin{cases} \mu_+(\theta) = \theta\,\alpha + (1 - \theta)\,\beta\,, \\ \dfrac{1}{\mu_-(\theta)} = \dfrac{\theta}{\alpha} + \dfrac{1 - \theta}{\beta}\,. \end{cases} \qquad (32)$$

Then $A_* \in M_\theta$ if and only if $(\mu_1, \ldots, \mu_N) \in K_\theta$ where K_θ is defined by:

$$\begin{cases} \mu_-(\theta) \leq \mu_j \leq \mu_+(\theta)\,, \ j = 1, \ldots, N\,, \\ \displaystyle\sum_{j=1}^{N} \dfrac{1}{\mu_j - \alpha} \leq \dfrac{1}{\mu_-(\theta) - \alpha} + \dfrac{N-1}{\mu_+(\theta) - \alpha}\,, \\ \displaystyle\sum_{j=1}^{N} \dfrac{1}{\beta - \mu_j} \leq \dfrac{1}{\beta - \mu_-(\theta)} + \dfrac{N-1}{\beta - \mu_+(\theta)}\,. \end{cases} \qquad (33)$$

Conversely, if $A_*(x)$ is any matrix satisfying (31) for a function θ satisfying $0 \leq \theta(x) \leq 1$, there exists a sequence $a_\varepsilon(x)$ that takes only the values α and β and satisfies (30) as well as $a_\varepsilon I \xrightarrow{H} A_*$.

Remark 19.

In the case where A_* is an isotropic material (i.e. $A_* = \mu I$), a tedious computation shows that the inequalities (33) are equivalent to:

$$\nu(\alpha\,; \beta\,; 1 - \theta) \leq \mu \leq \nu(\beta\,; \alpha\,; \theta)\,, \qquad (34)$$

where $\nu\left(\beta\,;\alpha\,;\theta\right)$ is given by:

$$\frac{\nu-\beta}{\nu+(N-1)\beta}=\theta\frac{\alpha-\beta}{\alpha+(N-1)\beta}\,. \tag{35}$$

Formula (35) corresponds to an explicit computation known as "coated spheres": a kernel made of a ball of material α and coated by a spherical shell of material β is embedded in this material ν. Formula (35) is known in physics under the names of Maxwell-Garnett, Lorentz-Lorentz, or Clausius-Mossoti. The bounds (34) appeared for the first time in the work of Hashin and Shtrikman (1962).

Remark 20.
If a_ε is assumed to take only the values α and β and to satisfy

$$\int_\Omega a_\varepsilon\left(x\right)dx=\gamma\,meas(\Omega)\,, \tag{36}$$

(with $\alpha<\gamma<\beta$), then θ satisfies:

$$\int_\Omega\theta\left(x\right)dx=\frac{\beta-\gamma}{\beta-\alpha}\,meas(\Omega)\,. \tag{36a}$$

From a variant of Proposition 10 we infer that if A_* satisfies (31) with θ satisfying (36a), there exists a sequence a_ε, taking only the values α and β, which satisfies (36) and $a_\varepsilon\,I\xrightarrow{\,H\,}A_*$.

Remark 21.
It is beyond the aim of this paper to explain in detail the proof of Proposition 10, and the reader is referred to Tartar (1985) for this proof. For a large part of the problems we are interested in, such a detailed description of M_θ will turn out to be superfluous and Proposition 9, which asserts that $\mu_-\left(\theta\left(x\right)\right)\le\mu_j\left(x\right)\le\mu_+\left(\theta\left(x\right)\right)$ a.e., $1\le j\le N$, will be sufficient. This remark is important since there is almost no complete description analogous to Proposition 10 in other situations.

Remark 22.
For a given matrix A_* satisfying $A_*\left(x\right)\in M_{\theta\left(x\right)}$, there are infinitely many sequences of matrices $A_\varepsilon=a_\varepsilon(x)\,I$ (with coefficients a_ε that only take the values α and β with the given volume fractions $\theta\left(x\right)$ and $1-\theta\left(x\right)$) that H-converge to A_*. It is not realistic to try to give a unique construction of the matrices A_ε for a given matrix A_*. From another point

of view, several procedures are known to describe explicitly (in a way that can be more or less complicated) matrices A_ε that $H-$converge to A_*. One of them generalizes to the case of confocal ellipsoids the computations which in the case of balls give formula (35) (see Tartar (1985) Section 4 for details). Another one consists of producing A_* as the limit of stratified materials, with layers having various directions and various scales; see Braidy and Pouilloux (1982), Tartar (1985), Section 5, and Francfort and Murat (1986), Section 4. Finally let us note that a given material A_* corresponds to various volume fractions that take their values in an interval. This interval is reduced to a point in the (limit) case where the matrix A_* has one eigenvalue equal to $\mu_-(\theta)$ and the $N - 1$ other eigenvalues equal to $\mu_+(\theta)$. This case is easy to obtain by using stratified materials; if e is an unit vector and $a_\varepsilon(x) = \tilde{a}_\varepsilon((x, e))$ with $\tilde{a}_\varepsilon(t) \rightharpoonup \theta(t)\alpha + (1 - \theta(t))\beta$ weakly $*$ in L^∞, then e is an eigenvector for $A_*(x)$ corresponding to the eigenvalue $\mu_{(\theta)((x,e))}$, and the orthogonal subspace is generated by the eigenvectors corresponding to the multiple eigenvalue $\mu_+(\theta)((x, e))$. The explicit formula for A_* can indeed be easily written in the case $A_\varepsilon((x, e))$.

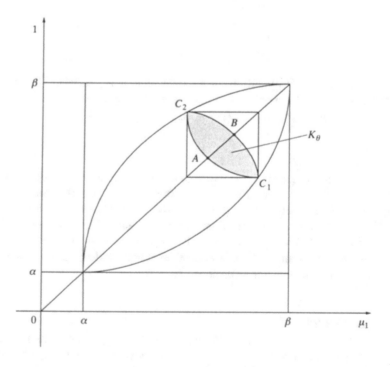

Figure 1.

Remark 23.

The case $N = 1$ (which is not completely useless) corresponds to $A_* = \mu_-(\theta)$. The case $N = 2$ corresponds to Figure 1. The equation of $C_1 \, A \, C_2$ is $\dfrac{1}{\mu_1 - \alpha} + \dfrac{1}{\mu_2 - \alpha} = \dfrac{1}{\mu_-(\theta) - \alpha} + \dfrac{1}{\mu_+(\theta) - \alpha}$; the equation of $C_1 \, B \, C_2$ is $\dfrac{1}{\beta - \mu_1} + \dfrac{1}{\beta - \mu_2} = \dfrac{1}{\beta \mu_-(\theta)} + \dfrac{1}{\beta - \mu_+(\theta)}$. The point C_1 has $\mu_1 = \mu_-(\theta)$, $\mu_2 = \mu_+(\theta)$ as coordinates and C_2 is the symmetric point of C_1 relative to the line $\mu_1 = \mu_2$.

PART III. GENERALIZED DOMAINS AND
NECESSARY CONDITIONS OF OPTIMALITY

Let Ω be an open bounded subset of $\mathbb{R}^N$. We want to find a measurable subset ω of Ω that satisfies

$$meas\,(\omega) = \gamma', \tag{37}$$

and that minimizes the functional

$$F_0(\omega) = \int_\omega g\,(x, u\,(x))\,dx + \int_{\Omega \backslash \omega} h\,(x, u\,(x))\,dx, \tag{38}$$

where u is the solution of:

$$-div\,(a\,(x)\,grad\,u) = f \ \text{ in } \Omega,\ u \in V = H_0^1\,(\Omega), \tag{39}$$

with $a\,(x) = \alpha$ on ω, β on $\Omega \setminus \omega$. Here f is given in V'. The hypotheses on g and h are specified later. Using the results of Part II we construct a relaxed problem and write the conditions of optimality.

For the first step we need the applications $u \longrightarrow g\,(\cdot, u\,(\cdot))$ and $u \longrightarrow h\,(\cdot, u\,(\cdot))$ to be continuous from V weak (we consider only bounded sets so we need only the applications to be sequentially continuous) in $L^1\,(\Omega)$ strong. Therefore we suppose, for example, that

$$\begin{cases} g \text{ and } h \text{ satisfy the Carathéodory's conditions} \\ \text{(measurability in } x, \text{ continuity in } \lambda \text{) with} \\ |g\,(x, \lambda)| + |h\,(x, \lambda)| \leq k_0\,(x) + c_0|\lambda|^2 \text{ with } k_0 \in L^1(\Omega). \end{cases} \tag{40}$$

Proposition 11. *A relaxed problem (in the sense of Section I.a.) of the preceding problem is as follows. Find θ and A_* satisfying*

$$\begin{cases} 0 \leq \theta\,(x) \leq 1 \ a.e. \qquad \displaystyle\int_\Omega \theta\,(x)\,dx = \gamma' \\ A_*\,(x) \in M_{\theta(x)} \ a.e. \ x \in \Omega \end{cases} \tag{37a}$$

(where $M_{\theta(x)}$ is defined in Proposition 10) that minimizes the functional:

$$F(\theta, A_*) = \int_\Omega \left[\theta(x)\, g(x, u(x)) + (1 - \theta(x))\, h(x, u(x)) \right] dx\,, \qquad (38a)$$

where u is the solution of:

$$-\operatorname{div}(A_* \operatorname{grad} u) = f \ \text{in}\ \Omega\,, \quad u \in V\,. \qquad (39a)$$

Remark 24.

The initial problem corresponds to the case in which θ is taken as the characteristic function of ω in the relaxed problem. It is indeed enough to note that $A \in M_1$ is equivalent to $A = \alpha I$ and $A \in M_0$ to $A = \beta I$. The topology on the space X described by (37a) is the weak $*$ topology of $L^\infty(\Omega)$ for θ and the $H-$convergence topology for A_*. The proof of Proposition 11 immediately follows from Proposition 10 where all the technical difficulties are solved.

Remark 25.

The generalized domains are made here of pairs (θ, A_*). The natural idea that is hidden behind Proposition 11 is the following. If there is no domain ω that minimizes the functional F_0, the minimizing sequence will, for example, be made of the material β in which inclusions of material α will be distributed with smaller and smaller sizes. This corresponds to some heterogeneous material, the structure of which is described by (θ, A_*) at the macroscopic level.

Remark 26.

Even though the definition of the set M_θ is quite complicated, we have the following property.

For every $\theta \in [0, 1]\,, M_\theta$ is a convex set of symmetric matrices. $\qquad (41)$

Indeed if $\varphi : \mathbb{R} \longrightarrow \mathbb{R}$ is a convex function, for any symmetric matrix A with eigenvalues $\mu_1, \ldots, \mu_N$ and for any orthonormal basis we have the inequality $\sum_i \varphi(A_{ii}) \leq \sum_i \varphi(\mu_i)$, which easily follows from the existence of an orthogonal matrix P such that $A_{ij} = \sum_k P_{ik}\mu_k P_{jk}$. Therefore, when φ is a convex function, the sets $\{A : \sum_i \varphi(\mu_i) \leq c\}$ and $\{A : \sum_i \varphi(A_{ii}) \leq c$

for any orthonormal basis} coincide and then define a convex set of matrices since the second one is a convex set. The set M_θ is defined as the intersection of two sets of this type: one corresponds to $\varphi(\lambda) = \dfrac{1}{\lambda\alpha}$ for $\lambda > \alpha$, $\varphi(\lambda) = +\infty$ if $\lambda \leq \alpha$ and the other corresponds to $\varphi(\lambda) = \dfrac{1}{\beta - \lambda}$ for $\lambda < \beta$, $\varphi(\lambda) = +\infty$ if $\lambda \geq \beta$. Finally, using again the fact that for some orthogonal matrix P one has $A_{ij} = \sum_k P_{ik}\mu_k P_{jk}$, it is easy to see that the two sets

$$\{A : \mu_-(\theta) \leq \mu_i \leq \mu_+(\theta) \ \forall i = 1, \ldots, N\}$$

$$\{A : \mu_-(\theta) \leq A_{ii} \leq \mu_+(\theta) \ \forall i = 1, \ldots, N \text{ for any orthonormal basis}\}$$

coincide and define a convex set of matrices since the second one is a convex set. Hence M_θ is a convex set.

In the sequel we compute the derivatives on X, that is, with respect to (θ, A_*), and then write the necessary conditions of optimality for a generalized domain. These conditions remain valid particularly if there exists a domain ω that is an optimal solution of the initial problem.

In order to be able to compute the derivatives we assume more hypotheses on g and on h than those assumed in (40) to obtain the relaxed problem. We assume that $u \longrightarrow g(\cdot, u(\cdot))$ and $u \longrightarrow h(\cdot, u(\cdot))$ are Gâteaux differentiable and locally Lipschitz continuous from V into $L^1(\Omega)$. This holds true if we impose that

$$\begin{cases} \dfrac{\partial g}{\partial \lambda} \text{ and } \dfrac{\partial h}{\partial \lambda} \text{ satisfy the Carathéodory's conditions with} \\ |\dfrac{\partial g}{\partial \lambda}(x, \lambda)| + |\dfrac{\partial h}{\partial \lambda}(x, \lambda)| \leq k_1(x) + c_1|\lambda| \text{ with } k_1 \in L^2(\Omega). \end{cases} \quad (40\text{a})$$

Then the map $(\theta, A_*) \longrightarrow F(\theta, A_*)$ is differentiable and to any increment $(\delta\theta, \delta A_*)$ corresponds the increment δF given by

$$\delta F = \int_\Omega \{ [g(x, u) - h(x, u)] \, \delta\theta \quad (42\text{a})$$

$$+ [\theta \dfrac{\partial g}{\partial \lambda}(x, u) + (1 - \theta) \dfrac{\partial h}{\partial \lambda}(x, u)] \delta u \} \, dx,$$

where δu is defined by

$$- div(A_* \, grad \, \delta u + \delta A_* \, grad \, u) = 0 \text{ in } \Omega, \quad \delta u \in V. \quad (42\text{b})$$

In order to eliminate δu in the expression of δF we introduce the adjoint state p as the solution of

$$- div(A_* \, grad \, p) = \theta \dfrac{\partial g}{\partial \lambda}(x, u) + (1 - \theta) \dfrac{\partial h}{\partial \lambda}(x, u) \text{ in } \Omega, \quad p \in V. \quad (43)$$

Since using integrations by parts the term containing δu in (42a) is easily proved to coincide with $-(\delta A_* \, grad\, u, \, grad\, p)$, we have the following result.

Proposition 12. *The map* $(\theta, A_*) \longrightarrow F(\theta, A_*)$ *is Gâteaux differentiable and its derivative is given by*

$$\delta F = \int_\Omega \{ [g(x, u) - h(x, u)] \, \delta\theta \, (\delta A_* \, grad\, u, \, grad\, p)\} \, dx, \qquad (44)$$

where p is the solution of (43).

Remark 27.

A necessary condition in order for F to achieve its minimum at (θ, A_*) is the following,

$$\delta F \geq 0 \text{ for any admissible increments } (\delta\theta, \delta A_*). \qquad (44a)$$

An admissible increment is defined as the derivative at $t = 0$ of a curve $t \longrightarrow (\theta(x, t), \, A_*(x, t))$ that remains in the set of constraints (37a). The convexity result (41) obtained in Remark 26 will considerably simplify the analysis of these admissible directions.

It is useful to deduce the consequences of (44a) in two steps: taking $\delta\theta = 0$ and making variations of A_* in M_θ, making variations of θ. Remark 26 implies that in the case $\delta\theta = 0$ the increment $\delta A_* = B_* - A_*$ is admissible at the point A_* for any $B_* \in M_\theta$; $\delta F \geq 0$ at (θ, A_*) and then implies

$$\begin{cases} \displaystyle\int_\Omega (B_*(x) grad\, u, \, grad\, p) \, dx \leq \int_\Omega (A_*(x) grad\, u, \, grad\, p) \, dx \\ \forall B_*(x) \in M_{\theta(x)}. \end{cases} \qquad (45)$$

Hence A_* satisfies almost everywhere a pointwise optimality condition that follows from the maximization of $(B_* \, grad\, u\,(x), \, grad\, p\,(x))$ with respect to $B_* \in M_{\theta(x)}$.

This maximization problem is easily solved by applying the following.

Lemma 1. *Let C_θ be the set of symmetric $N \times N$ matrices with eigenvalues between $\mu_-(\theta)$ and $\mu_+(\theta)$. Let e and e' be two unit vectors of $\mathbb{R}^N$. Then the inequality:*

$$\begin{cases} (Ae, e') \geq (Be, e'), \quad \forall B \in C_\theta, \\ A \in C_\theta, \end{cases} \qquad (46)$$

is equivalent to

$$\begin{cases} A \in C_\theta, \\ Ae = \dfrac{1}{2}(\mu_+(\theta) + \mu_-(\theta)) e + \dfrac{1}{2}(\mu_+(\theta) - \mu_-(\theta)) e', \\ Ae' = \dfrac{1}{2}(\mu_+(\theta) - \mu_-(\theta)) e + \dfrac{1}{2}(\mu_+(\theta) + \mu_-(\theta)) e', \end{cases} \qquad (46a)$$

which is also equivalent to

$$\begin{cases} A \in C_\theta, \\ A\,(e + e') = \mu_+\,(\theta)\,(e + e'), \\ A\,(e - e') = \mu_-\,(\theta)\,(e - e'). \end{cases} \qquad (46b)$$

To prove Lemma 1, set $X = e + e'$, $X' = e - e'$. Then (46) reads as

$$\begin{cases} (A\,X,\,X) - (A\,X',\,X') \geq (B\,X,\,X) - (B\,X',\,X'),\; \forall B \in C_\theta \\ A \in C_\theta, \end{cases}$$

which using the Rayleigh quotients is equivalent to (46b).

Remark 28.

There exists a matrix $A \in M_\theta$ satisfying (46b), for example, that one corresponding to a stratified material perpendicularly to X', where X' is parallel to $e - e'$ and perpendicular to $e + e'$. (If $e \neq e'$ or if $e = e'$ and $N \leq 2$, this A is unique.)

Now using only the following information upon M_θ (see (33) for the definition of K_θ):

$$\begin{cases} (\mu_1, \ldots, \mu_N) \in K_\theta \text{ implies} \\ \qquad \mu_-\,(\theta) \leq \mu_j \leq \mu_+\,(\theta),\; j = 1, \ldots, N\,; \\ \text{if } N \geq 2, \text{ there exists } (\mu_1, \ldots, \mu_N) \in K_\theta \text{ with} \\ \mu_1 = \mu_+(\theta),\; \mu_2 = \mu_-(\theta), \end{cases} \qquad (47)$$

we deduce from Lemma 1 that any A that maximizes $(Be,\,e')$ on M_θ satisfies (46a). This method can be useful if we observe that the result of Proposition 10 is not known for general systems. In contrast, if one considers functionals F_0 different from those defined by (38), one may need all the information upon M_θ.

If e and e' are unit vectors with $(e,\,e') = cos\,\varphi$, the maximum of $(B\,e,\,e')$ for B in M_θ is equal to $cos^2 \dfrac{\varphi}{2}\,\mu_+(\theta) sin^2 \dfrac{\varphi}{2}\,\mu_-(\theta)$. We thus proved the following.

Proposition 13. *If (θ, A_*) is a minimum of F, then except for the set where $|grad\,u||grad\,p| = 0$ one has:*

$$\begin{cases} A_*\,grad\,u = & \dfrac{1}{2}\,(\mu_+(\theta) + \mu_-(\theta))\,grad\,u \\ & + \dfrac{1}{2}\,(\mu_+(\theta) - \mu_-(\theta))\,\dfrac{|grad\,u|}{|grad\,p|}\,grad\,p, \\ A_*\,grad\,p = & \dfrac{1}{2}\,(\mu_+(\theta) - \mu_{(\theta)})\,\dfrac{|grad\,p|}{|grad\,u|}\,grad\,u \\ & + \dfrac{1}{2}\,(\mu_+(\theta) + \mu_-(\theta))\,grad\,p. \end{cases} \qquad (48)$$

In particular since $(grad\,u,\ grad\,p) = |grad\,u||grad\,p|\cos\varphi$, *one has*

$$\begin{cases} (A_*\,grad\,u,\ grad\,p) = \\ |grad\,u||grad\,p|\{\cos^2\dfrac{\varphi}{2}\,\mu_+(\theta) - \sin^2\dfrac{\varphi}{2}\,\mu_-(\theta)\}. \end{cases} \tag{48a}$$

We now make variations of θ. More precisely, we consider a curve $(\theta\,(x,t),\ A_*\,(x,\,t))$ defined for $x \in \Omega$ and $t \in [0,1]$ such that:

$$\begin{aligned} A_*\,(t)\,grad\,u ={}& \frac{1}{2}\,(\mu_+(\theta(t)) + \mu_-(\theta(t)))\,grad\,u \\ &+ \frac{1}{2}\,(\mu_+(\theta(t)) - \mu_-(\theta(t)))\,\frac{|grad\,u|}{|grad\,p|}\,grad\,p, \end{aligned}$$

$$\begin{aligned} A_*\,(t)\,grad\,p ={}& \frac{1}{2}\,(\mu_+(\theta(t)) - \mu_-(\theta(t)))\,\frac{|grad\,p|}{|grad\,u|}\,grad\,u \\ &+ \frac{1}{2}\,(\mu_+(\theta(t)) + \mu_-(\theta(t)))\,grad\,p, \end{aligned}$$

where for the sake of simplicity, we did not write the variable x. Here u and p are the solutions of (39a) and (43), corresponding to $A_* = A_*\,(0)$. To construct such matrices $A_*\,(t)$ it is sufficient to consider stratified materials in a direction perpendicular to $\dfrac{grad\,u}{|grad\,u|} - \dfrac{grad\,p}{|grad\,p|}$ (see Remark 28). If $\theta\,(t)$ is admissible (i.e., verifies $0 \le \theta\,(x,t) \le 1$ a.e. and $\displaystyle\int_\Omega \theta\,(x,t)\,dx = \gamma'$), then the curve $(\theta\,(x,t),\ A_*\,(x,t))$ remains in the admissible set (37a) and by differentiation in t of the relation (48a) we obtain from (44):

Proposition 14. *If F achieves its minimum in $(\theta\,,A_*)$, then for any admissible variation of $\delta\theta$ we have*

$$\int_\Omega Q\,(x)\,\delta\theta\,(x)\,dx \ge 0 \tag{49}$$

where

$$\begin{cases} Q\,(x) &= [g\,(x,u) - h\,(x,u)] \\ &- |grad\,u||grad\,p|\{\cos^2\dfrac{\varphi}{2}\,\dfrac{d\mu_+}{d\theta}\,(\theta) - \sin^2\dfrac{\varphi}{2}\,\dfrac{d\mu_-}{d\theta}\,(\theta)\}. \end{cases} \tag{50}$$

Remark 29.

We wrote $(grad\,u,\ grad\,p) = |grad\,u||grad\,p|\cos\varphi$. Note that this does not define φ when $|grad\,u||grad\,p| = 0$ but in this case φ does not appear in the function Q.

Using definition (32) of $\mu_-(\theta)$ and $\mu_+(\theta)$ we write Q in the form:

$$
\left\{
\begin{aligned}
Q(x) &= [g(x,u) - h(x,u)] \\
&\quad + \frac{\beta - \alpha}{\alpha\beta} |grad\,u||grad\,p|\{\alpha\beta \cos^2\frac{\varphi}{2} - \mu_-^2(\theta)\sin^2\frac{\varphi}{2}\}.
\end{aligned}
\right.
\tag{50a}
$$

The interpretation of (49) is straightforward. A Lagrange multiplier C_0 appears because of the constraint (37a) (i.e., $\int_\Omega \theta(x)\,dx = \gamma'$) and we have the following.

Proposition 15. *If F achieves its minimum at (θ, A_*), then there exists $C_0 \in \mathbb{R}$ such that*

$$
\left\{
\begin{aligned}
\theta(x) &= 0 &\Rightarrow Q(x) &\geq C_0, \\
0 < \theta(x) &< 1 &\Rightarrow Q(x) &= C_0, \\
\theta(x) &= 1 &\Rightarrow Q(x) &\leq C_0.
\end{aligned}
\right.
\tag{51}
$$

Remark 30.

If $Q(x) > C_0$, then $\theta(x) = 0$, that is, $A_* = \beta I$. If $Q(x) < C_0$, then $\theta(x) = 1$, that is, $A_* = \alpha I$.

If we had not imposed the constraint $meas\,\omega = \gamma'$, the constraint $\int_\Omega \theta(x)\,dx = \gamma'$ would not appear in the relaxed problem and then we would have $C_0 = 0$ in (51).

Remark 31.

If there exists an homogenized region, that is, a region where $0 < \theta(x) < 1$, one has necessarily $Q(x) = C_0$ a.e. in this region. In particular the same constant C_0 appears in each subregion if the homogenized region has several connected components.

Remark 32.

The necessary conditions of optimality are made of two parts. The first one is (51) and the second one is (48) which makes precise the structure of A_* when $|grad\,u||grad\,p| \neq 0$; in the region where $|grad\,u||grad\,p| = 0$ we are not able to make precise the structure of A_*. In fact an optimal solution always exists that uses no other materials but the stratified ones, that is, a matrix A_* which has one eigenvalue equal to $\mu_-(\theta)$ and the other ones equal to $\mu_+ (\theta)$. Indeed, as observed by Raitum (1978), one can change A_* in the region where $|grad\,u||grad\,p| = 0$ without changing $A_*\,grad\,u$ and thus without changing the solution u. This is easy in the region ω_1 where $|grad\,u| = 0$ where we change A_* without changing $A_*\,grad\,u$ and maintain the same θ. In the region ω_2 where $|grad\,u| \neq 0$ and $|grad\,p| = 0$ this turns out to be more difficult since we have to change both θ and A_*. With this purpose we replace A_* by the matrix corresponding to a stratified material, without changing either $A_*\,grad\,u$ nor $\displaystyle\int_{\omega_2} \theta(x)\,dx$ while diminishing $\displaystyle\int_{\omega_2} [(g(x,u) - h(x,u))]\,\theta(x)\,dx$. We use the following result for this.

Lemma 2. *For e and $\bar{e}$ given in $\mathbb{R}^N$ there exists a $B \in C_\theta$ such that $B\,e = \bar{e}$ if and only if*

$$\|\bar{e}\frac{1}{2}(\mu_+(\theta) + \mu_-(\theta))\,e\| \leq \frac{1}{2}(\mu_+(\theta) - \mu_-(\theta))\|e\|.$$

If e and $\bar{e}$ are both different from 0, the preceding inequality is true if and only if θ varies in an interval. If we denote by $[\theta_-(e,\bar{e}), \theta_+(e,\bar{e})]$ the interval of $\theta \in [0,1]$ for which the inequality is valid, then we can construct a matrix B_- corresponding to a stratified material with the volume fraction θ_- and a matrix B_+ corresponding to the volume fraction θ_+.

To prove Lemma 2, define M by $M = B - \dfrac{1}{2}(\mu_+(\theta) + \mu_-(\theta))\,I$ and note that $Be = \bar{e}$ if and only if $Me = \bar{e}\dfrac{1}{2}(\mu_+(\theta) + \mu_-(\theta))\,e$, whereas $B \in C_\theta$ if and only if $\|M\| \leq \dfrac{1}{2}(\mu_+(\theta)\mu_-(\theta))$.

It is then sufficient to minimize $\displaystyle\int_{\omega_2} [g(x,u) - h(x,u)]\,\overline{\theta}(x)\,dx$ under the constraints $\displaystyle\int_{\omega_2} \overline{\theta}(x)\,dx$ given and

$$\theta_-\,(grad\,u, A_*\,grad\,u) \leq \overline{\theta} \leq \theta_+\,(grad\,u, A_*\,grad\,u).$$

This minimum is achieved for a function $\overline{\theta}$ that takes only the values $\theta_{(grad\,u, A_*\,grad\,u)}$ and $\theta_+\,(grad\,u, A_*\,grad\,u)$. We choose as a new θ this solution $\overline{\theta}$ and as a new A_* the matrix of the corresponding stratified material. (Let us notice that in the case $\theta = 0$ or $\theta = 1$ the word "stratified" is not correct since the matrix corresponds to βI or αI, that is, to a pure isotropic material.)

Remark 33.

By considering the regions where $|grad\, u|\, |grad\, p| \neq 0$ and where $|grad\, u|\, |grad\, p| = 0$, and using Remarks 32 and 28, we can replace A_* by the matrix of a stratified material without changing either $A_*\, grad\, u$ nor $A_*\, grad\, p$, thus without changing u and p.

Remark 34.

Even if we proved the existence of an optimal solution (θ, A_*) with A_* corresponding to a stratified material, we cannot consider the minimization problem restricted to the class of those special pairs (θ, A_*) as a relaxed problem in the sense of Section I.a. We are only able to say that, in view of the particular form of the function F_0, there exists a solution which has this simple form. (A similar situation appears when minimizing a linear functional on a compact convex set: even if the minimum is always achieved at an extremal point, the set of extremal points is not always closed.) Nevertheless there are situations where the analysis of the conditions of optimality leads to simpler relaxed problems; see the following two sections.

PART IV. EXAMPLE ONE

Let Ω be an open subset of $\mathbb{R}^N$. (The case is motivated by applications corresponds to $N = 2$; see Remark 42.) We look for a measurable subset ω of Ω with given measure γ' and define:

$$a\,(x) = \begin{cases} \alpha \text{ on } \omega\,, \\ \beta \text{ on } \Omega \setminus \omega\,. \end{cases} \tag{52}$$

Let u be the solution of

$$-div\,(a\,grad\,u) = 1 \text{ in } \Omega\,, \quad u \in V = H_0^1\,(\Omega)\,. \tag{53}$$

We want to maximize $\displaystyle\int_\Omega u\,dx = \int_\Omega a|grad\,u|^2\,dx$, that is, to minimize

$$J_0\,(\omega) = -\int_\Omega u\,(x)\,dx\,. \tag{54}$$

We call the problem defined by (52), (53), and (54), the *initial problem.*

Applying Proposition 11 we obtain a relaxed problem in the sense of Section I.a. which is defined by

$$A_*(x) \in M_{\theta(x)}\,; \quad 0 \leq \theta\,(x) \leq 1\,; \quad \int_\Omega \theta\,(x)\,dx = \gamma'\,, \tag{55}$$

$$-div\,(A_*\,grad\,u) = 1 \text{ in } \Omega\,, \quad u \in V\,, \tag{56}$$

and since $g(x, u) = h(x, u) = -u$ we here minimize:

$$J(\theta, A_*) = \int_\Omega u(x)\, dx.$$ (57)

From the definition (43) of the adjoint state p we get

$$-div(A_*\, grad\, p) = -1 \text{ in } \Omega, \ p \in V,$$

and therefore

$$p = -u.$$ (58)

We deduce from Proposition 13 that if (θ, A_*) realizes the minimum of $J(\theta, A_*)$, then, in the region where $|grad\, u| \neq 0$ we have $A_*\, grad\, u = \mu_-(\theta)\, grad\, u$, an equality which is still valid in the region where $|grad\, u| = 0$. Thus:

$$A_*\, grad\, u = \mu_-(\theta)\, grad\, u \qquad \text{a.e. in } \Omega.$$ (59)

As in Remark 28, in the region where $|grad\, u| \neq 0$, A_* has one eigenvalue equal to $\mu_-(\theta)$ and from (33) we get that the other eigenvalues are equal to $\mu_+(\theta)$. Thus A_* is a matrix that corresponds to a stratified material with layers perpendicular to $grad\, u$.

Since the value of φ is π, formula (50a) gives

$$Q(x) = \frac{\beta - \alpha}{\alpha\beta} \mu_-^2(\theta) |grad\, u|^2$$

and Proposition 15 gives the following necessary condition of optimality. There exists $C_1 \geq 0$ such that

$$\begin{cases} \theta(x) = 0 & \Rightarrow \mu_-(\theta)\, |grad\, u| \leq C_1, \\ 0 < \theta(x) < 1 & \Rightarrow \mu_-(\theta)\, |grad\, u| = C_1, \\ \theta(x) = 1 & \Rightarrow \mu_-(\theta)\, |grad\, u| \geq C_1. \end{cases}$$ (60)

Formula (60) allows one to write θ as a function of $|grad\, u|$ in the following way.

$$\begin{cases} \text{if} \quad 0 \leq |grad\, u(x)| \leq \dfrac{C_1}{\beta}, \text{ then } \theta(x) = 0, \\[2mm] \text{if} \quad \dfrac{C_1}{\beta} \leq |grad\, u(x)| \leq \dfrac{C_1}{\alpha}, \text{ then} \\[2mm] \qquad \theta(x) = [\dfrac{C_1}{\alpha} - \dfrac{C_1}{\beta}]^1 (|grad\, u(x)| - \dfrac{C_1}{\beta}), \\[2mm] \text{if} \quad \dfrac{C_1}{\alpha} \leq |grad\, u(x)|, \text{ then } \theta(x) = 1. \end{cases}$$ (60a)

Remark 35.
The case $C_1 = 0$ is obtained only if $\theta \equiv 1$, which means $\gamma' = meas\,\Omega$.

The necessary conditions of optimality for the relaxed problem (55), (56), and (57) led us to the two conditions (59) and (60). In what follows we prove that each of these conditions is capable of solving a minimization problem that is simpler than the relaxation problem itself.

The first problem consists of minimizing:

$$\tilde{J}(\theta) = -\int_\Omega u(x)\,dx\,,\tag{61}$$

where u is the solution of

$$-div\,(\mu_-(\theta)\,grad\,u) = 1 \text{ in } \Omega\,,\quad u \in V\,,\tag{62}$$

under the constraints:

$$0 \le \theta(x) \le 1 \text{ a.e. }\,;\,\int_\Omega \theta(x)\,dx = \gamma'\,.\tag{63}$$

The second one consists of minimizing:

$$\int_\Omega (\Phi_{C_1}(|grad\,v|) - v)\,dx\,,\tag{64}$$

where $\Phi_{C_1} : \mathbb{R} \longrightarrow \mathbb{R}$ is the convex function defined by $\Phi_{C_1}(0) = 0$ and

$$\frac{d\Phi_{C_1}}{d\lambda}(\lambda) = \begin{cases} \beta\lambda \text{ if } 0 < \lambda < \dfrac{C_1}{\beta}\,, \\[2mm] C_1 \text{ if } \dfrac{C_1}{\beta} < \lambda < \dfrac{C_1}{\alpha}\,, \\[2mm] \alpha\lambda \text{ if } \dfrac{C_1}{\alpha} < \lambda. \end{cases}\tag{65}$$

Condition (59) shows that any solution of the relaxed problem satisfies (62) and, since any function u which is a solution of (62) can be obtained as a solution of (56) for some matrix A_* which corresponds to a stratified material with layers perpendicular to $grad\,u$, the relaxed problem (55), (56), (57) is equivalent to problem (61), (62), (63).

Remark 36.

Problem (61), (62), (63) is also a relaxed problem in the sense of Section I.a. In order to prove this we remark that

$$\tilde{J}(\theta) = \min_{v \in H_0^1(\Omega)} \int_{\Omega} (\mu_-(\theta) |grad\,v|^2 - 2v)\,dx, \qquad (66)$$

and that (61), (62), (63) is equivalent to minimizing (in θ) the function $\tilde{J}$ defined by (66) under the constraint (63). The important fact is that $\tilde{J}$ is convex in θ since the function $\mu_-(\theta) w^2$ is convex in (θ, w) for $\theta \geq 0$ and w in $\mathbb{R}^N$. (The latest assertion results from the fact that $\frac{1}{t} w^2$ is convex in (t, w) for $t > 0$.) This gives the property (2b) if the space

$$X = \{\theta : 0 \leq \theta(x) \leq 1, \ \int_{\Omega} \theta(x)\,dx = \gamma'\}$$

is equipped with the weak $*$ topology of $L^\infty(\Omega)$, whereas X_0 corresponds to the characteristic functions of sets ω of measure γ'.

In order to satisfy the property (2c) we consider the matrix A_* which corresponds to a stratified material with layers perpendicular to $grad\,u$ (where u is the solution of (62)) with volume fraction θ of material α. We then construct a sequence $\varphi_n I$ that H-converges to A_* with $\varphi_n = \theta_n \alpha + (1\theta_n)\beta$ where θ_n is a characteristic function which converges weak $*$ in $L^\infty(\Omega)$ to θ. We get using the convergence of energies (Remark 16):

$$\lim \tilde{J}(\theta_n) = -\int_{\Omega} (A_*\,grad\,u,\ grad\,u)$$

$$= -\int_{\Omega} \mu_-(\theta) |grad\,u|^2\,dx = \tilde{J}(\theta).$$

As in Remark 28 we see that the precise knowledge of K_θ is not used here and that it is enough to know the way to treat the case of stratified materials.

Remark 37.

As said in the previous remark, the function

$$\int_{\Omega} (\mu_-(\theta) |grad\,v|^2 - 2v)\,dx$$

is convex in (θ, v). This shows that $\tilde{J}$ is convex in θ and permits us to avoid any use of homogenization since we can prove directly the existence of the solution for problem (61), (62), (63), whereas the initial problem corresponds to the case in which θ is a characteristic function.

The problem of minimizing in θ $\tilde{J}$ defined by (66) under the constraint (63) is also equivalent to minimizing in (θ, v) the function $\int (\mu_- (\theta) |grad\, v|^2 - 2v)\, dx$ for v in $H_0^1 (\Omega)$ and θ satisfying (63).

As for any convex minimization problem, the necessary conditions for this last problem are also sufficient: the variation in v gives equation (62), and the variation in θ gives, after introduction of a Lagrange multiplier, the inequalities (60). Hence the set of solutions (θ, v) is convex.

The function $\mu_- (\theta)\, w^2$ is not strictly convex in (θ, w) but it is not affine on any segment $[(\theta_1, w_1), (\theta_2, w_2)]$ unless $\mu_- (\theta_1)\, w_1 = \mu_- (\theta_2)\, w_2$. This gives the next result:

$$\mu_- (\theta)\, grad\, u \text{ is uniquely defined.} \tag{67}$$

It is then natural to introduce the dual problem of (66), which involves $q = \mu_- (\theta)\, grad\, u$:

$$\tilde{J} (\theta) = - \min_{-div\, q=1} \int_\Omega \frac{1}{\mu_- (\theta)} |q|^2\, dx. \tag{68}$$

The problem is thus to search $\max_\theta \min_q \int_\Omega \frac{1}{\mu (\theta)} |q|^2\, dx$ for θ satisfying (63) and $q \in (L^2 (\Omega))^N$ such that $- div\, q = 1$ in Ω. The integrand is convex in q and affine (therefore concave) in θ. The minimax theorem implies:

$$\min_\theta \tilde{J} (\theta) = - \min_{div\, q=1} \max_\theta \int_\Omega \frac{1}{\mu (\theta)} |q|^2\, dx. \tag{69}$$

The solution q_0 is unique and the solutions θ maximize

$$\int_\Omega \frac{1}{\mu_- (\theta)} |q_0|^2\, dx$$

under the constraints $0 \le \theta (x) \le 1$, $\int_\Omega \theta (x)\, dx = \gamma'$, which is equivalent to the maximization of $\int_\Omega \theta |q_0|^2\, dx$. Once again we find the condition that there exists $C_1 \ge 0$ such that $\theta = 1$ if $|q_0| > C_1$ and $\theta = 0$ if $|q_0| < C_1$, but we find also that C_1 is independent of the solution θ.

$$\text{Two distinct solutions } \theta \text{ correspond to the same value of } C_1. \tag{70}$$

The second problem (minimize the function Φ_{C_1} defined by (64) and (65)) can easily be associated with the relaxed problem since Euler's equation for this convex problem can be proved to give exactly the characterization (60).

Remark 38.

The conditions of optimality can also be viewed as Euler's equation for the dual minimization problem:

$$\min_{-div\, q=1} \int_{\Omega} \Phi^*_{C_1} (|q|)\, dx, \qquad (71)$$

where the dual convex function $\Phi^*_{C_1} : \mathbb{R} \longrightarrow \mathbb{R}$ is given by $\Phi^*_{C_1}(0) = 0$ and

$$\frac{d\,\Phi^*_{C_1}}{d\lambda}(\lambda) = \begin{cases} \dfrac{1}{\beta}\lambda \text{ if } 0 < \lambda < C_1, \\[2mm] \dfrac{1}{\alpha}\lambda \text{ if } C_1 < \lambda. \end{cases} \qquad (72)$$

The optimality conditions (60) (which are necessary and sufficient) permit us to describe the behavior of the solution in the homogenized regions, that is, in the regions where $0 < \theta(x) < 1$, as well as on the interfaces where θ is discontinuous.

If $0 < \theta(x) < 1$ in an open set Ω' then we have $\mu_-(\theta) = \dfrac{C_1}{|grad\, u|}$ in Ω' and therefore

$$-div\,(\frac{grad\, u}{|grad\, u|}) = \frac{1}{C_1} \text{ in the set where } 0 < \theta(x) < 1. \qquad (73)$$

If u is regular in Ω', a classical computation shows that the mean curvature of the equipotentials of u is given by $div\,(\dfrac{grad\, u}{|grad\, u|})$. In the two-dimensional case this is the usual curvature and we get

$$\begin{cases} \text{If } N = 2 \\ \text{and if the solution } u \text{ is smooth in the homogenized region,} \\ \text{the equipotentials of } u \text{ are pieces of circles of radius } C_1. \end{cases} \qquad (74)$$

Remark 39.

In the region where $0 < \theta(x) < 1$ we have $\dfrac{C_1}{\beta} \leq |grad\, u| \leq \dfrac{C_1}{\alpha}$ almost everywhere.

Even if there are several disjoint regions where $0 < \theta(x) < 1$, the radius of the equipotentials of u (which are pieces of a circle) is the same in the various regions.

Let us consider now a smooth interface on which θ is discontinuous (with θ and u smooth on both sides), and let $\theta_- < \theta_+$ be the two limits of θ. Then the conditions on the interface are the continuity of u and the continuity of $\mu_- (\theta) \dfrac{\partial u}{\partial n}$ (this follows from (62)), and also the inequalities (60):

$$|\mu_- (\theta) \, grad \, u| \geq C_1 \text{ on the side } \theta_- \, ,$$
$$|\mu_- (\theta) \, grad \, u| \leq C_1 \text{ on the side } \theta_+ \, .$$

Comparison of these inequalities on the normal and tangential derivatives show that the tangential derivatives vanish and that

$$\left\{ \begin{array}{l} \text{a regular interface where } \theta \text{ is discontinuous is an} \\ \text{equipotential of } u; \ \mu_- (\theta)|grad \, u| \text{ takes the value } C_1 \\ \text{on both sides of this interface.} \end{array} \right. \qquad (75)$$

Remark 40.
In the case where Ω is a ball or an annulus, the invariance by rotation and the convexity of $\bar{J}$ imply that the solution has radial symmetry. A computation in polar coordinates shows that $q(r) = \mu_- (\theta) \dfrac{\partial u}{\partial r}$ is of the form $-\dfrac{r}{N} + \dfrac{C}{r^{N-1}}$ which implies that there is no homogenized region in this case. Consequently the solution is unique.

In the case where Ω is a ball of radius R the solution can be easily computed: $C = 0$ and θ is equal to 0 in some ball of radius R_0 and to 1 in the annulus $R_0 \leq |x| \leq R$. Thus $\gamma' = \dfrac{S_N}{N}(R^N - R_0^N)$, where S_N is the surface of the sphere $\{x \in \mathbb{R}^N : |x| = 1\}$. The case where Ω is an annulus leads to more tedious computations.

Remark 41.
If the solution does not contain any homogenized region, if the interfaces between $\theta = 0$ and $\theta = 1$ are smooth, and if Ω is simply connected, then Ω is a ball. This results from the characterization (75) and from a theorem of Serrin (1971). Indeed we can set $grad \, w = \mu_- (\theta) \, grad \, u$ since by (75) the compatibility conditions for w are satisfied. Therefore w is the solution of $-\Delta \, w = 1$, $w \in H_0^1 (\Omega)$ and on some equipotential of w one has $|grad \, w| = C_1$ which cannot happen unless Ω is a ball.

In the case where Ω is not simply connected, w does not vanish on the whole boundary and Serrin's theorem cannot be used. It is probable that in this case the annulus is the only domain that gives a solution without an homogenized region.

Remark 42.

In the general case we are not able to describe the structure of the solution. The variation of parameters such as α, β, or the case of an equation $-div\,(\mu_-\,(\theta)\,grad\,u) = f$ with $f \not\equiv 1$ could generate strange phenomena. There may be some similarity here with "chaotic" phenomena: the appearance of the "chaos" could be nothing but a classical bifurcation for some homogenized problem.

The motivations of this first example are concerned with the case $N = 2$. One of the motivations (cf. Joseph, et al. (1984)) is to describe the flow of two nonmiscible fluids in a pipe. In this model the homogenized regions should describe some "turbulence." Unfortunately this "turbulence" does not appear in the case of a circular cylindrical pipe, as previously explained. Another motivation for this example (cf. Lurie et al. (1982)) is to try to maximize the torsional rigidity of a cylindrical bar made of two different materials with given volume fractions. The reader is also referred to the papers of Kohn and Strang (1986), Goodman, Kohn and Reyna (1986), and Kawohl et al. (1991) which appeared after the present one.

PART V. EXAMPLE TWO

We keep the notation of the first example but we now minimize

$$J_0\,(\omega) = \int_\Omega u\,(x)\,dx\,, \qquad (76)$$

with u given by (52) and (53).

The relaxed problem is then given by (55), (56), and

$$J\,(\theta, A_*) = \int_\Omega u\,(x)\,dx\,, \qquad (77)$$

and the adjoint state is

$$p = u\,. \qquad (78)$$

Proposition 13 shows that at the optimum one has:

$$A_*\,grad\,u = \mu_+\,(\theta)\,grad\,u \qquad \text{a.e.} \qquad (79)$$

In formula (50a) we have to take $\varphi = 0$, therefore

$$Q\,(x) = (\beta - \alpha)|grad\,u|^2$$

and Proposition 15 proves that there exists $C_1 \geq 0$ such that

$$\begin{cases} \theta\,(x) = 0 & \Rightarrow |grad\,u| \geq C_1\,, \\ 0 < \theta\,(x) < 1 & \Rightarrow |grad\,u| = C_1\,, \\ \theta\,(x) = 1 & \Rightarrow |grad\,u| \leq C_1\,. \end{cases} \qquad (80)$$

Hence the relaxed problem consists of minimizing $\int_\Omega u\,(x)\,dx$, where u is the solution of

$$- div\,(\mu_+\,(\theta)\,grad\,u) = 1 \text{ in } \Omega\,, \quad u \in H_0^1\,(\Omega) \tag{81}$$

under the constraint

$$0 \leq \theta\,(x) \leq 1 \text{ a.e. ; } \int_\Omega \theta\,(x)\,dx = \gamma'\,. \tag{82}$$

This problem is equivalent to:

$$\min_\theta \max_v \int (-\mu_+\,(\theta)\,|grad\,v|^2 + 2v)\,dx \tag{83}$$

for v in $H_0^1\,(\Omega)$ and θ satisfying (82). The minimax theorem can be applied to this problem since the integrand is concave in v and affine (thus convex) in θ. Moreover there is uniqueness of the v and two distinct solutions correspond to the same value of C_1.

The optimality conditions (80) can also be interpreted as the Euler equation corresponding to the minimization of

$$\min_v \int (\Psi_{C_1}\,(|grad\,v|) - v)\,dx\,, \tag{84}$$

where $\Psi_{C_1}\,(0) = 0$ and:

$$\frac{d\,\Psi_{C_1}}{d\lambda}\,(\lambda) = \begin{cases} \alpha\lambda \text{ if } 0 < \lambda < C_1\,, \\ \beta\lambda \text{ if } C_1 < \lambda. \end{cases} \tag{85}$$

Since this function is convex, the conditions of optimality are necessary and sufficient.

Similarly to (75) we obtain:

$$\begin{cases} \text{On a smooth interface on which } \theta \text{ is discontinuous one has} \\ \dfrac{\partial u}{\partial n} = 0 \text{ and } |grad\,u| = C_1 \text{ on both sides of the interface.} \end{cases} \tag{86}$$

The case where Ω is a ball is instructive. We deduce from (84) that the solution is radial and from (86) that the function θ is continuous. Since the equation gives $\mu_+\,(\theta)\dfrac{\partial u}{\partial r} = -\dfrac{r}{N}$, there always exists an homogenized region in which θ varies linearly in r. According to the value of γ' (which can be explicitly related to C_1) we obtain one of the next two possibilities:
(i) if the radius R of the ball satisfies $R \geq N\beta C_1$, then

$$\begin{cases} \dfrac{du}{dr}\,(r) = \begin{cases} -r/N\alpha & \text{if} \quad 0 \leq r < N\alpha C_1\,, \\ -C_1 & \text{if} \quad N\alpha C_1 < r < N\beta C_1\,, \\ -r/N\beta & \text{if} \quad N\beta C_1 < r < R\,, \end{cases} \\ u\,(R) = 0\,. \end{cases} \tag{87}$$

(ii) if the radius R satisfies $N\alpha C_1 \leq R \leq N\beta C_1$, then

$$\begin{cases} \dfrac{du}{dr}(x) = \begin{cases} -r/N\alpha & \text{if} \quad 0 < r < N\alpha C_1\,, \\ -C_1 & \text{if} \quad N\alpha C_1 < r < R\,, \end{cases} \\ u(R) = 0\,. \end{cases} \qquad (88)$$

Note that in order to find the optimal solution one has first to break the symmetry and to approximate an homogenized domain. The radial symmetry is recovered for the homogenized problem.

Acknowledgment. The authors are grateful to Anca-Maria Toader who translated this paper from the original French version, which was published in D. Bergman et. al., *Les Méthodes de l'Homogénéisation: Théorie et Applications en Physique,* Collection de la Direction des Etudes et Recherches d'Electricité de France, **57**, Eyrolles, Paris, (1985), pp. 319–369.

REFERENCES

– Bensoussan A., Lions J.L., Papanicolaou G., (, 1978), *Asymptotic analysis for periodic structures,* Studies in mathematics and its applications, **5**, North-Holland, Amsterdam..

– Braidy P. and Pouilloux D., (1982), *Mémoire d'option,* Ecole Polytechnique, unpublished.

– De Giorgi E. and Spagnolo S., (1973), Sulla convergenza degli integrali dell'energia per operatori ellitici del secondo ordine, *Bull. U.M.I.* **8, 391–411.**.

– Hadamard J., (1907), *Mémoire sur le problème d'analyse relatif à l'équilibre des plaques élastiques encastrées* in *Oeuvres choisies, Tome II,* Editions du CNRS, Paris, (1968), 515–641..

– Hashin Z. and Shtrikman S., (1962), *A variational approach to the theory of effective magnetic permeability of multiphase materials,* J. Applied Phys. **33**, 3125–3131..

– Joseph D., Renardy M., and Renardy Y., (1984), *Instability of the flow of immiscible liquids with different viscosities in a pipe,* J. Fluid Mech. **141**, 309–317..

– Lions J.L.. (1968), *Sur le contrôle optimal de systèmes gouvernés par des équations aux dérivées partielles,* Dunod - Gauthier Villars, Paris..

– Lurie K.A. Cherkaev A.V., and Fedorov A.V., (1982), *Regularization of optimal design problems for bars and plates,* J. Optim. Th. Appl. **37**, 499–543..

– Marino A. and Spagnolo S., (1969), *Un tipo di approssimazione dell'operatore* $\Sigma D_i\,(a_{ij}\,D_j)$ *con operatori* $\Sigma D_j\,(b\,D_j)$, Ann. Sc. Norm. Sup. Pisa **23**, 657–673..

– Meyers N.G., (1963), *An L^p-estimate for the gradient of solutions of second order elliptic divergence equations,* Ann. Sc. Norm. Sup. Pisa **17**, 189–206..

– Murat F., (1977), *Contre exemples pour divers problèmes où le contrôle intervient dans les coefficients,* Ann. Mat. Pura Appl. **112**, 49–68..

– Murat F., (1978a), *H-convergence,* Séminaire d'analyse fonctionnelle et numérique de l'Université d'Alger 1977-78, mimeographed. Also translated in the present book..

– Murat F.. (1978b), *Compacité par compensation,* Ann. Sc. Norm. Pisa **5**, 489–507..

– Raitum U.E.. (1978) *The extension of extremal problems connected with a linear elliptic equation,* Soviet Math. Dokl. **19**, 1342–1345..

– Sanchez-Palencia E., (1980), *Non homogeneous media and vibration theory,* Lecture Notes in Physics **127**, Springer Verlag, Berlin..

– Serrin J., (1971), *A symmetry problem in potential theory,* Arch. Rat. Mech. Anal. **43**, 304–318..

- Spagnolo S., (1968), *Sulla convergenza di soluzioni di equazioni paraboliche ed ellitiche*, Ann. Sc. Norm. Sup. Pisa **22**, 571–597..
- Tartar L., (1974), *Problèmes de contrôle des coefficients dans des équations aux dérivées partielles*, in *Control theory, numerical methods and computer systems modelling (Proceedings Iria 1974)*, ed. by Bensoussan A. & Lions J.L., Lecture Notes in Economics and Mathematical Systems **107**, Springer Verlag, Berlin, 420 426. Also translated in the present book..
- Tartar L., (1977a), *Cours Peccot, Collège de France*, unpublished. Partially written in Murat F. (1978a),.
- Tartar L., (1977b), *Estimation de coefficients homogénéisés*, in *Computing methods in applied sciences and engineering (Proceedins Iria 1977), Vol. I*, ed. by Glowinski R. & Lions J.L., Lecture Notes in Mathematics **704**, Springer Verlag, Berlin, 364-373. Also translated in the present book..
- Tartar L., (1979), *Compensated compactness and applications to partial differential equations*, in *Non linear analysis and mechanics: Heriot Watt Symposium, Vol. IV*, ed. by Knops R.J., Research Notes in Mathematics **39**, Pitman, London, 136–212..
- Tartar L., (1983), *Compacité par compensation: résultats et perspectives*, in *Non linear partial differential equations and their applications, Collège de France Seminar, Vol. IV*, ed. by Brezis H. & Lions J.L., Research Notes in Mathematics **84**, Pitman, London, 350–369..
- Young L.C., (1969), *Lectures on the calculus of variations and optimal control theory*, W.B. Saunders, Philadelphia..

REFERENCES ADDED

AT THE TIME OF THE TRANSLATION

- Francfort G. and Murat F., (1986), *Homogenization and optimal bounds in linear elasticity*, Arch. Rat. Mech. Anal. **94**, 307–334.
- Goodman J., Kohn R.V., and Reyna L., (1986), *Numerical study of a relaxed variational problem from optimal design*, Comp. Meth. Appl. Math. Eng. **57**, 107–127..
- Kawohl B., Stara J., and Wittum G., (1991), *Analysis and numerical studies of a problem of shape design*, Arch. Rat. Mech. Anal. **114**, 349–363..
- Kohn R.V. and Strang G., (1986), *Optimal design and relaxation of variational problems I, II & III*, Comm. Pure Appl. Math **39**, 113-137, 139-182 & 353–377..
- Murat F. and Tartar L., (1985), *Optimality conditions and homogenization*, in *Non-linear variational problems*, ed. by Marino A., Modica L., Spagnolo S. & Degiovanni M., Research Notes in Mathematics **127**, Pitman, London, 1–8..
- Spagnolo S., (1976), *Convergence in energy for elliptic operators*, in *Proceedings of the Third Symposium on Numerical Solutions of Partial Differential Equations (College Park, 1975)*, ed. by Hubbard B., Academic Press, New York, 469–498.
- Tartar L., (1985), *Estimations fines de coefficients homogénéisés*, in *Ennio De Giorgi Colloquium*, ed. by Kree P., Research Notes in Mathematics **125**, Pitman, London, 168–187.

Effective Characteristics of Composite Materials and the Optimal Design of Structural Elements*

K.A. Lurie and A.V. Cherkaev

Contents

---

*The present article is a translation of an article originally written in Russian and published in *Uspekhi Mekhaniki (Advances in Mechanics)* vol. *9, no. 2, 1986, 3–81*. Parts 6.2 and 6.3 of the original paper are omitted here since they overlap with the paper translated in Chapter 5 of this volume.

1. Introduction

This paper is concerned with structural optimization problems related
to a design of inhomogeneous continuous media. Many natural formations,
such as the trunk of a tree, a leaf, or bone tissue, have sharply defined
internal inhomogeneity leading to the anisotropy of their physical properties varying from one point to another. Since these formations are highly
expedient from the point of view of structural mechanics, it would be rea-
sonable to expect that the requirement of optimality for a construction that
is artificially designed from a given set of materials would by itself bring
into existence the composite media having the best microstructure.

It has been just this conclusion to which mathematical research of optimal design problems has led. This circumstance generates specific difficulties connected with the correct formulation of the problem that must allow
for a possibility of merging a composite material assembled from initial
phases. Thus, the description of effective properties of composite materials
that may be built from a given set of materials becomes a necessary part
of the process. This problem has a long history [51, 72, 86, 96] and still
evokes continuous interest among the mechanical community [15, 32, 42,
60].

These considerations have predetermined the structure of the current
paper. In Section 2 typical statements of optimal design problems are discussed and reasons leading to the concept of microstructure are set forth.
Using the terminology of optimization theory we can say that we are dealing here with space dependent chattering regimes of control, that is, with
the multidimensional analogue of a phenomenon appearing very often in
optimal problems with one independent variable.

A definition of a basic concept of *G*-closure of the control set is given in
Section 3. For the problems under consideration the controls are physical
characteristics of initial materials, such as conductivity, elastic moduli, and
so on. The *G*-closure is defined as a set of physical characteristics of all the

possible composite materials that can be built from given materials used as initial phases.

In the same section a special class of layered composites of kth rank is described. The $k = 2$ class is used for a description of G-closure of an arbitrary set of isotropic conductive media in the case of two spatial variables.

Section 4 contains information on quasiconvex functions. This concept makes it possible to design, in a regular way, the minimal extensions of initial sets of materials and in particular, create their G-closures. Examples of such extensions are concentrated in section 5. Namely, the microstructures of composite materials are determined that produce constructions of minimal heat conductivity or of minimal rigidity in two-dimensional elasticity. The set of effective constants for binary mixtures of conducting materials with given volume fractions of phases is constructed. In this way formulae generalizing known Hashin-Shtrikman estimates for the anisotropic composites are obtained. The same problem for a set of elastic composites produced by isotropic phases with the same bulk modulus and different shear moduli is solved in the context of the bending of thin plates. The same is done for a two-dimensional polycrystal produced by a single anisotropic conducting phase. A problem of a G-closure of a random set of anisotropic conductors in the two-dimensional case is also solved.

The last, sixth, section is devoted to problems of optimal design; some results from the previous sections are used for their solution.

2. On the specific features of optimal design problems for inhomogeneous bodies

2.1. *On the formulation of basic optimal design problems.*

The optimal design problems had arisen almost at the same time as the calculus of variations (Newton, Lagrange, Euler). However, they did not form an independent theoretical field in mechanics before the middle the 1930s, despite the appearance of a number of outstanding papers (Clausen [53], Michell [73], and Nikolai [30]). This circumstance is amazing because at the same time engineering methods of the design of complicated structures and constructions developed rapidly. Therefore, by the last decade when interest in these problems began increasing quickly, engineers had already had a great deal of experience in design and many intuitive ideas about the optimal structures based on this experience.

Development of modern optimal structural theory had begun with the research of relatively simple objects (beams, bars), and the results often seemed obvious for experienced engineers. This fact has generated a certain scepticism towards the practical significance of the theory. However, as the methods of optimal control and variational analysis improved, the range of objects that allowed for the theoretical research of their limit properties has expanded. Papers have started appearing that are devoted to the optimal properties of more difficult multidimensional structures (such as plates, shells, trusses, etc.) for which intuitive notions of optimal behavior were much less developed. As a result, theoretical methods of optimal design have obtained a practical importance. An indisputable advantage of these methods is the fact that due to them the problem of design has become a regular mathematical problem.

The problems we discuss in this paper are formulated in the following way. One has to design a structural element (a plate, a bar, a three-dimensional elastic body, etc.) that is under the action of a given load and is working under certain conditions on the boundary. The aim of the design is to obtain a construction with the minimal value of a certain functional characteristic (such as weight, natural oscillation period, etc.) under additional constraints on geometric, kinematic, or stiffness properties (such as deflection at some point, work of applied force).

To achieve this aim one can make use of the problem's parameters such as thickness of plates (shells), elastic moduli, material density, and the like. These characteristics are defined at every point of the construction and called control functions (controls). The controls take values in some sets U that characterize resources at the designer's disposal.

The typical problem of this type is the optimal distribution problem of force materials in the volume of construction. The set U for this problem consists of initially given materials from which one has to assemble the optimal construction [1, 2, 11, 17, 21, 31, 38, 41, 54, 55, 57, 79, 82, 84, 87, 89, 97]. Of course, numerous modifications of the just-formulated basic problem are possible.[1]

Complexity of design objects has initially led to the assumption that the optimal project is obtained as a result of relatively small modifications of some, prompted by experiment, basic construction. This idea has been realized mainly numerically where nothing could be said *a priori* about the convergence of the process and on how the dimensionality of a finite-dimensional approximating problem influenced this convergence. Meanwhile this influence turned out to be essential because the increasing precision of computations leads as a rule to a discovery of extreme variability of properties of the optimal construction throughout its volume [31, 45, 80].

The main weakness of the indicated approach is in the fact that the analogy with the discrete problem is incorrect. Unfortunately, the un-

---

[1] In this paper we do not touch problems related to the optimal shape of a domain occupied by a medium, although these problems can formally be included in a general scheme under the assumption that among permissible media there are materials with "ideal" properties (ideal conductors or isolators, absolutely solid bodies, vacuum, etc.).

derestimation of this circumstance has become the source of an array of mistakes. A project obtained as a result of a "naive" numerical solution can turn out to be better than the initial one but does not have to be necessarily the best one. Small modifications of parameters of a numerical scheme or changes in the method of calculation or in the character of discretization lead, generally speaking, to a totally different "even better" project that can be preferable from the viewpoint of practical realization [46].

A similar instability was in fact discovered in [45, 80]. This demonstrates that the selection of a numerical procedure must be founded on concepts that guarantee the existence of optimal control.

2.2. *Necessary conditions for optimality.*

The instability of the numerical scheme can be foreseen during the investigation of necessary optimality conditions, including the condition of stationarity and the Weierstrass inequality [21]. For the problems considered here it often turns out that none of the projects satisfying the conditions of stationarity complies with the Weierstrass condition and so is not optimal. Numerous examples of this phenomenon are given in [21-23, 25, 39, 67]. We cite one of them that is analyzed in detail later.

Let us consider the problem of a thin prismatic bar under torsion with a constant section S. The torsional rigidity I is expressed by the integral

$$I = 2 \int_S w \, dx, \quad x = (x^1, x^2), \tag{2.1}$$

where w is the Prandtl function, that is, the solution of the boundary value problem

$$\nabla \cdot \mathbf{D}(x) \cdot \nabla w = -2, \quad w|_{\partial S} = 0, \tag{2.2}$$

$$\mathbf{D}(x) = u(x)\mathbf{E}.$$

The scalar elastic compliance of the bar material at the point $x \in S$ is denoted by $u(x)$, the unit tensor is denoted by $\mathbf{E}$.

The function $u(x)$ should be chosen so as to maximize the rigidity I of the bar, with the average value of elastic compliance u_0 and interval $[u_1, u_2]$ of variation of compliance given by

$$u_1 \leq u(x) \leq u_2, \quad \int_S u(x)dx = u_0 \text{ mes } S, \tag{2.3}$$

$$0 < u_1 < u_0 < u_2 < \infty.$$

As stated here the problem was considered in [59] where it is shown that the Weierstrass necessary condition calls for the optimal project to be assembled from materials with the limit values of the compliances $u = u_1$ or $u = u_2$, in accordance with a rule [59]:

$$
\begin{aligned}
u = u_1, & \quad \text{if} \quad (\nabla w)^2 \geq \varkappa \frac{u_2}{u_1} \\
u = u_2, & \quad \text{if} \quad (\nabla w)^2 \leq \varkappa \frac{u_1}{u_2}
\end{aligned}
\tag{2.4}
$$

where $\varkappa = const > 0$ is the Lagrange multiplier responsible for the integral restriction (2.3).

Since $u_1/u_2 < 1$, (2.4) implies that none of the stationary regimes is optimal if values $(\nabla w)^2$ belong to a "forbidden" interval $(\varkappa u_1/u_2, \varkappa u_2/u_1)$.

On the other hand, the zones occupied by the materials with compliances u_1 and u_2 are divided by a line Γ with the normal $\mathbf{n}$ and the tangent $\mathbf{t}$, along which the value $(\nabla w)^2$ undergoes a jump. The value of this jump is defined by the condition of continuity

$$
[\nabla w \cdot \mathbf{t}]_1^2 = 0, \qquad [u \nabla w \cdot \mathbf{n}]_1^2 = 0,
\tag{2.5}
$$

where a symbol $[\bullet]_1^2 = (\bullet)_2 - (\bullet)_1$ denotes the jump of the value in square brackets. From the first inequality (2.4), which holds near the switching line Γ from the side where $u = u_1$, we obtain the following, taking into account (2.5):

$$
\varkappa \leq \frac{u_1}{u_2} [(\nabla w)^2]_1 = \frac{u_2}{u_1} [(\nabla w)^2]_2 + [\nabla w \cdot \mathbf{t}]^2 \left(\frac{u_1}{u_2} - \frac{u_2}{u_1} \right).
\tag{2.6}
$$

The last condition is consistent with the second inequality (2.4) only if

$$
\nabla w \cdot \mathbf{t} = 0,
\tag{2.7}
$$

both expressions (2.4) then hold as equalities on the opposite sides of the curve Γ.

Thus the curve Γ should satisfy simultaneously two conditions (2.4) and (2.7). This overdetermines the problem of finding Γ. Formally the situation is as if the position of the curve and the slope of its tangent were defined by two independent conditions. Clearly the problem is unsolvable *a fortiori* in the class of smooth curves. It is expected that the solution will exist among generalized curves, whose singularities are everywhere dense in a set of a nonzero measure.

2.3. *On the existence of optimal control.*

The problem of existence is not new in control theory. In optimization problems with one independent variable it was solved after the significance of "chattering"regimes of control was understood and methods for their description were developed [40, 9, 44, 8].

Chattering regimes are characterized by the fact that the main interval of variation of the independent variable is subdivided into an infinite number of subintervals on which the control assumes various admissible values. The boundary points of these subintervals ("switching points") follow infinitely often one after another, filling out everywhere densely a set of complete measure. A similar distribution of values of the control cannot be characterized in the usual sense by any of the limit functions. It makes sense to speak only about a minimizing sequence of controls, but not about the limit control function.

A well-known example of the optimal problem [44]

$$\dot{x} = u, \qquad u \in U, \qquad U = \{u : -1, 1\}, \qquad x(0) = x(T) = 0, \qquad (2.8)$$

$$I[x] = \int_0^T x^2 \, dt = \min$$

illustrates this phenomenon. For the sequence of controls $\{u_n(t)\}$,

$$u_n(t) = \begin{cases} 1, & t \in \left(\dfrac{2i-1}{n}, \dfrac{2i}{n} \right) \\ -1, & t \in \left(\dfrac{2i}{n}, \dfrac{2i+1}{n} \right), \end{cases} \qquad i = 1, ..., n-1,$$

there is a sequence of values $I[x]$ approaching zero as $n \to \infty$. At the same time, however, the number of subintervals where the function $u_n(t)$ is constant on the interval $[0, T]$ tends to infinity as $n \to \infty$, and this behavior cannot be associated with any kind of limit function. In other words, in this problem the functional $I[x]$ never reaches its (zero) lower bound. Other examples of optimization problems for which there is no optimal control can be found in papers [106, 108].

In spite of the fact that optimal chattering regimes were initially perceived as exceptional, later it became clear that they were in fact a rule, whereas "good," that is, smooth or piecewise smooth, controls were exceptions that were agreed upon individual properties of optimal problems [44].

It is not surprising that a similar conclusion extends to the problem with many independent variables, too. Besides, in distributed systems chattering regimes can exist that are essentially of multidimensional origin. Such regimes disappear when the system degenerates into the one dimensional. A large class of problems, in which the appearance of the multidimensional chattering regimes is almost inevitable, constitute the problems of the optimal distribution of material characteristics of continuous media. It was precisely for these problems that this effect was originally discovered theoretically [20].

Let us illustrate it by an example of the optimal problem related to the distribution of temperature in a heat-conducting medium. Let us suppose that we have two isotropic media characterized by constant heat conductances u_1 and u_2. These media should be located in a given domain V so as to minimize some cost functional that is connected with the distribution of temperature, for a known source of heat and specified boundary conditions. For example, a mean square difference between the acting temperature T and its desirable value T_0 may be considered as a cost

$$I[T] = \int_V [T - T_0(x)]^2 dx. \tag{2.9}$$

It is clear that for obtaining the required distribution of heat it is necessary to facilitate, to the best possible extent, conditions of heat flow in a certain favorable direction and to hamper, to the maximal extent, such flow in the perpendicular direction. And this must be done at every point of the domain. Now it is obvious that the heat conductance of a desired medium should depend on the direction at every point. In other words, it must be a tensor function of coordinates. But how can we create the required anisotropic medium from the materials at our disposal? The difficulty resides in the fact that we do not have materials with a necessary degree of anisotropy. Available materials are isotropic by themselves. The only solution is to construct the required medium artificially, assembling from given materials a composite of a more or less complicated interior microstructure. The simplest example is the laminated composite. Its effective characteristics (heat conductances) evidently are different in the directions along and across layers. It is also possible to consider much more complicated microstructures. However, we show that in our case this is unnecessary. It is important to note that the problem under consideration (as many other similar problems) contains no parameter that might set up lower bounds for the size of microinclusions of given materials, in particular, the width of layers in the laminated composite. Therefore, one should expect that the optimal value of the functional in these problems is reached under some infinitely frequent subdivision of the initially given domain into parts occupied by different components. In what follows, references to multidimensional chattering regimes and media with microstructure are understood in this sense.

Looking for a solution of the problem as the curve (surface) separating media with $u = u_1$ and $u = u_2$, we obtain the result described earlier (cf.

[21]): smooth curves (surfaces) do not satisfy the Weierstrass condition, so they are not optimal. Emerging generalized curves (surfaces) completely correspond to the intuitive idea about microstructures arising in optimal constructions.

Although the given illustration belongs to a particular problem, there is no doubt that the described phenomenon has a completely general character. It arises in the case when the initial set of controls does not possess a certain property of completeness. In the following it is shown that for the problems of building continuous media the completeness of a set of admissible materials takes place if this set contains the initial media as well as all the composites assembled from them.

How can we correctly formulate the optimal problem taking into consideration a possibility of chattering regimes? For the problems with one independent variable the answer to this question is given by a known Filippov's lemma [40].

Consider the controlled system

$$\frac{dx}{dt} = X(t, x, u), \qquad x = (x^1, \ldots, x^n), \qquad u \in U, \qquad (2.10)$$

$$X = (X^1, \ldots, X^n),$$

and let the set U of admissible values of the control u consist of two points $u = u_1$ and $u = u_2$. Then, according to Filippov's lemma, one introduces a "relaxed" problem connected with the system

$$\frac{dx}{dt} = \langle X(t, x, u) \rangle \triangleq m X(t, x, u_1) + (1 - m) X(t, x, u_2), \qquad (2.11)$$

where the concentration $m = m(t)$ is to be determined along with $x(t)$. In the general case of an arbitrary set U the right-hand side $\langle X(t, x, u) \rangle$ of the relaxed system is restructured to yield the least convex hull $\operatorname*{conv}_{u \in U} X(t, x, u)$ of the set $\{X(t, x, u)\}$, $u \in U$ for each fixed pair of (t, x). For problem (2.8) we have:

$$\operatorname*{conv}_{u \in U} X(t, x, u) = \langle u \rangle = -m + 1 - m = 1 - 2m,$$

where m is the concentration of parts $u = -1$ in the chattering regime. This concentration becomes an additional control. Its optimal value, evidently, is equal to $1/2$. The problem has a solution, because the functional $I[x]$ attains its lower bound, which equals zero.

In the problem of relaxation all the weak limits of solutions $x_{(n)}$, corresponding to all possible sequences $\{u_n\}$, are generated by a certain function

$m(t)$ (i.e., by the convex hull of the right-hand sides), and no other weak limits are produced [8]. In this sense the problem of relaxation represents the minimal extension of the initial problem [8, 43].

Thus in the one-dimensional case the relaxation is the averaging of the right-hand sides of the equations with a suitable concentration function. For the multidimensional case there is no similar general procedure of relaxation, and special rules must be derived for the various types of problems. Such solution is given in the following for the problems connected with the operator $\nabla \cdot \mathbf{D} \cdot \nabla$ (heat conduction, bars under torsion) and with the operator $\nabla\nabla \cdot \cdot \mathbf{D} \cdot \cdot \nabla\nabla$ (plane problem of elasticity, bending of plates).

The Filippov's techniques do not immediately transfer to the relaxation in the multidimensional case. The new element that now arises is the dependence of effective properties of a composite not only upon the concentration, but also upon the shape of inclusions of the initially given components.

The effective properties of any composite are characterized by the tensor $\mathbf{D}_0$ connecting the values of various state variables averaged over some elementary volume of the composite medium, a so-called physically small volume. This volume is taken to be sufficiently small in comparison to some length scale of the problem "in the large," defined by the distribution of the external sources, the form of the domain, and so on. At the same time, the volume is large enough to include a great number of parts occupied by different compounds belonging to the admissible set U.

The problem of evaluation of tensor $\mathbf{D}_0$ for the composites of a given microstructure has generated many publications [4, 5, 37, 65]. There is an asymptotic procedure that allows us to calculate the effective properties of a composite possessing some given microstructure. This procedure (called homogenization) requires the solution of certain accessory boundary value problems that turn out to be fairly complicated and may be solved only numerically for more or less nontrivial microstructures. An analytical solution takes place only for laminated composites (one-dimensional averaging).

3. G-closed sets: definitions and general properties

3.1. G-closure of the set of admissible controls.

We possess a certain set U of initially given components at our disposal, these components generally being anisotropic materials themselves. We are interested in the determination of effective material properties of the whole set of composites that may be obtained from the given ingredients with the aid of the process of mixing. Speaking about the material properties we, of course, bear in mind a description of the mentioned set in terms of the invariants of the corresponding tensor $\mathbf{D}_0$.

The necessity of obtaining an invariant description of the whole set of possible composites arises from the fact that we do not know in advance which element of this set will actually participate in the optimal distribution of materials at any point of the region. In order to illustrate this approach mathematically, we introduce definitions.

Let a process in the media occupying the volume $V \subset R^n$ be described by the differential equation

$$L(\mathbf{D}(x))w = f, \qquad x = (x^1, \ldots, x^n), \tag{3.1}$$

where $L(\mathbf{D}(x))$ is a linear elliptic coercive operator depending on the tensor $\mathbf{D}(x)$ which characterizes material properties of the medium. The following can serve as examples: the operator of the heat conduction theory

$$L(\mathbf{D}) \triangleq \nabla \cdot \mathbf{D} \cdot \nabla, \qquad \nabla \triangleq (\partial/\partial x^1, \ldots, \partial/\partial x^n) \tag{3.2}$$

containing the tensor $\mathbf{D}$ of coefficients of the heat conduction, a symmetric, positive definite tensor of the second rank, and the operator of the plane elasticity

$$L(\mathbf{D}) \triangleq \nabla\nabla \cdot\cdot \mathbf{D} \cdot\cdot \nabla\nabla, \qquad \nabla = (\partial/\partial x^1, \partial/\partial x^2) \tag{3.3}$$

that depends on the tensor $\mathbf{D}$ of the elastic compliances, a self-adjoint [36] positive definite tensor of the fourth rank.

Let us assume that set U of the values $\mathbf{D}(x)$ consists of a finite number of points[2]

$$U = \{\mathbf{D}_1, \ldots, \mathbf{D}_N; \ \mathbf{D}_i > 0, \qquad \|\mathbf{D}_i\|_{L_\infty} \leq C \leq \infty \qquad \forall i \in (\overline{1, N})\},$$

corresponding to the values $\mathbf{D}(x)$ in the subdomains $V_1, \ldots, V_N$ that the domain $V : V = \bigcup_i V_i$ is divided into

$$\mathbf{D}(x) = \sum_{i=1}^{N} \chi_i(x)\mathbf{D}_i; \tag{3.4}$$

here $\chi_i(x)$ is the characteristic function of the ith subdomain

$$\chi_i(x) = \begin{cases} 1, & \text{if} \quad x \in V_i, \\ 0, & \text{if} \quad x \notin V_i. \end{cases} \tag{3.5}$$

---

[2]The inequality $\mathbf{D} > 0$ means the positiveness of a quadratic form

$$e^T \cdot \mathbf{D} \cdot e > 0 \qquad \forall e \in R^n.$$

Consider a sequence $\{V^S\}$ of subdivisions of the domain V into parts V_i^S and the corresponding sequences of controls $\{\mathbf{D}^S(x)\}$,

$$\mathbf{D}^S(x) = \sum_{i=1}^{N} \chi_i^S(x)\mathbf{D}_i$$

and the solutions $\{w^S\}$ of equation (3.1). Let us assume that the sequence $\{w^S\}$ weakly converges in a certain (energetic) Sobolev's space to a limit w^0

$$w^S \rightharpoonup w^0$$

(the mark $\rightharpoonup$ denotes convergence in weak topology); that is, the following equality holds (for the operator (3.2))

$$\lim_{S \to \infty} \int_V (\nabla w^S \cdot \nabla \eta + c w^S \eta)\, dx = \int_V (\nabla w^0 \cdot \nabla \eta + c w^0 \eta)\, dx$$

$$\forall \eta \in \overset{\circ}{W}_2^1(V), \qquad c > 0,$$

or the equality (for the operator (2.3))

$$\lim_{S \to \infty} \int_V (\nabla\nabla w^S \cdot\cdot \nabla\nabla\eta + c_1 \nabla w^S \cdot \nabla\eta + c_0 w^S \eta)\, dx =$$
$$\int_V (\nabla\nabla w^0 \cdot\cdot \nabla\nabla\eta + c_1 \nabla w^0 \cdot \nabla\eta + c_0 w^0 \eta)\, dx,$$

$$\forall \eta \in \overset{\circ}{W}_2^2(V), \qquad c_1 > 0,\ c_0 > 0.$$

The element w^0, however, may not satisfy (3.1) for any tensor $\mathbf{D}(x)$ of type (3.4); that is, this element cannot be the solution of the initial problem with the tensor $\mathbf{D}$ from the initial set U. On the other hand, it is known [13, 33, 71] that under the formulated conditions there is a symmetric positive definite tensor $\mathbf{D}_0 = \mathbf{D}_0(x)$, with the components belonging to the space $L_\infty(V)$, that satisfies the equation

$$L(\mathbf{D}_0(x))w^0 = f. \tag{3.6}$$

In other words, the following limit equality holds

$$w^S \triangleq L^{-1}(\mathbf{D}^S)f \underset{W_2^p(V)}{\rightharpoonup} w^0 \triangleq L^{-1}(\mathbf{D}_0)f \qquad \forall f \in L_2(V).$$

This definition is kept in mind [71] when one says [13, 109] that the sequence of operators $\{L(\mathbf{D}^S)\}$ G-converges to an operator $L(\mathbf{D}_0)$

$$L(\mathbf{D}^S) \underset{G}{\rightarrow} L(\mathbf{D}_0).$$

Since for the linear elliptic operators the type of the G-limit operator remains the same [13, 71, 109] and it is only the tensor $\mathbf{D}$ that changes, one can speak about G-convergence of a sequence of tensors $\mathbf{D}^S(x)$ themselves [13]:

$$\mathbf{D}^S \underset{G}{\rightarrow} \mathbf{D}_0;$$

the tensor $\mathbf{D}_0$ is called a G-limit of the sequence $\{\mathbf{D}^S\}$ with respect to the operator L.

Let us note that in the problems that are of interest to us about inhomogeneous bodies (composites) the weak-limit value of the solution w^0 may be interpreted as a result of the averaging of the solution w in a "physically small volume" (cf. Section 2.3). This interpretation, accepted in the homogenization theory [13, 65, 37], is justified by the fact that subdivisions forming a sequence $\{\chi^S\}$ tend to become almost periodic as s increases.

The tensor $\mathbf{D}_0$ is called an effective tensor of material characteristics of the composite. It defines the behavior of the composite in any strictly interior subdomain of the domain V. This tensor depends only on the set U of the initially given materials and the sequence $\{\chi^S\}$ of subdivisions of the domain. It does not depend on the sources f and the boundary conditions:

$$\mathbf{D}_0 = \mathbf{D}_0(U, \{\chi^S\}).$$

We have named the range of the tensor $\mathbf{D}_0$, corresponding to all the weak limits of solutions w^0 (or all the sequences $\{\mathbf{D}^S\}$) the G-closure of the set U. This set is denoted by GU:

$$GU = \{\mathbf{D}_0 : \mathbf{D}^S = \sum \chi_i^S \mathbf{D}_i \underset{G}{\rightarrow} \mathbf{D}_0, \qquad \mathbf{D}_i \in U \qquad \forall \{\chi_i^S\}\}. \qquad (3.7)$$

Evidently one has $U \in GU$.

In what follows, when it does not lead to confusion, we apply the same symbols to the set of materials and the set of invariants of the tensors of their material constants.

In most of the design problems the initial sets of materials U are not G-closed, that is, they do not coincide with their G-closure. We cite a few examples:

1) a discrete set that consists of several materials (it is possible to obtain materials with intermediate properties assembling composites);

2) an arbitrary set of isotropic media (laminated composites take us out of this class; note that these are specifically the composites that often prove to be optimal);

3) a set of anisotropic media that differ only by the orientation of their principal axes (assembling of polycrystal structures leads to the changing of the principal rigidities; in particular, it is possible to obtain an isotropic medium).

Thus, in order to formulate the optimization problem correctly one has to construct the G-closure of the initial set of controls. Although at first it seems that this problem presents essential difficulties, it has been solved explicitly in many important cases.

Remark.

In one-dimensional problems the role of GU is played by the right-hand sides of the initial differential equations (3.11) averaged according to the Filippov's lemma.

In applications one often deals with problems in which the microstructure is subject to conditions that specify relative amounts of each initial material. This makes it necessary to describe the set of all composites containing given initial materials in the given volume fractions in the microstructure. We call such a problem G_mU — the G_m closure problem for the initial set U, denoting by a symbol m the volume fractions (concentrations) of the initial phases. In order to obtain the exact definition of the set G_mU, let us assume that

$$\chi_i^S(x) \underset{L_\infty(V)}{\overset{*}{\rightharpoonup}} m_i(x). \tag{3.8}$$

The value $m_i(x)$ has the meaning of concentration of the ith phase at the point x. It is clear that

$$0 \le m_i(x) \le 1, \qquad i = 1, \ldots, N,$$

and

$$\sum_{i=1}^{N} m_i = 1.$$

The set G_mU now is described by

$$G_mU = \{\mathbf{D}_0 : \mathbf{D}^S \underset{G}{\rightarrow} \mathbf{D}_0, \qquad \mathbf{D}_i \in U \qquad \forall \{\chi_i^S\} : \chi_i^S \overset{*}{\rightharpoonup} m_i\}. \tag{3.9}$$

A one-dimensional analogy of the G_m-closure problem is trivial: fixing the concentrations of the phase we obtain the only one right side of the relaxed system. In a multidimensional case, due to a variety of possible microstructures, one obtains entire sets of composites satisfying the formulated conditions.

It is obvious that the the union of G_m-closures corresponding to all possible m represents the G-closure of the initial set of materials:

$$GU = \bigcup_m G_m U. \tag{3.10}$$

The following property is implied directly from the definition of the set $G_m U$.

Let $\mathbf{D}'$ be a G-limit tensor of properties of a mixture of materials with tensors $\mathbf{D}_1$, $\mathbf{D}_2$ of material characteristics taken in concentrations m_1', m_2' and let $\mathbf{D}''$ be a G-limit tensor for a mixture of the same materials taken in concentrations m_1'', m_2''. Invariants Λ', Λ'' of the tensors $\mathbf{D}'$, $\mathbf{D}''$ belong, by the definition, to sets

$$\Lambda' \in G(m_1', m_2', \mathbf{D}_1, \mathbf{D}_2), \qquad \Lambda'' \in G(m_1'', m_2'', \mathbf{D}_1, \mathbf{D}_2). \tag{3.11}$$

Make up a "composite of the second rank," that is, a mixture of the materials $\mathbf{D}'$ and $\mathbf{D}''$, and denote by ν_1 and ν_2 concentrations of the phases $\mathbf{D}'$ and $\mathbf{D}''$, respectively. Invariants Λ''' of the tensor of properties of such a composite belong to the set

$$\Lambda''' \in G(\nu_1, \nu_2, \mathbf{D}', \mathbf{D}'').$$

On the other hand, any composite in $G(\nu_1, \nu_2, \mathbf{D}', \mathbf{D}'')$ is composed of the materials $\mathbf{D}_1$, $\mathbf{D}_2$ in concentrations respectively equal to $\nu_1 m_1' + \nu_2 m_1''$, $\nu_1 m_2' + \nu_2 m_2''$. So the following inclusion holds

$$G(\nu_1, \nu_2, \mathbf{D}', \mathbf{D}'') \subset G(\nu_1 m_1' + \nu_2 m_1'', \ \nu_1 m_2' + \nu_2 m_2'', \mathbf{D}_1, \mathbf{D}_2),$$

where $\mathbf{D}'$, $\mathbf{D}''$ are tensors with the invariants that satisfy (3.11).

Corollary. *The set $G_m U$ is G-closed; that is, it coincides with its G-closure.*

3.2. *Laminated composites of the kth rank.*

As noted, the problem of computing an effective tensor $\mathbf{D}_0$ that corresponds to a particular microstructure, as a rule, can be solved only numerically. We now describe a special class of microstructures, called layers of the kth rank [24,25,68]; this class is important for our future purposes. Earlier such microstructures were considered in [90]. An effective tensor of these microstructures can be calculated analytically.

The simplest representatives of the named class are composites formed by layers of initial materials (layers of the first rank). The tensor of the material properties associated with any such medium changes only along the normal $\mathbf{n}$ to layers: $\mathbf{D}(x) = \mathbf{D}(\mathbf{n})$.

In order to calculate the effective tensor of this composite it is sufficient to represent the initial equation in the form of a system of the first order, solvable with respect to the derivatives along the direction $\mathbf{n}$, and then to apply the procedure of one-dimensional averaging in this direction.

A tensor $\mathbf{D}_l$ of the conductance of a laminated composite for the system (3.1), (3.2) has the following form [68]

$$\mathbf{D}_l = \mathbf{D}_l(\mathbf{n}, m) = m_1\mathbf{D}_1 + m_2\mathbf{D}_2 - \frac{m_1 m_2}{\widetilde{D}}(\mathbf{D}_1 - \mathbf{D}_2) \cdot \mathbf{nn} \cdot (\mathbf{D}_1 - \mathbf{D}_2), \quad (3.12)$$

where $\mathbf{D}_1$, $\mathbf{D}_2$ are conductance tensors of phases, m_1, $m_2 = 1 - m_1$ are the concentrations of these phases, and the symbol $\widetilde{D}$ is defined as $\widetilde{D} = \mathbf{n} \cdot (m_1\mathbf{D}_2 + m_2\mathbf{D}_1) \cdot \mathbf{n}$.

Similarly, the effective tensor of elastic moduli of a laminated composite for the system (3.1), (3.3) is computed as

$$\mathbf{D}_l = \mathbf{D}_l(\mathbf{n}, m) =$$
$$m_1\mathbf{D}_1 + m_2\mathbf{D}_2 - \frac{m_1 m_2}{\widetilde{D}}(\mathbf{D}_1 - \mathbf{D}_2) \cdot \cdot \mathbf{nnnn} \cdot \cdot (\mathbf{D}_1 - \mathbf{D}_2), \quad (3.13)$$

where

$$\widetilde{D} = \mathbf{nn} \cdot \cdot (m_1\mathbf{D}_2 + m_2\mathbf{D}_1) \cdot \cdot \mathbf{nn}.$$

In order to construct layers of the second rank, introduce two different composites of the first rank corresponding to two different concentrations m and directions of the normal $\mathbf{n}_1$; these composites will be considered as the initial phases. Construct from these phases a layer medium of the first rank directing the normal along the axis $\mathbf{n}_2$ (Figure 1). The tensor of effective characteristics $\mathbf{D}_l^{(2)}$ is defined for the equations (3.1) and (3.2) by the formula (3.12) where instead of $\mathbf{D}_1$ and $\mathbf{D}_2$ one should take two different values of $\mathbf{D}_l(\mathbf{n}_1, m)$ and make the basis vector $\mathbf{n}$ equal with $\mathbf{n}_2$. Repeating the described procedure k times we obtain the structure of the laminated composite of the kth rank.

The concept of the layers of the kth rank corresponds to an idea of a hierarchy of lengthscales for the inclusions forming the microstructure: the width ε_1 of layers of the first rank tends to zero faster than the width

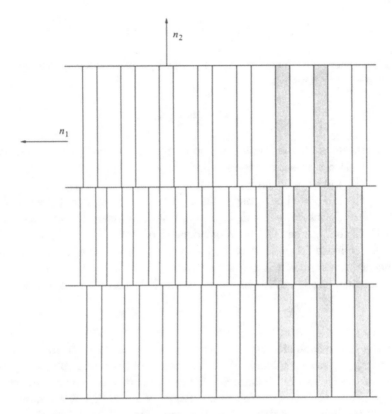

Figure 1. Example of a layer microstructure of rank 2; n_1 and n_2 are the normals to layers of rank 1 and 2.

ε_2 of layers of the second rank, and so on. For this reason the laminated composite of the ith rank can serve as an homogeneous initial material for the construction of a composite of the $(i+1)$th rank if $\varepsilon_i/\varepsilon_{i+1} \to 0$. Consequently, the composite of the kth rank is a result of k consecutive operations of one-dimensional averaging.

The effective properties of composites of the kth rank depend on the latent parameters of the structure, such as the normals $n_1, \ldots, n_k$ and the concentrations of materials in layers of each rank.

A similar procedure can be accomplished for systems described by any stationary elliptic operators, in particular, for stationary Maxwell's equations in an inhomogeneous medium, for equations of linear elasticity, and so on. The process of formation of layers of the kth rank is monotone; on each step it extends a set $\{\Lambda_k\}$ of invariants of the composites of the kth rank:

$$\{\Lambda_k\} \subset \{\Lambda_{k+1}\}, \qquad k = 1, 2, \ldots .$$

On the other side, the invariants of the composites evidently are bounded, so there is the limit set

$$\{\Lambda\} = \lim_{k \to \infty} \{\Lambda_k\}$$

of invariants of all the laminated composites of any rank, which is defined as the least set possessing the properties:

1) the initial set belongs to $\{\Lambda\}$; and

2) the laminated composite that is composed from any elements of the set $\{\Lambda\}$ and that corresponds to any normal to layers $\mathbf{n}$ belongs to $\{\Lambda\}$.

Note that the set $\{\Lambda\}$ does not depend on the method of construction of the sequence $\{\Lambda_k\}$; that is, it does not depend on the normals $\mathbf{n}_k$ and concentrations m_k of the previous ranks.

The set $\{\Lambda\}$ is not larger than the set GU of all the composites

$$\{\Lambda\} \subset GU.$$

On the other hand, it turns out that for the examples considered in the following, these sets coincide. In other words, in these examples the procedure of the "multidimensional averaging," that is, the construction of GU $(G_m U)$, and the construction of sequences of finite or infinite number of one-dimensional averaging lead to the same results ([63,105,107]).

As an example consider the process of the construction of the second rank layers for equations (3.1) and (3.2) in the plane (x^1, x^2). Replace this equation by an equivalent system of the first order [21], resolved with respect to the derivatives of the phase variables in a certain direction $\mathbf{n}$:

$$z_{\mathbf{n}}^1 = (-z_t^2 + f)u^{-1}, \qquad z_{\mathbf{n}}^2 = uz_t^1, \qquad z^1 = w, \qquad \tilde{f}_{\mathbf{n}} = f; \qquad (3.14)$$

the direction $\mathbf{t}$ is chosen to be orthogonal to $\mathbf{n}$.

Assuming that $u = u(\mathbf{n})$ (layers with the normal $\mathbf{n}$), apply the averaging operation (2.11) to the system (3.14). The result will depend on $\mathbf{n}$:

$$z_{\mathbf{n}}^1 = (-z_t^2 + \tilde{f})\langle u^{-1}\rangle_1, \qquad z_{\mathbf{n}}^2 = z_t^1 \langle u \rangle_1. \qquad (3.15)$$

If the set U consists of two elements, $U = (u_1, u_2)$, then the following holds

$$\langle u \rangle_1 = \text{conv } u(x) \triangleq m_{11}u_1 + m_{12}u_2, \quad m_{11} + m_{12} = 1, \quad m_{11} \geq 0, \quad m_{12} \geq 0.$$

Going over to the next step we consider the system (3.15) as the initial one. We resolve it with respect to the derivatives in the direction $\mathbf{t}$:

$$z_t^1 = (\langle u \rangle_1)^{-1} z_\mathbf{n}^2, \qquad z_t^2 = \tilde{f} - (\langle u^{-1} \rangle_1)^{-1} z_\mathbf{n}^1.$$

Now assume that in this system the coefficients $(\langle u \rangle_1)^{-1}$, $(\langle u^{-1} \rangle_1)^{-1}$ depend on the $\mathbf{t}$ direction (layers of the second rank) which at this step plays the same role as the direction $\mathbf{n}$ did at the previous step. Repeating the operation (2.11) of one-dimensional averaging we obtain the system

$$z_t^1 = \left\langle \langle u \rangle_1^{-1} \right\rangle_2 z_\mathbf{n}^2, \qquad z_t^2 = \tilde{f} - \left\langle \langle u^{-1} \rangle_1^{-1} \right\rangle_2 z_\mathbf{n}^1, \qquad (3.16)$$

where

$$\langle u \rangle_2 = \operatorname{conv} u \triangleq \sum_{(i)} m_{2i} u_i, \qquad \sum_{(i)} m_{2i} = 1, \qquad m_{2i} \geq 0.$$

The microstructure obtained in this way (layers of the second rank) is represented on Figure 1. The second step of the described procedure obviously consists of rotating the composite of the first rank about the angle of $90°$ and then assembling a new laminated composite.

Returning to the initial notation, one represents the system (3.16) in the following form ($z^1 = w$),

$$\frac{\partial}{\partial t} \frac{1}{\left\langle \frac{1}{\langle u \rangle_1} \right\rangle_2} \frac{\partial w}{\partial t} + \frac{\partial}{\partial n} \left\langle \frac{1}{\langle \frac{1}{u} \rangle_1} \right\rangle_2 \frac{\partial w}{\partial n} \triangleq \nabla \cdot \mathbf{D}_0 \cdot \nabla w = f,$$

where the tensor $\mathbf{D}_0$ has the dyadic representation:

$$\mathbf{D}_0 = \lambda_1 \mathbf{tt} + \lambda_2 \mathbf{nn},$$

$$\lambda_1 = \frac{1}{\left\langle \frac{1}{\langle u \rangle_1} \right\rangle_2}, \qquad \lambda_2 = \left\langle \frac{1}{\langle \frac{1}{u} \rangle_1} \right\rangle_2.$$

The double averaging performed here leads to the set $\mathbf{D}_0$ of tensors, whose eigenvalues depend on the latent parameters of the microstructure, that is, the concentrations m_{11}, m_{12}, m_{2i}. The layers of the second rank consist of phases that are themselves composites of the first rank and have different volume fractions $m_{11}^{(i)}$ of the material u_1. The total concentration m_1 of this material in the composite is obviously equal to

$$m_1 = \sum_{(i)} m_{2i} m_{11}^{(i)}.$$

The remaining free parameters $m_{11}^{(i)}$, m_{2i} characterize a set of points of the plane (λ_1, λ_2) corresponding to various composites with the fixed concentration m_1 of the first phase. It is checked directly that the following curves serve as the boundary of this set (Figure 2):

$$1 + \frac{u_1}{\lambda_1 - u_1} + \frac{u_1}{\lambda_2 - u_1} = \frac{1}{1 - m_1}\left(1 + \frac{2u_1}{u_2 - u_1}\right) \qquad (3.17)$$

and

$$1 + \frac{u_2}{\lambda_1 - u_2} + \frac{u_2}{\lambda_2 - u_2} = \frac{1}{m_1}\left(1 + \frac{2u_2}{u_1 - u_2}\right); \qquad (3.18)$$

these curves correspond to matrix composites (Figure 3). In these composites material u_1 or u_2 forms inclusions in the matrix of the remaining material.

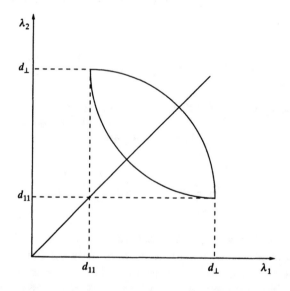

Figure 2. The set of eigenvalues (λ_1, λ_2) of the effective tensor of all laminates of rank 2, composed from two isotropic phases u_1 and u_2 taken in concentrations m_1 and m_2 $(m_1 + m_2 = 1)$; the following notations are used: $d_\perp = m_1 u_1 + m_2 u_2$, $d_\parallel = (m_1 u_1^{-1} + m_2 u_2^{-1})^{-1}$.

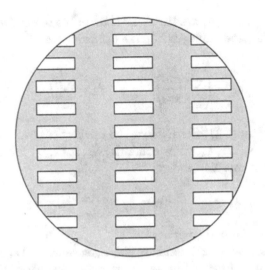

Figure 3. Microstructure of the matrix composite.

At the same time one can see that the further increase in the rank of layers does not lead to the extension of the obtained set. As shown in the following, the curves (3.17) and (3.18) provide bounds for the set of invariants of all possible microstructures [26], and not only for laminates.

3.3. *Reuss–Voigt estimates.*

The simplest estimates of the set $G_m U$ for the operator (3.2) are given by the well-known Reuss–Voigt inequalities [86, 96]

$$(\langle \mathbf{D}^{-1} \rangle)^{-1} \leq \mathbf{D}_0 \leq \langle \mathbf{D} \rangle . \tag{3.19}$$

Here $\langle \mathbf{D} \rangle \triangleq \sum_{i=1}^{N} m_i \mathbf{D}_i$ is the average value of the tensor (3.4). It is assumed that in the neighborhood of a point x the initial phases have concentrations m_i (3.8).

Consider the process described by equation (3.2) or by the equivalent system

$$\nabla \cdot \mathbf{j} = f, \qquad \mathbf{j} = \mathbf{D} \cdot \nabla w, \qquad f \in L_2(V). \tag{3.20}$$

If $U = \{\mathbf{D}_i = u_i \mathbf{E}; \ i = 1, 2, \ldots, N, \ 0 < u_1 \leq u_2 \leq \cdots \leq u_N < \infty\}$, where $\mathbf{E}$ is the unit tensor, then (3.19) is equivalent to the following inequalities

$$d_{\parallel} \leq \lambda_1 \leq \lambda_2 \leq \cdots \leq \lambda_n, \tag{3.21}_1$$

$$\lambda_1 \leq \lambda_2 \leq \cdots \leq \lambda_n \leq d_\perp, \qquad (3.21)_2$$

where λ_i $(i = 1, 2, \ldots, n)$ are the eigenvalues of the tensor $\mathbf{D}_0$ numbered in increasing order, and values $d_\|$, $d_\perp$ are defined as

$$d_\| = \left(\sum_{i=1}^{N} \frac{m_i}{u_i} \right)^{-1}, \qquad d_\perp = \sum_{i=1}^{N} m_i u_i. \qquad (3.22)$$

The estimates (3.21) characterize the smallest cube in the space of the eigenvalues $\lambda_1, \lambda_2, \ldots, \lambda_n$ which contains the set $G_m U$. In fact, the vertex of the cube with coordinates

$$\lambda_1 = d_\|, \qquad \lambda_2 = \lambda_3 = \cdots = \lambda_n = d_\perp \qquad (3.23)$$

corresponds to a layered composite.

The inequalities (3.21) estimate the eigenvalues of $\mathbf{D}_0$ independently of one another. These estimates are sharp in the sense that there are composites for which one of the eigenvalues λ_i coincides with any point of the interval (3.21).

The layers of the second rank formed by the initial phases are examples of such composites.

On the other hand, it is evident that not all the points (3.21) are attainable; that is, not all of them correspond to real microstructures. For example, there is no isotropic composite with the eigenvalues

$$\lambda_1 = \lambda_2 = \cdots = \lambda_n = d_\|.$$

Thus the cube (3.21) bounds the set greater than the set $G_m U$. In spite of this fact, in the plane case $(n = 2)$ the union of the estimates (3.21) corresponding to all of the values m_i coincides with the set GU.

3.4. *G-closure of the set of isotropically conducting phases (plane case).*

First consider the case of two phases $(N = 2)$. For a layered rank one structure the tensor $\mathbf{D}_0$ is characterized by the eigenvalues (3.23) with $n = 2$. The matrix of this tensor has the maximal range (i.e. has the maximal difference $\lambda_2 - \lambda_1$ of the eigenvalues) among all the matrices satisfying (3.21).

Eliminating the parameters m_1, $m_2 = 1 - m_1$ $(0 \leq m_1 \leq 1)$ from (3.22) and (3.23), we obtain the relationship

$$u_1 \leq \lambda_1 = \frac{u_1 u_2}{u_1 + u_2 - \lambda_2} \leq u_2, \qquad (3.24)$$

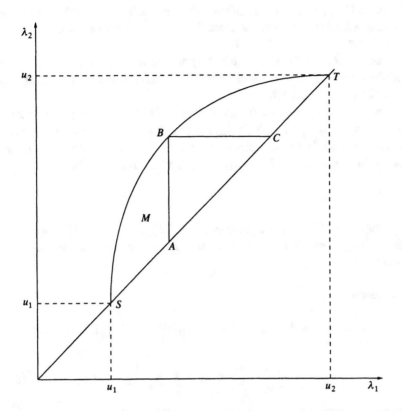

Figure 4. Set GU generated by two isotropic phases u_1 and u_2 (the plane case).

defining the curve SBT that is shown in Figure 4.

This curve represents the locus of the vertices B of triangles ABC that are the sets of values λ_1, λ_2 characterized by the inequalities (3.21). Every such triangle contains the set $G_m U$ according to (3.21). The union of the triangles ABC for all the values m $(0 \leq m \leq 1)$ represents a set containing GU. This set is depicted by a configuration $SBTCAS$. The initial materials u_1 and u_2 correspond to the vertices S and T, respectively.

The estimate of the set GU given by the configuration $SBTCAS$ is nevertheless sharp, because the laminated composites correspond to the boundary SBT and layers of the second rank correspond to the interior points and the diagonal $SACT$. Thus the eigenvalues λ_1, λ_2 of the tensors $\mathbf{D}_0$ forming the set GU satisfy the following inequalities [33, 93]

$$u_1 \leq \frac{u_1 u_2}{u_1 + u_2 - \lambda_2} \leq \lambda_1 \leq \lambda_2 \leq u_2. \tag{3.25}$$

Note that adding to the initial set U materials whose eigenvalues are represented by the points M from GU does not change the set GU: if $M \in GU$, then $G(U + M) = GU$.

Indeed one can associate with the point M a composite assembled from the phases u_1, u_2. This composite will be indistinguishable in its properties from the added material.

In particular, the union of the sets (3.21) for all the parameters m_i $(0 \le m_i \le 1, i = 1, 2, \ldots, N, \sum_{i=1}^{N} m_i = 1)$ leads to the inequalities (3.25) in which u_2 needs to be replaced by u_N. These inequalities characterize the G-closure of the set $U = (u_1, \ldots, u_N)$ of isotropic media.

Remark 1.

The curve SBT in coordinates $\lambda_2 = \|\mathbf{D}\|$, $\lambda_1^{-1} = \|\mathbf{D}^{-1}\|$ is a straight line passing through the points S and T that correspond to the initial phases.

Remark 2.

In the case of n-dimensional space $(\lambda_1.\lambda_2, \ldots, \lambda_n)$ the set GU satisfies the inequalities:

$$u_1 \le \frac{u_1 u_2}{u_1 + u_2 - \lambda_n} \le \lambda_1 \le \lambda_2 \le \cdots \le \lambda_n \le u_2. \qquad (3.26)$$

They are obtained by analogy with (3.25) [33]. Unlike the case $n = 2$, the set (3.26) is larger than GU.

The description of the set GU for this case is given in the Section 5.2.

3.5. *Minimal extensions and G-closures of sets of admissible controls.*

In Section 3.1 the G-closure of the set U of initial phases was defined as the set GU of tensors of effective constants of all the composites assembled from the initial materials.

Knowledge of the sets GU and $G_m U$ is useful for a wide class of problems in mechanics of inhomogeneous bodies. One often has to deal with mixtures in which a shape of inclusions is either unknown or it changes from one point to another in a random way. Under these conditions prediction of the effective properties of a composite can be based only on estimates for the sets GU and $G_m U$. Besides, knowledge of the limit values of the effective moduli and microstructures realizing these values give an idea about the best geometric forms of reinforcing inclusions.

The sets GU and $G_m U$ play a distinctive role in optimal design. They allow us to formulate a wide class of optimization problems for the multidimensional objects that are described by linear elliptic equations. In fact, all the controls corresponding to the extreme points of all weakly semi-continuous functionals of solutions w belong to the set GU $(G_m U)$ by virtue of its construction. Recall that weakly (lower) semicontinuous functionals are functionals $I[w]$ for which the following limit inequality holds

$$\lim_{w^s \to w^0} I[w^s] \geq I[w^0].$$

The following can serve as an example of a semicontinuous functional

$$I[w] = \int_V f(w)\, dx,$$

where $f(w)$ is a convex function.

The minimum of a semicontinuous functional is reached for a weakly limit element w^0 and, consequently, for a control $\mathbf{D}^0 \in GU$. For such functionals the following equality holds

$$\inf_{\mathbf{D} \in U} I[w(\mathbf{D})] = \min_{\mathbf{D} \in GU} I[w(\mathbf{D})].$$

Thus the control, solving the optimal problem, belongs *a fortiori* to the set GU. However, in order to guarantee the existence of optimal control it is unnecessary to know the set GU as a whole. For example, one can determine that the optimal control belongs only to a certain part of the boundary of the set GU. In [34] it is shown that for weakly semicontinuous functionals of solutions of equations (3.1) and (3.2) the extreme points are reached when the control belongs to the set of composites of the first rank in a general multidimensional case $(n \geq 2)$; this set, as shown in Section 3.4, forms a part of the boundary of the set GU if $n > 2$ (cf. (3.26)) and the whole boundary if $n = 2$ (cf. (3.24)). Introducing the symbol MU to denote this part of the boundary we have:

$$\inf_{\mathbf{D} \in U} I[w(\mathbf{D})] = \min_{\mathbf{D} \in MU} I[w(\mathbf{D})] = \min_{\mathbf{D} \in GU} I[w(\mathbf{D})]. \tag{3.27}$$

On the other hand, it is possible *to define* the set MU by the equality (3.27). We call any such set the minimal extension of the set U [43]. The set GU itself is, certainly, the minimal extension. However, in Sections 5.1 and 5.2, examples of sets $MU \neq GU$ will be given, these examples using a particular kind of the minimized functional. A description of the sets GU $(G_m U)$ gives the maximal information about the effective properties of all the composites. This information can be used for solving a particular optimal design problem only partially . For this purpose it is sufficient to know how to construct some of the minimal extensions MU. Often this represents a simpler problem.

It is also possible to indicate design problems for which the correct statement necessarily requires the knowledge of the entire set GU.

Remark.

In variational problems, and in particular, in design problems it is customary to prove the existence of the solution on the basis of the compactness of the set U. Even a weak compactness of this set often turns out to be sufficient , for example, when controls are the right-hand sides of linear differential equations expressing the density of distributed sources. In all these cases one has the same initially given set of controls U.

However, the set U is not compact in many optimal design problems. At the same time the weak compactness of these sets is insufficient for the existence of the optimal element in these sets (see examples previously mentioned). Under these conditions an alternative arises: we have either to find the optimal control on a compact subset of the set U or to complement this set by new elements extending it to the set MU. It is often hard to justify physically the first way for problems of structural design, because the requirement of the compactness should follow from the mechanical formulation of the problem and it should not be caused only by wanting the existence of the optimal control. More specifically, when physical and technological reasons dictate a certain smoothness of control, for example, the function of thickness of a thin-shelled construction, this approach has a successful application [6, 18, 55, 78].

By contrast, the extension of the initial sets U to the sets MU or GU turns in some sense the set of admissible controls into a compactum. At the same time this extension corresponds to an intuitively expected pattern showing a frequent alternation of the initial materials. In the microscale, the composites that make up the G-closures are the same initial phases belonging to the set U. In the macroscale, the composites have new characteristics that the initial phases do not have. These new characteristics are agreed upon the intrinsic microstructure of composites, which, therefore, plays the role of an additional control factor.

4. Quasiconvex functions and extensions of sets of admissible controls

4.1. *Some information on quasiconvex functions.*

The construction of the minimal extensions of sets of admissible controls in multidimensional problems and also the description of the sets GU and $G_m U$ are based on using some properties of quasiconvex functions. These functions, originally introduced by Morrey [74], were investigated in several papers [47, 48, 54, 75, 83, 91, 94, 98] and also in papers of other authors [24–26, 28, 68–70], where, however, a different terminology was used.

Let $\mathbf{v} = (v_1, \ldots, v_m)$ be an m-dimensional vector, with components belonging to a space $L_p(\overline{V})$:

$$v_i \in L_p(\overline{V}), \qquad i = 1, \ldots, m,$$

where $\overline{V}$ is a domain in R^n: $\overline{V} \subset R^n$.

Consider a sequence $\{\mathbf{v}^s\}$ converging weakly to a vector $\mathbf{v}^0$: $\{v_i^0 \in L_p(\overline{V}), i = 1, \ldots, m\}$

$$\mathbf{v}^s \underset{L_p(\overline{V})}{\rightharpoonup} \mathbf{v}^0. \tag{4.1}$$

Let us introduce a scalar function $f : R^m \to R^1$ of the argument $\mathbf{v}$ and a sequence $\{f(\mathbf{v}^s)\}$ that we assume to be convergent weakly to a limit f^0:

$$f(\mathbf{v}^s) \rightharpoonup f^0; \tag{4.2}$$

at the same time let us consider a functional

$$I[\mathbf{v}] \triangleq \int_V f(\mathbf{v}) \, dx, \tag{4.3}$$

where V is an arbitrary subdomain in $\overline{V}$.

Note that the value f^0 depends not only on $\mathbf{v}^0$ but also on the character of the passage to the limit (4.1), that is, on the sequence $\{\mathbf{v}^s\}$ itself.

In addition, let us calculate the value $f(\mathbf{v}^0)$. We are interested in a relation between $f(\mathbf{v}^0)$ and f^0, and also in the value $\inf_{\mathbf{v}^s \to \mathbf{v}^0} I[\mathbf{v}^s]$.

It is known [35] that if $\{\mathbf{v}^s\}$ is an arbitrary sequence converging weakly to $\mathbf{v}^0$ and $f(\mathbf{v})$ is a convex function, then

$$f^0 \geq f(\mathbf{v}^0), \tag{4.4}$$

$$\min_{\mathbf{v}^s \to \mathbf{v}^0} I[\mathbf{v}^s] = \int_V f(\mathbf{v}^0) \, dx. \tag{4.5}$$

If $f(\mathbf{v})$ is an affine function, then

$$f^0 = f(\mathbf{v}^0). \tag{4.6}$$

Thus for an affine function the limit f^0 is uniquely determined by the value $\mathbf{v}^0$. In other cases a problem of the estimation of the value f^0 arises.

If f is convex, then the estimation (4.4) holds. It is sharp in the sense that there is a sequence $\{\mathbf{v}^s\}$ satisfying (4.1) for which (4.4) holds as an identity. The existence of the mentioned sequence is equivalent to the relation (4.5).

If the function $f(\mathbf{v})$ is not convex then a function $f_c(\mathbf{v})$ is used for the estimation of the value f^0. The function $f_c(\mathbf{v})$ is the greatest convex function that does not exceed $f(\mathbf{v})$:

$$f_c(\mathbf{v}) \triangleq \sup_{\omega \in \Omega} \omega(\mathbf{v});$$

where Ω denotes a set of all the convex functions of the argument $\mathbf{v}$ such that $\omega(\mathbf{v}) \leq f(\mathbf{v}) \; \forall \mathbf{v}$.

This definition and (4.4) imply

$$f^0 \geq (f_c)^0 \geq f_c(\mathbf{v}^0). \tag{4.7}$$

This inequality also gives the sharp estimation of the value f^0. As to the lower bound of the functional $I[\mathbf{v}^s]$ with respect to elements $\mathbf{v}^s$, $\mathbf{v}^s \rightharpoonup \mathbf{v}^0$, it is equal to

$$\inf_{\mathbf{v}^s \to \mathbf{v}^0} I[\mathbf{v}^s] = \int_V f_c(\mathbf{v}^0) \, dx. \tag{4.8}$$

A construction of the function $f_c \triangleq \langle\langle f \rangle\rangle$ [35] is realized with the help of the double Young transform of the function f. In this way we can justify the procedure of the relaxation (2.11) suggested by Filippov for the system of the ordinary differential equations (2.10). In this system the state variables x strongly converge in L_p and the corresponding controls u converge only weakly. In this case the vector X plays the role of $\mathbf{v}$ and the relaxation (2.11) is realized by introducing the vector $f_c = \langle\langle X \rangle\rangle$, the double Young transform of the function X with respect to controls u and with fixed (t, x). Thus a procedure of the convexification (building of f_c) extends the set of controls U to the set MU (see (3.27)) by introducing new controls — concentrations m (cf. (2.10) and (2.11)).

Problems with partial derivatives containing a control in the main part of a differential operator are distinguished by the fact that both the controls and the state variables (called parametric variables [21]), generally speaking, converge only weakly. So generalization of one-dimensional representations in this case requires the participation of parametric variables in the procedure of the minimal extension.

However, here it is necessary to take into consideration that parametric variables, such as stresses and strains in elasticity, current density, and gradient of a potential, and the like, are subject to differential constraints of the following type

$$\sum_{j=1}^{n}\sum_{i=1}^{m} a_{ijk}\frac{\partial v_i}{\partial x^j} \text{ is bounded in } L_p, \qquad k = 1, 2, \ldots \qquad (4.9)$$

where a_{ijk} are certain real coefficients.
For example, the gradient of a potential $\mathbf{v} = \nabla\omega$ has the zero curl

$$\nabla \times \mathbf{v} = 0, \qquad (4.10)$$

and a vector of current density $\mathbf{j} = \mathbf{D} \cdot \nabla w$ (see (3.20)) has a bounded divergence:

$$\nabla \cdot \mathbf{j} \text{ is bounded in } L_2. \qquad (4.11)$$

The following constraints imposed on the tensor of strain $\mathbf{e} = \nabla\nabla w$ and on the tensor of bending moments $\mathbf{M} = \mathbf{D} \cdot \cdot \nabla\nabla w$ are essential for problems of bending of plates connected with an operator $\nabla\nabla \cdot \cdot \mathbf{D} \cdot \cdot \nabla\nabla$ (cf. (3.3))

$$\text{Ink } \mathbf{e} = 0, \qquad (4.12)$$

$$\nabla\nabla \cdot \cdot \mathbf{M} \text{ is bounded in } L_2. \qquad (4.13)$$

In elasticity problems the strain tensor $\mathbf{e}$ is subject to a compatibility condition [19]

$$\text{Ink } \mathbf{e} \triangleq \nabla \times (\nabla \times \mathbf{e})^{\top} = 0,$$

which is similar to (4.9), and the symmetric stress tensor σ is subject to a boundedness condition for the divergence: $\nabla \cdot \sigma$ is bounded in L_2. Note that the order of derivatives in the linear form of type (4.9) is not important for the sequel.

The formulated constraints relate to invariant linear forms of derivatives of parametric variables. They lead to the fact that certain combinations of these variables remain continuous on an arbitrary line where the variables themselves have a jump. Equation (4.10) implies continuity of the component of $\mathbf{v}$ tangent to the jump line and (4.11) implies continuity of the normal component of $\mathbf{j}$.

Returning to the main problem of this section, we are interested in a relation between $f(\mathbf{v}^0)$ and f^0 when elements of the sequence $\{\mathbf{v}^s\}$ are not arbitrary but satisfy additional constraints (4.9). It is clear that under such circumstances a class of functions satisfying (4.4) and (4.6) can only get extended. The functions satisfying (4.4) under conditions (4.9) are called quasiconvex and the functions satisfying (4.6) under the same conditions are called quasi-affine. (Note that in paper [54] the constraint (4.9) is connected with a term "A–quasi-convexity.") An equivalent definition of a quasi-convex function consists in the following [54]: a function $f : R^m \to R^1$ is called quasiconvex if the inequality

$$\int_V f(\mu + \xi(x))\, dx \geq \int_V f(\mu)\, dx = f(\mu)\ \text{mes}\ V$$

holds $\forall \mu \in R^m$ for any hypercube $V \subset R^n$ and for any function $\xi \in L(V)$, where $\xi = (\xi_1, \xi_2, \ldots, \xi_m)$, with

$$L(V) \triangleq \left\{ \xi \in L_\infty(V),\quad \int_V \xi(x)\, dx = 0,\quad \sum_{j=1}^{n}\sum_{i=1}^{m} a_{ijk}\frac{\partial \xi_i}{\partial x^j} = 0 \right\}.$$

The given definition is equivalent to the ordinary convexity if the last condition in this characteristic of a set is absent. Thus the main difference is the fact that the test functions should belong to a subset for which the following holds

$$\sum_{j=1}^{n}\sum_{i=1}^{m} a_{ijk}\frac{\partial \xi_i}{\partial x^j} = 0.$$

Let us denote by $f_q(\mathbf{v})$ the greatest quasi convex function that does not exceed $f(\mathbf{v})$. The function $f_q(\mathbf{v})$, as shown in [74], has the property that for $f : 0 \leq f(\mathbf{v}) \leq a + b|\mathbf{v}|^p$, $p > 1$ the following equality holds (cf. (4.8))

$$\inf_{\substack{\mathbf{v}^s \to \mathbf{v}^0 \\ \mathbf{v}^s \in (4.9)}} I[\mathbf{v}^s] = \int_V f_q(\mathbf{v}^0)\, dx. \tag{4.14}$$

Thus, the multidimensional analogy of Filippov's lemma includes the procedure of constructing a quasiconvex hull of differential constraint of the problem. The constructing of the convex hull is reduced to the Young transform, whereas the formation of the quasiconvex envelope represents a more difficult problem which is considered in Section 4.3.

Note that all the convex (affine) functions are quasi-convex (quasi-affine). On the other hand, it is possible to give important examples of functions that are not convex but are quasiconvex.

4.2. *Examples of quasi-convex but not convex functions.*

Example 1.

Let $\mathbf{v}$ be a rectangular matrix of dimension $(n \times k)$:

$$\mathbf{v} = (\nabla w_1, \nabla w_2, \ldots, \nabla w_k), \qquad w_s \in W_1^p(V).$$

Then any minor of this matrix

$$\varphi_1 = \text{sub det } \mathbf{v}$$

is a quasi-affine function [47].

Example 2.

Let $n = 2$ and let $\mathbf{v}$ be a (2×2)-matrix

$$\mathbf{v} = (\mathbf{j}_1, \mathbf{j}_2),$$

where vectors $\mathbf{j}_1$, $\mathbf{j}_2$ satisfy the condition (4.11). Then $\varphi_2 = \text{det } \mathbf{v}$ is a quasi-affine function [24].

Example 3.

Let $n > 2$ and let $\mathbf{v}$ be $(n \times n)$-matrix

$$\mathbf{v} = (\mathbf{j}_1, \ldots, \mathbf{j}_n),$$

where vectors $\mathbf{j}_1, \ldots, \mathbf{j}_n$ satisfy (4.11). Then the function

$$\varphi_3 = (n - 1) \text{tr } \mathbf{v}^2 - (\text{tr } \mathbf{v})^2$$

is quasiconvex [95].

Example 4.

Let $\mathbf{v} = (\mathbf{v}_1, \mathbf{v}_2)$, where vector $\mathbf{v}_1$ satisfies the condition (4.10) and vector $\mathbf{v}_2$ satisfies the condition (4.11). Then the function

$$\varphi_4 = \mathbf{v}_1 \cdot \mathbf{v}_2$$

is quasi-affine [75, 94].

Example 5.

Let $n = 2$, $\mathbf{e} = \nabla\nabla w$, $w \in W_2^2(V)$. Then the function $\varphi_5 = \det \mathbf{e}$ is quasi-affine [25, 48].

Example 6.

Let $n = 2$, $\mathbf{M}$ be a symmetric (2×2)-tensor satisfying the condition (4.13). Then the function $\varphi_6 = -\det \mathbf{M}$ is quasi-convex [10].

Example 7.

Let $n = 2$, $\mathbf{v} = (\mathbf{v}_1, \mathbf{v}_2)$, where symmetric (2×2)-tensor $\mathbf{v}_1$ satisfies the condition (4.12) and a symmetric (2×2)-tensor $\mathbf{v}_2$ satisfies the condition (4.13). Then the function $\varphi_7 = \mathbf{v}_1 \cdot \cdot \mathbf{v}_2$ is quasi-affine [25].

Other examples of quasi-affine (quasiconvex) functions of variables that are related to the plane problem of elasticity are given in Section 5 in the discussion on particular problems.

In all of the examples given here the properties of quasi-convexity (quasi-affinity) can be proved with the help of the method presented in [54]. These examples were constructed in the cited original papers.

For the mentioned examples of quasi-affine functions it is possible to indicate the following characteristic properties uncovering their physical sense. First, these functions are invariant with respect to the choice of a basis in R^n. Second, on any line of the discontinuity of $\mathbf{v}$ a jump of the function $f(\mathbf{v})$ is linear with respect to jumps of the argument. Since the passage to the weak limit represents a linear operation, such behavior of $f(\mathbf{v})$ explains the quasi-affinity [68]. For Example 2 we have:

$$f(\mathbf{v}) = \det \mathbf{v} = j_{1n} j_{2t} - j_{1t} j_{2n},$$

where $\mathbf{n}$, $\mathbf{t}$ are the normal and tangential directions to the discontinuity line. Since the normal components of the vectors $\mathbf{j}_1, \mathbf{j}_2$ are continuous, the jump of $f(\mathbf{v})$ depends linearly on the jumps $\mathbf{j}_1, \mathbf{j}_2$.

The examples previously mentioned give us the possibility of constructing the G-closures and the minimal extensions of the sets of admissible controls for the problems formulated in Section 2.

4.3.   *Method of construction of the minimal extensions of sets of admissible controls.*

Properties of quasiconvex functions make it possible to give a procedure of quasiconvexification, which is similar to Filippov's procedure for ordinary differential equations.

Consider a scalar function

$$h^s = h(\chi^s, \mathbf{v}^s),$$

depending on two arguments. The first of them represents characteristic vector–function $\chi^s = (\chi_1^s, \ldots, \chi_N^s)$, the second one is a vector of parametric variables satisfying condition (4.9). Let vectors χ^s, $\mathbf{v}^s$ form sequences weakly converging to their limits $m = (m_1, \ldots, m_N)$ and $\mathbf{v}^0$, respectively, that is, conditions (3.8) and (4.1) hold. Let us estimate the weak limit of the sequence $\{h^s\}$ by a function depending only on the values m and $\mathbf{v}^0$; that is, let us find an inequality similar to (4.4)

$$\lim_{\text{weak}} h(\chi^s, \mathbf{v}^s) \geq \hat{h}(m, \mathbf{v}^0), \quad \text{if } \chi^s \overset{*}{\rightharpoonup} m, \ \mathbf{v}^s \underset{L_p}{\rightharpoonup} \mathbf{v}^0.$$

The estimation will be sharp if there exist sequences χ^s, $\mathbf{v}^s$ for which it holds as an equality. Let us denote by h_q a function $\hat{h}$ realizing such an estimation:

$$h_q(m, \mathbf{v}^0) = \inf_{\chi^s, \mathbf{v}^s} \lim_{\text{weak}} h(\chi^s, \mathbf{v}^s), \quad \forall \chi^s \in (3.8), \ \forall \mathbf{v}^s \in (4.1), (4.9). \quad (4.15)$$

If the differential constraints (4.9) are not taken into account, then in the capacity of $\hat{h}(m, \mathbf{v}^0)$ one can take the greatest convex function h_c not exceeding h:

$$\lim_{\text{weak}} h(\chi^s, \mathbf{v}^s) \geq h_c(m, \mathbf{v}^0). \quad (4.16)$$

It is clear that this inequality always holds, but the estimation that it gives will not be, generally speaking, sharp (the extension of the initial problem will not be minimal) because of the fact that the constraints (4.9) were not taken into account:

$$h_q(m, \mathbf{v}^0) \geq h_c(m, \mathbf{v}^0).$$

Our aim is to find h_q. For this function we obtain double-sided estimations, that is, we will construct functions h_l and h_p such that

$$h \geq h_l \geq h_q \geq h_p \geq h_c. \quad (4.17)$$

If it proves to be $h_l = h_p$, then the desired function h_q is obtained.

The necessity of using double-sided estimations for the value h_q is connected with the fact that the direct test on quasiconvexity is difficult because the definition of quasiconvexity is nonconstructive: it includes the integration over an arbitrary hypercube and the differential constraints (4.9) should be taken into account.

In order to construct the function h_p let us consider a linear combination of the following type,

$$g(\chi, \mathbf{v}, \alpha, \beta) \triangleq h(\chi, \mathbf{v}) + \sum_{(i)} \alpha_i \varphi_i(\mathbf{v}) + \sum_{(j)} \beta_j \psi_j(\mathbf{v}), \qquad (4.18)$$

where φ_i, ψ_j are, respectively, some quasi-affine (but not affine) and quasiconvex (but not convex) functions of $\mathbf{v}$, and α_i, β_j are real constants with

$$-\infty \leq \alpha_i \leq \infty, \qquad -\infty \leq \beta_j \leq 0. \qquad (4.19)$$

According to (4.7) we have

$$\lim_{\text{weak}} g(\chi^s, \mathbf{v}^s, \alpha, \beta) \geq g_c(m, \mathbf{v}^0, \alpha, \beta),$$

where $g_c \triangleq \langle\langle g \rangle\rangle$ is the greatest convex function of variables χ and $\mathbf{v}$ not exceeding g.

On the other hand, by virtue of properties of the functions φ_i, ψ_j and the conditions (4.19), the following inequality holds,

$$\lim_{\text{weak}} h(\chi^s, \mathbf{v}^s) + \sum_{(i)} \alpha_i \varphi_i(\mathbf{v}^0) + \sum_{(j)} \beta_j \psi_j(\mathbf{v}^0) \geq \lim_{\text{weak}} g(\chi^s, \mathbf{v}^s, \alpha, \beta).$$

Combining the last two inequalities we obtain the following estimations,

$$\lim_{\text{weak}} h(\chi^s, \mathbf{v}^s) \geq$$
$$\max_{\alpha, \beta \in (4.19)} \{ g_c(m, \mathbf{v}^0, \alpha, \beta) - \sum_{(i)} \alpha_i \varphi_i(\mathbf{v}^0) - \sum_{(j)} \beta_j \psi_j(\mathbf{v}^0) \} \triangleq h_p(m, \mathbf{v}^0). \qquad (4.20)$$

Referring to the definition (4.15) we obtain the inequality (4.17) on h_q and h_p. The estimation that holds for all the admissible values of the parameters α, β can be sharpened with the help of the operation max in the right-hand side (4.20). When $\alpha = \beta = 0$ the estimation coincides with (4.16); for other values of parameters the right-hand side (4.20) represents a quasiconvex, but, generally speaking, not convex function of arguments m, $\mathbf{v}^0$. The effectiveness of the estimation obtained in this way is defined by a number of quasiconvex (but not convex) functions φ_i, ψ_j that were used to obtain it. The estimation h_p uses algebraic corollaries of quasiconvexity of the selected functions φ_i, ψ_j. As shown in the following, the estimations of such kind prove for an array of examples to be sharp, that is, $h_p = h_q$. This indicates the "completeness" of the system of the selected functions.

Remark.

The procedure described here was proposed in [47, 54, 91, 63] for the functions of form $\varphi = \operatorname{sub} \det \nabla w$. It was called a polyconvexification because of the fact that the function φ turns out to be polylinear; these functions were sufficient for the problems considered in the indicated papers. The generalization suggested here also uses quasiconvex functions ψ; the necessity of considering these functions is defined by a more general constraint (4.9). The symbol h_p introduced in [91] nevertheless remains in this case, too.

As to the function h_l estimating h_q from below (see (4.17)), the construction of layers of the kth rank described in Section 3.2 are used for its construction. By its very definition, this construction gives examples of weakly convergent sequences $\{\chi^s, \mathbf{v}^s\}$ such that the weak limit

$$\lim_{\text{weak}} h(\chi^s, \mathbf{v}^s) = h(m, \mathbf{v}^0, \mu_1, \ldots, \mu_k, \mathbf{n}_1, \ldots, \mathbf{n}_k) \qquad (4.21)$$

depends on intrinsic parameters of a microstructure, that is, concentrations $\mu_1, \ldots, \mu_k$ of phases in layers of various ranks and vectors $\mathbf{n}_1, \ldots, \mathbf{n}_k$ of normals to these layers. As shown in Section 3.2, the limit value $\lim_{\text{weak}} h(\chi^s, \mathbf{v}^s)$ in (4.21) is calculated analytically for the layered microstructures of the kth rank. The minimal value $h_l^{(k)}$ of the right-hand side (4.21) for the varying intrinsic parameters of the microstructure $\mu_1, \ldots, \mu_k$ and $\mathbf{n}_1, \ldots, \mathbf{n}_k$ estimates h_q from above:

$$h_l^{(k)} = \min_{\substack{\mu_1, \ldots, \mu_k \\ \mathbf{n}_1, \ldots, \mathbf{n}_k}} h(m, \mathbf{v}^0, \mu_1, \ldots, \mu_k, \mathbf{n}_1, \ldots, \mathbf{n}_k) \geq h_q.$$

In fact, layered microstructures correspond to specific sequences and so the minimum $h(m, \mathbf{v}^0, \mu_1, \ldots, \mu_k, \mathbf{n}_1, \ldots, \mathbf{n}_k)$ with respect to the parameters of these sequences is not smaller than the value h_q which is the minimum $h(\chi^s, \mathbf{v}^s)$ with respect to all the possible sequences $\{\chi^s, \mathbf{v}^s\}$.

It is obvious that the value $h_l^{(k)}$ does not increase as the rank k does. However, the sequence $\{h_l^{(k)}\}$, $k = 1, 2, \ldots$ is bounded below by the value h_p. So there exists some finite or infinite rank of layers giving the best estimation h_l, if only the value h_p is not equal to $-\infty$.

Let us emphasize that the given scheme for constructing the estimation h_l is connected with the passages to the limit of a particular kind, such that on every step the weak limit is taken with respect to a sequence, whose components vary only along the normal **n** to the layers. Such one-dimensional weak passages to the limit, performed repeatedly for various **n**, approximate the limit h_q of an arbitrary multidimensional weakly convergent sequence of subdivisions $\{\chi^s\}$.

The laminated composites of the kth rank are not the only representatives of microstructures, properties of which can be calculated analytically. One could use constructions of the "embedded spheres" type [56, 15] or "embedded ellipsoids" type [95]. (L. Tartar, a private communication by R. Kohn).

5. Constructing G-closures for certain sets

5.1. *Estimate for density of energy of composites [68, 10].*

1. A process in a medium is described by equations (3.1) and (3.2). Density of the energy can be represented in the two equivalent forms:

$$N = \nabla w \cdot \mathbf{j} = \nabla w \cdot \mathbf{D} \cdot \nabla w \tag{5.1}$$

or

$$N = \nabla w \cdot \mathbf{j} = \mathbf{j} \cdot \mathbf{D}^{-1} \cdot \mathbf{j}. \tag{5.2}$$

Here $\mathbf{j} = \mathbf{D} \cdot \nabla w$ $(\nabla \cdot \mathbf{j} = f)$ denotes the current density (heat flux).

Consider a sequence of subdivisions of a domain V, occupied by the medium, into parts such that the conductivity $\mathbf{D} = \mathbf{D}^s(x) > 0$ satisfies (3.4) and (3.8). In other words, the medium is a mixture of initial materials $\mathbf{D}_i$ with concentrations m_i. Let us find the range of density of the energy of such a medium depending on its microstructure.

The quasi-affinity of the function N (Example 4, Section 4.2) implies that density of the energy N^0 of the limit medium is equal to

$$N^0 = \lim_{\text{weak}} N^s = \nabla w^0 \cdot \mathbf{j}^0,$$

where

$$w^s \underset{W_2^1(V)}{\longrightarrow} w^0, \qquad \mathbf{j}^s \underset{L_2(V)}{\longrightarrow} \mathbf{j}^0.$$

The energy determines the G-limit state law

$$\mathbf{j}^0 = \mathbf{D}_0 \cdot \nabla w^0$$

and can be presented in the form

$$N^0 = \nabla w^0 \cdot \mathbf{D}_0 \cdot \nabla w^0 = \mathbf{j}^0 \cdot \mathbf{D}_0^{-1} \cdot \mathbf{j}^0. \tag{5.3}$$

For the estimate of density of the energy one should estimate the function N^0 by a value depending only on average fields ∇w^0, $\mathbf{j}^0$, on the conductivity of phases $\mathbf{D}_1, \ldots, \mathbf{D}_N$, and on the concentrations $m_1, \ldots, m_N$.

Since $N^s = \nabla w^s \cdot \mathbf{D}^s \cdot \nabla w^s$ is a convex function of the arguments ∇w^s, $(\mathbf{D}^s)^{-1}$, the estimate (4.16) implies that

$$N^0 \geq \nabla w^0 \cdot \left(\sum_{i=1}^{N} m_i \mathbf{D}_i^{-1} \right)^{-1} \cdot \nabla w^0. \tag{5.4}$$

The tensor $\sum_{i=1}^{N} m_i \mathbf{D}_i^{-1}$ which is a part of the estimate represents a weak limit of the sequence $\{(\mathbf{D}^s)^{-1}\}$, where

$$(\mathbf{D}^s)^{-1} = \sum_{i=1}^{N} \chi_i^s \mathbf{D}_i^{-1}.$$

The right-hand side of (5.4) is the greatest convex function of the arguments m, ∇w^0, not exceeding $\lim_{\text{weak}} N^s = N^0$.

Using the representation for the density N in the form (5.2) we obtain another estimate

$$N^0 = \mathbf{j}^0 \cdot \mathbf{D}_0^{-1} \cdot \mathbf{j}^0 \geq \mathbf{j}^0 \cdot \left(\sum_{i=1}^{N} m_i \mathbf{D}_i \right)^{-1} \cdot \mathbf{j}^0. \tag{5.5}$$

The inequalities (5.4) and (5.5) estimate the minimal value of the energy in the given fields ∇w^0 and $\mathbf{j}^0$ respectively. At the same time they estimate the maximal λ_n and the minimal λ_1 eigenvalues of the tensor $\mathbf{D}_0$. Comparing (5.3) with (5.4) and (5.5) we obtain

$$\lambda_n \leq \sum_{i=1}^{N} m_i d_n(i), \qquad \lambda_1 \geq \left(\sum_{i=1}^{N} m_i d_i^{-1}(i) \right)^{-1}, \tag{5.6}$$

where $d_n(i)$, $d_1(i)$ are the maximal and the minimal eigenvalues of the tensor of the conductivity of the ith phase.

The inequalities (5.6) coincide with the Reuss–Voigt estimates (3.21). As it was noted, these estimates are sharp since they correspond to layered microstructures oriented either along or across the vector ∇w^0.

In this example only the property of the quasi-affinity of the estimated function N was used. The coefficients α_i, β_j were equal to zero.

2. Now consider a problem of estimating density of the energy N for the elastic plate investigated in [10]. The bending of the plate is described the

equations (3.1), (3.3) or by an equivalent system

$$\mathbf{e} = \nabla\nabla w, \quad \mathbf{M} = \mathbf{D} \cdot \cdot \mathbf{e}, \quad \nabla\nabla \cdot \cdot \mathbf{M} = f, \tag{5.7}$$

where $\mathbf{e}$ is the (2×2)–stress tensor $(e_{ig} \in L_2(S))$, $\mathbf{M}$ is the (2×2)–tensor of bending moments $(M_{ij} \in L_2(S))$, and $\mathbf{D}$ is a self-adjoint positive definite elastic tensor of the plate. Let the initial phases be isotropic and characterized by tensors $\mathbf{D}_1$ and $\mathbf{D}_2$ of type

$$\mathbf{D}_1 = k_1 \mathbf{a}_1 \mathbf{a}_1 + \mu_1 (\mathbf{a}_2 \mathbf{a}_2 + \mathbf{a}_3 \mathbf{a}_3), \tag{5.8}$$

$$\mathbf{D}_2 = k_2 \mathbf{a}_1 \mathbf{a}_1 + \mu_2 (\mathbf{a}_2 \mathbf{a}_2 + \mathbf{a}_3 \mathbf{a}_3), \tag{5.9}$$

where k_1, μ_1 $(k_2$, $\mu_2)$ are the bulk and shear moduli of the first (second) phase, respectively. The eigentensors $\mathbf{a}_1$ (spherical) and $\mathbf{a}_2$, $\mathbf{a}_3$ (deviators) are orthonormalized (in the sense of convolution over two indices) and have the following form [36]

$$\mathbf{a}_1 = \frac{1}{\sqrt{2}}(\mathbf{ii} + \mathbf{jj}), \ \mathbf{a}_2 = \frac{1}{\sqrt{2}}(\mathbf{ii} - \mathbf{jj}), \ \mathbf{a}_3 = \frac{1}{\sqrt{2}}(\mathbf{ij} + \mathbf{ji}), \tag{5.10}$$

$(\mathbf{i}, \mathbf{j}$ are two orts on a plane S occupied by the plate). Assume also that $k_1 < k_2$, $\mu_1 < \mu_2$. The density of elastic energy N is expressed as

$$N = \mathbf{e} \cdot \cdot \mathbf{D} \cdot \cdot \mathbf{e} = \mathbf{M} \cdot \cdot \mathbf{D}^{-1} \cdot \cdot \mathbf{M}.$$

Consider again the process of forming a composite from the phases and (5.8), (5.9); that is, assume that $\mathbf{D} = \mathbf{D}^s(x)$ satisfies (3.4) and (3.8); the concentrations m_1 and $m_2 = 1 - m_1$ of the phases $\mathbf{D}_1$ and $\mathbf{D}_2$ in the composite are considered to be fixed.

In order to apply the estimate we assume that

$$h = \mathbf{e} \cdot \cdot \mathbf{D} \cdot \cdot \mathbf{e}, \qquad \varphi = 2 \det \mathbf{e} \qquad \text{(Example 5)}.$$

We obtain

$$g^s = h^s + \alpha\varphi^s = \mathbf{e} \cdot \cdot (\mathbf{D}^s + \alpha\mathbf{D}_{I_2}) \cdot \cdot \mathbf{e}^s,$$

where

$$\mathbf{D}_{I_2} \triangleq \mathbf{a}_1 \mathbf{a}_1 - \mathbf{a}_2 \mathbf{a}_2 - \mathbf{a}_3 \mathbf{a}_3;$$

the tensors $\mathbf{a}_1$, $\mathbf{a}_2$, $\mathbf{a}_3$ are defined by the formulae (5.10). Here the following representation is used

$$2 \det \mathbf{e} = \mathbf{e} \cdot \cdot \mathbf{D}_{I_2} \cdot \cdot \mathbf{e}. \tag{5.12}$$

Using the quasi-affinity of the function h

$$\lim_{\text{weak}} h^s = \nabla\nabla w^0 \cdot \cdot \mathbf{D}_0 \cdot \cdot \nabla\nabla w^0$$

and referring to (3.8) we obtain an estimate

$$\nabla\nabla w^0 \cdot \cdot \mathbf{D}_0 \cdot \cdot \nabla\nabla w^0 \geq \max_{\alpha} \nabla\nabla w^0 \cdot \cdot \{[m_1(\mathbf{D}_1 + \alpha\mathbf{D}_{I_2})^{-1} +$$
$$m_2(\mathbf{D}_2 + \alpha\mathbf{D}_{I_2})^{-1}]^{-1} - \alpha\mathbf{D}_{I_2}\} \cdot \cdot \nabla\nabla w^0 \triangleq$$
$$\max_{\alpha} \mathbf{e}^0 \cdot \cdot \overline{\mathbf{D}} \cdot \cdot \mathbf{e}^0 \triangleq \overline{\Pi}, \tag{5.13}$$

where the parameter α should be chosen in such a way as to keep tensors $\mathbf{D}_1 + \alpha\mathbf{D}_{I_2}$, $\mathbf{D}_2 + \alpha\mathbf{D}_{I_2}$ positive semidefinite (this follows from the construction of the convex hull g_c).
The tensor $\overline{\mathbf{D}}$ has dyadic representation [36]

$$\overline{\mathbf{D}} = \bar{k}\mathbf{a}_1\mathbf{a}_1 + \bar{\mu}(\mathbf{a}_2\mathbf{a}_2 + \mathbf{a}_3\mathbf{a}_3), \tag{5.14}$$

where

$$\bar{k} = \left(\frac{m_1}{k_1 + \alpha} + \frac{m_2}{k_2 + \alpha}\right)^{-1} - \alpha,$$

$$\bar{\mu} = \left(\frac{m_1}{\mu_1 - \alpha} + \frac{m_2}{\mu_2 - \alpha}\right)^1 + \alpha.$$

The range of α is defined by inequalities[3]

$$-k_1 \leq \alpha \leq \mu_1.$$

On the other hand, representing the density of energy in the form of $h = \mathbf{M} \cdot \cdot \mathbf{D}^{-1} \cdot \cdot \mathbf{M}$ and taking into account the quasiconvexity of a function

$$\psi = -2 \det \mathbf{M} = -\mathbf{M} \cdot \cdot \mathbf{D}_{I_2} \cdot \cdot \mathbf{M}$$

---

[3]For other values of α the function g_c equals ∞.

(Example 6), we obtain yet one more estimate

$$\mathbf{M}^0 \cdot \cdot \mathbf{D}_0^{-1} \cdot \cdot \mathbf{M}^0 \geq$$

$$\max_{\mu_2^{-1} \geq \beta \geq 0} \mathbf{M}^0 \cdot \cdot \{[m_1(\mathbf{D}_1^{-1} + \beta \mathbf{D}_{I_2})^{-1} + m_2(\mathbf{D}_2^{-1} + \beta \mathbf{D}_{I_2})^{-1}]^{-1} - \beta \mathbf{D}_{I_2}\} \cdot \cdot \mathbf{M}^0 =$$

$$\max_{\mu_2^{-1} \geq \beta \geq 0} \mathbf{M}^0 \cdot \cdot \underline{\mathbf{C}} \cdot \cdot \mathbf{M}^0 \triangleq \underline{\Pi}, \quad (5.15)$$

where the tensor $\underline{\mathbf{C}}$ equals

$$\underline{\mathbf{C}} = \underline{k}^{-1} \mathbf{a}_1 \mathbf{a}_1 + \underline{\mu}^{-1}(\mathbf{a}_2 \mathbf{a}_2 + \mathbf{a}_3 \mathbf{a}_3), \quad (5.16)$$

$$\underline{k}^{-1} = \left(\frac{m_1}{k_1^{-1} + \beta} + \frac{m_2}{k_2^{-1} + \beta}\right)^{-1} - \beta,$$

$$\underline{\mu}^{-1} = \left(\frac{m_1}{\mu_1^{-1} - \beta} + \frac{m_2}{\mu_2^{-1} - \beta}\right)^{-1} + \beta,$$

and admissible values of the parameter β (when the function g_c is not equal to ∞) belong to the interval $0 \leq \beta \leq \mu_2^{-1}$.

In order to obtain the explicit formulae for the estimates let us compute the values α_0 and β_0 of the parameters α and β maximizing the right-hand sides of the inequalities (5.13) and (5.15). These values are given in Tables 1 and 2 with the values of the estimates for the elastic energy $\overline{\Pi}$ and $\underline{\Pi}$. The value α_0 (β_0) depends on a ratio $\xi(\zeta)$ of the absolute values of deviatoric and spherical parts of the strain tensor (tensor of moments); that is, on

$$\xi = |\mathbf{e} \cdot \cdot \mathbf{a}_2'| / |\mathbf{e} \cdot \cdot \mathbf{a}_1|, \qquad \zeta = |\mathbf{M} \cdot \cdot \mathbf{a}_2''| / |\mathbf{M} \cdot \cdot \mathbf{a}_1|. \quad (5.17)$$

Here $\mathbf{a}_2'$ and $\mathbf{a}_2''$ denote the deviators coaxial with the tensors $\mathbf{e}$ and $\mathbf{M}$, respectively.

The dependence of the estimates on the invariant characteristics of stress and strain distinguishes them from the Reuss–Voigt estimates of the energy. The Reuss–Voigt inequalities ("Hill fork" [58]) are an example of estimates that can be obtained without fixing the strain states. In fact, as one can see from the tables, there exist states for which $\alpha = 0$ or $\beta = 0$ and the estimate coincides with its value according to Reuss–Voigt.

TABLE 1. Estimation of the energy of the "soft" plate

$\xi \geq \xi_1$	$\xi_2 \leq \xi \leq \xi_1$	$\xi \leq \xi_2$
$\alpha_0 = -k_1$	$\alpha_0 = \dfrac{\Delta k \tilde{\mu} - \Delta \mu \tilde{k}\xi}{\Delta k + \Delta \mu \xi}$	$\alpha_0 = \mu_1$
$\overline{\Pi} = k_1 e_1^2 + \left[\left(\dfrac{m_1}{k_1+\mu_1} + \dfrac{m_2}{k_1+\mu_2}\right)^{-1} - k_1\right] e_2^2$	$\overline{\Pi} = \langle k\rangle e_1^2 + \langle \mu\rangle e_2^2 - \dfrac{m_1 m_2}{k+\tilde{\mu}}(\Delta k e_1 + \Delta \mu e_2)^2$	$\overline{\Pi} = \mu_1 e_2^2 + \left[\left(\dfrac{m_1}{k_1+\mu_1} + \dfrac{m_2}{k_2+\mu_1}\right)^{-1} - \mu_1\right] e_1^2$

Notations: $\Delta a = a_2 - a_1$; $\langle a\rangle = m_1 a_1 + m_2 a_2$; $\tilde{a} = m_1 a_2 + m_2 a_1$, $\xi_1 = \dfrac{\tilde{\mu}+k_1}{m_1 \Delta \mu}$, $\xi_2 = \dfrac{m_1 \Delta k}{k+\mu_1}$.

TABLE 2. Estimate of the energy of the "rigid" plate

$\zeta \geq \zeta_1$	$\zeta_2 \leq \zeta \leq \zeta_1$	$\zeta \leq \zeta_2$
$\beta_0 = 0$	$\beta_0 = \dfrac{\Delta(k^{-1})(\widetilde{\mu^{-1}}) - \Delta(\mu^{-1})(\widetilde{k^{-1}})\zeta}{\Delta(k^{-1}) + \Delta(\mu^{-1})\zeta}$	$\beta_0 = \mu_2^{-1}$
$\underline{\Pi} = \langle k\rangle^{-1} M_1^2 + \langle \mu\rangle^{-1} M_2^2$	$\underline{\Pi} = \langle k^{-1}\rangle M_1^2 + \langle \mu^{-1}\rangle M_2^2 - \dfrac{m_1 m_2}{k^{-1}+\mu^{-1}}\left[\Delta(k^{-1})M_1 + \Delta(\mu^{-1})M_2\right]^2$	$\underline{\Pi} = \mu_2^{-1} M_2^2 + \left[\left(\dfrac{m_1}{k_1^{-1}+\mu_2^{-1}} + \dfrac{m_2}{k_2^{-1}+\mu_2^{-1}}\right)^{-1} - \dfrac{1}{\mu_2}\right] M_1^2$

Notations: $\zeta_1 = \dfrac{(\widetilde{\mu^{-1}})\Delta(k^{-1})}{(\widetilde{k^{-1}})\Delta(\mu^{-1})}$, $\zeta_2 = -\dfrac{m_2\Delta(k^{-1})}{(\widetilde{k^{-1}})+\mu_2^{-1}}$.

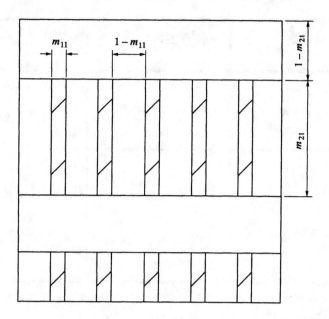

Figure 5. Matrix composite realizing extremal values of energy of elastic strain.

The obtained estimates of the energy are sharp as they are realized by layers of the first and second ranks shown in Figure 5 (matrix structures). The elastic moduli of such composites are computed according to the formulae in Section 3.3. They depend both on moduli of initial phases and on parameters of the structure — concentrations m_{11} and m_{21} of layers of the first and second ranks (see Figure 5). These structures can degenerate into the layers of the first rank. The matrix structure is orthotropic. Therefore the density of the elastic energy for such a composite depends, besides parameters m_{11} and m_{21} connected by the condition $m_{11}m_{21} = m_1$, also on the angle φ between principal axes of the elastic tensor and the stress tensor: $\Pi = \Pi(m_1, m_{11}, \varphi)$. The estimates given in Tables 3 and 4 are reached for stationary values of parameters m_{11}, φ [39] which together with other characteristics of the optimal structures are given in the Tables.

It should be emphasized that the quadratic form $N = \mathbf{e} \cdot\cdot \mathbf{D} \cdot\cdot \mathbf{e} = \mathbf{M} \cdot\cdot \mathbf{C} \cdot\cdot \mathbf{M}$ determines the tensors $\mathbf{D}$ and $\mathbf{C}$ not uniquely but up to additive terms that are orthogonal to the tensors $\mathbf{e}$ and $\mathbf{M}$, respectively. So in order to prove the sharpness of the estimates it is sufficient to find at least one representative of the described equivalence class, that is, to find a tensor $\widetilde{\mathbf{D}}$ or $\widetilde{\mathbf{C}}$ for which the following holds

$$\mathbf{e} \cdot\cdot (\widetilde{\mathbf{D}} - \mathbf{D}) \cdot\cdot \mathbf{e} = 0 \quad \text{or} \quad \mathbf{M} \cdot\cdot (\widetilde{\mathbf{C}} - \mathbf{C}) \cdot\cdot \mathbf{M} = 0.$$

TABLE 3. Parameters of the optimal "soft" structures

$\xi \geq \xi_1$	$\xi_2 \leq \xi \leq \xi_1$	$\xi \leq \xi_2$
$m_{11} =$ $P_a(m_1, m_2, \mathbf{D}_1, \mathbf{D}_2, \xi)$	$m_{11} = 1$	$m_{11} =$ $P_b(m_1, m_2, \mathbf{D}_1, \mathbf{D}_2, \xi)$
$\varphi = 0$	$\varphi = 0$	$\varphi = 0$

Notations:

$$P_a(m_1, m_2, \mathbf{D}_1, \mathbf{D}_2, \xi) = \frac{1+m_1}{2} + \frac{1}{2\xi}\left(\frac{k_2 + \mu_2}{\mu_2 - \mu_1} - m_2\right);$$

$$P_b(m_1, m_2, \mathbf{D}_1, \mathbf{D}_2, \xi) = \frac{1+m_1}{2} - \frac{\xi}{2}\left(\frac{k_2 + \mu_2}{k_2 - k_1} - m_2\right).$$

TABLE 4. Parameters of the optimal "rigid" structures

$\zeta \geq \zeta_1$	$\zeta_2 \leq \zeta \leq \zeta_1$	$\zeta < \zeta_2$
$m_{11} = 1$	$m_{11} = 1$	$m_{11} =$ $P_b(m_1, m_2, \mathbf{C}_1, \mathbf{C}_2, \zeta)$
$\cos 2\varphi =$ $-\dfrac{\Delta(k^{-1})(\widetilde{\mu^{-1}})}{\Delta(\mu^{-1})(k^{-1})}$	$\varphi = \frac{\pi}{2}$	$\varphi = 0$

The tensors of the effective constants of the microstructures given in the tables can serve as examples of such representatives. In particular, the microstructure in Figure 5 is orthotropic, but the corresponding estimate of the energy has such a form as if it were generated by an isotropic tensor (5.14).

5.2. G_m-closure of binary mixture of isotropic conductive phases [28].

The Reuss–Voigt estimates describe the least cube in the space of eigenvalues $(\lambda_1, \ldots, \lambda_n)$ of the tensor $\mathbf{D}_0$ inside which the set $G_m U$ is disposed. In order to give the exact description of this set it is necessary to have linked estimates of the eigenvalues which determine, for fixed concentration m, the range of one eigenvalue if the values of other eigenvalues are given. For this purpose one should subject a material to n different independent "tests", that is, to consider n solutions $w(i)$ $(i = 1, \ldots, n)$ (3.1) and (3.2), corresponding either to n linearly independent right-hand sides $f(i)$ or to n different boundary conditions. Since the tensor $\mathbf{D}_0$ of the effective characteristics is defined exclusively by a medium itself, these n tests are subject only to the condition of linear independence (nondegeneracy).

The necessity of considering n tests is connected with the fact that we determine the tensor $\mathbf{D}_0$ from values of the quadratic form associated with it. In order to find the tensor $\mathbf{D}_0$ completely it is necessary to know the values of this form generated by n linearly independent vectors.

This circumstance distinguishes the problem of finding G_m-closure from that of the previous section, where the quadratic form was estimated for a fixed value of the vector argument.

So, consider a quadratic form h^s defined as the sum of the strain energy densities of a medium subject to n independent tests:

$$h^s = \sum_{i=1}^{n} \nabla w^s(i) \cdot \mathbf{D}^s(x) \cdot \nabla w^s(i); \qquad (5.18)$$

tensor of conductivity $\mathbf{D}^s(x)$ equals

$$\mathbf{D}^s(x) = \chi_1^s \mathbf{D}_1 + \chi_2^s \mathbf{D}_2, \ \mathbf{D}_1 = u_1 \mathbf{E}, \ \mathbf{D}_2 = u_2 \mathbf{E}, \ 0 < u_1 \le u_2 < \infty,$$

where $\chi_1^s(x)$, $\chi_2^s(x)$ are characteristic functions of subdomains occupied by materials $\mathbf{D}_1$ and $\mathbf{D}_2$ with isotropic conductivities u_1 and u_2. The sequences $\{\chi_1^s\}$, $\{\chi_2^s\}$ are assumed to be weakly convergent to the limits $m_1(x)$ and $m_2(x)$ $(m_1 + m_2 = 1)$ respectively (cf. (3.8)).

The sequence $\{\chi^s\}$ converges weakly, by virtue of quasi-affinity of each term, to the function

$$h^0 = \sum_{i=1}^{n} \nabla w^0(i) \cdot \mathbf{D}_0(x) \cdot \nabla w^0(i).$$

Let us estimate h^0 from below by a function depending only on $\nabla w^0(i)$, m_1, m_2, $\mathbf{D}_1$, $\mathbf{D}_2$. Let us make use of the formula (4.20). For quasi-affine functions let us take minors of the second rank of the matrix $\|\nabla w(1), \nabla w(2), \ldots, \nabla w(n)\|$ (Example 1), admitting the following representation

$$\varphi_{rq} = \nabla w(r) \cdot \mathbf{O}_{rq} \cdot \nabla w(q),$$

where $\mathbf{O}_{rq}$ is a skew-symmetric $(n \times n)$-matrix of projection onto a plane defined by orts $\mathbf{e}_r$, $\mathbf{e}_q$ and the subsequent rotation about an angle of $90°$ in this plane:

$$\mathbf{O}_{rq} = \mathbf{e}_r \mathbf{e}_q - \mathbf{e}_q \mathbf{e}_r.$$

A function

$$g = \sum_{i=1}^{n} \nabla w(i) \cdot \mathbf{D} \cdot \nabla w(i) + \sum_{i=1}^{n} \sum_{j=1}^{n} \alpha_{ij} \nabla w(i) \cdot \mathbf{O}_{ij} \cdot \nabla w(j) \triangleq$$

$$\varepsilon \cdot \mathbf{A}(\mathbf{D}, \alpha_{ij}) \cdot \varepsilon, \qquad \alpha_{ij} = \alpha_{ji}$$

represents a quadratic form over n^2-dimensional vector

$$\varepsilon = (\nabla w(1), \ldots, \nabla w(n)),$$

generated by $n^2 \times n^2$-matrix $\mathbf{A}(\mathbf{D}, \alpha_{ij})$ that has a block representation

$$\mathbf{A}(\mathbf{D}, \alpha_{ij}) = \begin{pmatrix} \mathbf{D} & \alpha_{12}\mathbf{O}_{12} & \cdots & \alpha_{1n}\mathbf{O}_{1n} \\ -\alpha_{12}\mathbf{O}_{12} & \mathbf{D} & \cdots & \alpha_{2n}\mathbf{O}_{2n} \\ \vdots & \vdots & \ddots & \vdots \\ -\alpha_{1n}\mathbf{O}_{1n} & \cdots & \cdots & \mathbf{D} \end{pmatrix}. \qquad (5.19)$$

A function g_c equals

$$g_c = \begin{cases} \varepsilon^0 \cdot [m_1 \mathbf{A}^{-1}(\mathbf{D}_1, \alpha_{ij}) + m_2 \mathbf{A}^{-1}(\mathbf{D}_2, \alpha_{ij})]^{-1} \cdot \varepsilon^0, & \text{if } |\alpha_{ij}| \leq u_1 \, \forall i, j, \\ \\ -\infty, & \text{if } \exists i, j : |\alpha_{ij}| > u_1, \end{cases}$$

where

$$\varepsilon^s \xrightarrow[L_2(V)]{} \varepsilon^0 = \left(\nabla w^0(1), \ldots, \nabla w^0(n)\right).$$

If $|\alpha_{ij}| \leq u_1$, $\forall i, j$, then the estimate (4.20) has the form:

$$h^0 = \sum_{i=1}^{n} \nabla w^0(i) \cdot \mathbf{D}_0 \cdot \nabla w^0(i) \geq$$

$$\varepsilon^0 \cdot [m_1 \mathbf{A}^{-1}(\mathbf{D}_1, \alpha_{ij}) + m_2 \mathbf{A}^1(\mathbf{D}_2, \alpha_{ij})]^{-1} \cdot \varepsilon^0 -$$

$$\sum_{i=1}^{n} \sum_{j=1}^{n} \alpha_{ij} \nabla w^0(i) \cdot \mathbf{O}_{ij} \cdot \nabla w^0(j)$$

or

$$\varepsilon^0 \cdot \{\mathbf{A}(\mathbf{D}_0, \alpha_{ij}) - [m_1 \mathbf{A}^1(\mathbf{D}_1, \alpha_{ij}) + m_2 \mathbf{A}^{-1}(\mathbf{D}_2, \alpha_{ij})]^{-1}\} \cdot \varepsilon^0 \geq 0. \quad (5.21)$$

Taking into consideration the arbitrariness of the vector ε^0 and the positive definiteness of the matrices $\mathbf{A}(\mathbf{D}_0, \alpha_{ij})$, $\mathbf{A}(\mathbf{D}_1, \alpha_{ij})$, and $\mathbf{A}(\mathbf{D}_2, \alpha_{ij})$, we obtain a matrix inequality bounding the invariants of the tensor $\mathbf{D}_0$,

$$\mathbf{A}^{-1}(\mathbf{D}_0, \alpha_{ij}) \leq m_1 \mathbf{A}^{-1}(\mathbf{D}_1, \alpha_{ij}) + m_2 \mathbf{A}^{-1}(\mathbf{D}_2, \alpha_{ij}). \quad (5.22)$$

With the aim of transforming this inequality let us note that in the principal axes of the tensor $\mathbf{D}_0$ the matrices $\mathbf{A}(\mathbf{D}_k, \alpha_{ij})$ $(k = 0, 1, 2)$ represent the direct sum of the block

$$\mathbf{A}_1(\mathbf{D}_k, \alpha_{ij}) = \mathbf{D}_k + \mathcal{A}, \quad (5.23)$$

where $(n \times n)$-matrix $\mathcal{A} = \|\alpha_{ij}\|$ is such that $\alpha_{ii} = 0$ $(i = 1, \ldots, n)$, and (2×2)-blocks

$$\mathbf{A}_{ij}(\mathbf{D}_k, \alpha_{ij}) = \begin{pmatrix} \lambda_i^{(k)} & -\alpha_{ij} \\ -\alpha_{ij} & \lambda_j^{(k)} \end{pmatrix}, \quad i, j = 1, \ldots, n; \ i \neq j,$$

where $\lambda_i^{(k)}$, $\lambda_j^{(k)}$ $(i, j = 1, 2, \ldots, n)$ are eigenvalues of the tensor $\mathbf{D}_k$. Let us assume that $\alpha_{ij} = u_1$ $\forall i, j$ and apply the inequality (5.22) to the matrix $\mathbf{A}_1(\mathbf{D}_k, \alpha_{ij})$. It is necessary to take into account that for the given values of parameters α_{ij} the matrix $\mathbf{A}_1(\mathbf{D}_1, \alpha_{ij})$ becomes the following dyad

$$\mathbf{A}_1(\mathbf{D}_1, \alpha_{ij}) = nu_1 \mathbf{e}'^{\mathsf{T}} \mathbf{e}',$$

where

$$\mathbf{e}' = \frac{1}{\sqrt{n}} \underbrace{(1, 1, \ldots, 1)}_{n \text{ times}}.$$

The projection of the inequality (5.22) for the block $\mathbf{A}_1(\mathbf{D}_k, u_1)$ onto the "direction" $\mathbf{e}'^{\mathsf{T}} \mathbf{e}'$ gives after transformations a scalar inequality estimating the eigenvalues λ_i of the tensor $\mathbf{D}_0$ [27, 28]:

$$1 + \sum_{i=1}^{n} \frac{u_1}{\lambda_i - u_1} \le \frac{1}{m_2} \left(1 + \frac{n u_1}{u_2 - u_1} \right). \tag{5.24}$$

Similar inequalities for the blocks $\mathbf{A}_{ij}$ turn to be weaker than (5.24).

Another estimate of the eigenvalues λ_i can be obtained considering a quadratic form $(\mathbf{j} = \mathbf{D} \cdot \nabla w)$:

$$h = \sum_{i=1}^{n} \mathbf{j}(i) \cdot \mathbf{D}^{-1} \cdot \mathbf{j}(i)$$

and a quasiconvex function (Example 3)

$$\psi_3 = (n-1)(\operatorname{tr} \mathbf{J})^2 - \operatorname{tr} \mathbf{J}^2, \qquad \mathbf{J} = (\mathbf{j}(1), \ldots, \mathbf{j}(n)).$$

Operating as in the previous case we obtain the estimate

$$1 + \sum_{i=1}^{n} \frac{u_2}{\lambda_i - u_2} \ge \frac{1}{m_1} \left(1 + \frac{n u_2}{u_1 - u_2} \right). \tag{5.25}$$

The inequalities (5.24) and (5.25) (see [27, 28]) together with the Reuss–Voigt estimates $(3.21)_1$ and $(3.21)_2$ bound the set $G_m U$.[4] It is easy to verify that $(3.21)_1$ is a corollary of $(3.21)_2$, (5.24) and (5.25). Hence, the set $G_m U$ is bounded only by the last three inequalities [28].

The estimates $(3.21)_2$, (5.24), and (5.25) are sharp since they correspond to orthogonal matrix structures of the nth rank described in Section 3.2 (cf. (3.17) and (3.18)). A component of the boundary of the set $G_m U$ on which (5.24) holds as an equality corresponds to structures where the material $\mathbf{D}_1$ plays the role of the matrix and $\mathbf{D}_2$ plays the role of inclusions (see Figure 5); the materials change their roles on the component (5.25) of the boundary. The equality $\lambda_n = d_\perp$ (see (3.22)) holds for the structures with inclusions in form of cylinders, that is, on the layers of the smaller rank.

For the case $n = 2$ the domain $G_m U$ is shown in Figure 2; for the case $n = 3$, in Figure 6. Let us note that the range of the conductivity

---

[4]Independently of papers [27, 28] (see also [102, 103]) the inequalities (5.24) and (5.25) were given in [76, 77] without proof; an independent proof of these inequalities and construction of the sharp estimates $G_m U$ taking into account the inequality $(3.21)_2$ was given in [95].

λ of isotropic composites $\mathbf{D}_0 = \lambda \mathbf{E}$ for these cases coincides with Hashin-Shtrikman estimates [56].

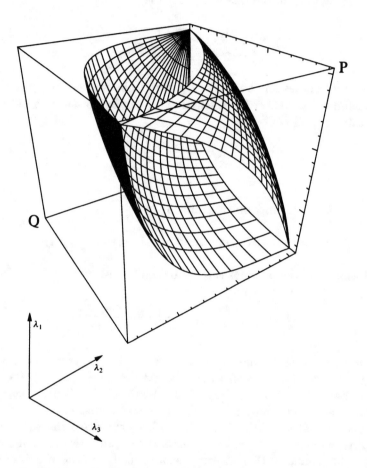

Figure 6. The set $G_m U$ generated by two isotropic phases u_1 and u_2 (three-dimensional case); point P has coordinates $(d_\perp, d_\perp, d_\perp)$ and point Q – coordinates $(d_\parallel, d_\parallel, d_\parallel)$.

The results obtained here also allow us to describe the set GU of invariants of all composites assembled from the materials with conductivities u_1 and u_2. For this purpose, according to (3.10), it is necessary to construct the union of the sets $G_m U$ with respect to the parameter m.

It is easy to see that boundary elements of the set GU have the maximal value of the eigenvalue λ_n ($\lambda_n \geq \lambda_{n-1} \geq \cdots \geq \lambda_1$) for all m: $\lambda_n = m_1 u_1 + m_2 u_2$.

Eliminating the parameters m_1 and $m_2 = 1 - m_1$ from the last equality and from (5.24) and (5.25), we see that the boundary of the set GU is described by the relations:

$$1 + \sum_{i=1}^{n-1} \frac{u_1}{\lambda_i - u_1} = \frac{u_2 + (n-2)u_1}{\lambda_n - u_1},$$

$$u_1 \le \lambda_n \le u_2, \; \lambda_i \le \lambda_n, \; i = 1, 2, \ldots, n - 1.$$

Let us note that for $n = 2$ these relations turn into (3.24).

The components of the boundary of the set GU correspond to matrix structures of the $(n-1)$th rank with cylindric inclusions from the material u_1 or u_2.

5.3. G_m-*closure of a set that consists of two isotropic phases with respect to an operator of plane elastic problem; the case with partially known* $\mathbf{D}_0$.

This problem can be solved similarly to the one considered earlier; here we restrict ourselves to the case when initial phases have equal bulk moduli $k > 0$ and different shear moduli μ_1 and μ_2 $(0 < \mu_1 \le \mu_2 < \infty)$:

$$\mathbf{D}_1 = k\mathbf{a}_1\mathbf{a}_1 + \mu_1(\mathbf{a}_2\mathbf{a}_2 + \mathbf{a}_3\mathbf{a}_3), \; \mathbf{D}_2 = k\mathbf{a}_1\mathbf{a}_1 + \mu_2(\mathbf{a}_2\mathbf{a}_2 + \mathbf{a}_3\mathbf{a}_3). \quad (5.26)$$

Here $\mathbf{a}_1$, $\mathbf{a}_2$, $\mathbf{a}_3$ are eigentensors determined by the formulae (5.10).

For the problems of this type the following general statement holds [68]. Let components of a composite of phase $\mathbf{D}_i$ be such that their expansions over the eigenbases

$$\mathbf{D}_i = d_1\mathbf{c}_1\mathbf{c}_1 + d_{2i}\mathbf{c}_{2i}\mathbf{c}_{2i} + d_{3i}\mathbf{c}_{3i}\mathbf{c}_{3i}$$

contain the term $d_1\mathbf{c}_1\mathbf{c}_1$, common for all phases.[5] Then the effective tensor $\mathbf{D}_0$ of the mixture is characterized by the decomposition

$$\mathbf{D}_0 = d_1\mathbf{c}_1\mathbf{c}_1 + d_{20}\mathbf{c}_{20}\mathbf{c}_{20} + d_{30}\mathbf{c}_{30}\mathbf{c}_{30},$$

that includes the same term $d_1\mathbf{c}_1\mathbf{c}_1$. In other words, under the formulated conditions d_1 remains the eigenvalue and $\mathbf{c}_1$ the corresponding eigentensor of the tensor $\mathbf{D}_0$.[6]

Indeed, consider a weakly convergent sequence of tensors $\mathbf{e} \triangleq \nabla\nabla w$

$$\mathbf{e}^s \underset{L_2(S)}{\rightharpoonup} \mathbf{e}^0$$

---

[5]Here $\mathbf{c}_1$, $\mathbf{c}_{2i}$, $\mathbf{c}_{3i}$ are eigentensors of a random type.

[6]This property reflects a particular role of representations of the type (5.26) (see [36]) in describing the effective constants of composites.

and the corresponding sequence

$$\mathbf{M}^s \triangleq \mathbf{D}^s \cdot \cdot \mathbf{e}^s \underset{L_2(S)}{\rightharpoonup} \mathbf{D}_0 \cdot \cdot \mathbf{e}^0 \triangleq \mathbf{M}^0,$$

where

$$\mathbf{M}^s = d_1 e_1^s \mathbf{c}_1 + d_2^s e_2^s \mathbf{c}_2^s + d_3^s e_3^s \mathbf{c}_3^s, \ e_i^s \triangleq \mathbf{e}^s \cdot \cdot \mathbf{c}_i^s, \ \mathbf{c}_1^s \triangleq \mathbf{c}_1.$$

Contracting the last equation with the constant tensor $\mathbf{c}_1$ we obtain

$$\mathbf{M}^s \cdot \cdot \mathbf{c}_1 = d_1 e_1^s \underset{L_2(S)}{\rightharpoonup} d_1 e_1^0 = \mathbf{M}^0 \cdot \cdot \mathbf{c}_1, \ e_1^0 \triangleq \mathbf{e}^0 \cdot \cdot \mathbf{c}_1.$$

We now expand the tensor $\mathbf{D}_0$ over the orthogonal basis $(\mathbf{c}_1, \mathbf{c}_2, \mathbf{c}_3)$ and arrive at the relationship for $\mathbf{M}^0$

$$\mathbf{M}^0 = \sum_{j=1}^{3} d_{ij}^0 \mathbf{c}_i \mathbf{c}_j \cdot \cdot \mathbf{e}^0 \ (d_{ij}^0 = d_{ji}^0, \ i, j = 1, 2, 3).$$

Contracting this expression with the tensor $\mathbf{c}_1$ and using the aforementioned, we get

$$d_1 e_1^0 = \sum_{j=1}^{3} d_{1j}^0 e_j^0,$$

from which, in view of the arbitrariness of $\mathbf{e}^0$, the given statement implies.

In particular, the mixture of the phases (5.26) has the tensor $\mathbf{D}_0$ of the following type

$$\mathbf{D}_0 = k \mathbf{a}_1 \mathbf{a}_1 + \mu_{20} \mathbf{a}_{20} \mathbf{a}_{20} + \mu_{30} \mathbf{a}_{30} \mathbf{a}_{30},$$

where $\mathbf{a}_{20}$, $\mathbf{a}_{30}$ are certain orthogonal deviators. The tensor $\mathbf{D}_0$ describes a medium with cubic symmetry. The Young moduli of such a medium are equal to each other in two mutually perpendicular directions [15]. For the construction of the set $G_m U$ it remains to find a set that is formed by the coefficients μ_{20}, μ_{30}.

Consider a function h of the type

$$h = \nabla\nabla w(1) \cdot \cdot \mathbf{D}(x) \cdot \cdot \nabla\nabla w(1) + \nabla\nabla w(2) \cdot \cdot \mathbf{D}(x) \cdot \cdot \nabla\nabla w(2),$$

representing a sum of densities of elastic energy of the plate for two tests $w(1)$ and $w(2)$-deflections, corresponding to different independent loadings.

Let us use quasi-affine functions of the following type

$$\varphi_5 = \nabla\nabla w(i) \cdot \cdot \mathbf{D}_{I_2}(x) \cdot \cdot \nabla\nabla w(i), \ i = 1,2 \ (\text{see } (5.12) \text{ and Example 5)}$$

$$\varphi_8 = \nabla\nabla w(1) \cdot \cdot \mathbf{O} \cdot \cdot \nabla\nabla w(2),$$

where $\mathbf{O} = \mathbf{a}_2\mathbf{a}_3 - \mathbf{a}_3\mathbf{a}_2$ (see (5.10)) is a tensor of the fourth rank projecting on a deviator subspace and realizing a rotation of a plane about an angle of $\pi/4$. The fact that the function φ is quasi-affine can be confirmed directly using the results of [48].

Indeed, the quadratic form φ_8 has the following type

$$\varphi_8 = [w_{xx}(1) - w_{yy}(1)]w_{xy}(2) - [w_{xx}(2) - w_{yy}(2)]w_{xy}(1)$$

and represents "null-Lagrangian" [48]; integrating by parts the relationship $\int_s \varphi_8 \, dxdy$ one can note that the result does not have surface terms.

Repeating the procedure of the previous section we arrive at the tensor inequality (5.22), where the block matrix $\mathbf{A}(\mathbf{D}, \alpha_{ij})$ has the type:

$$\mathbf{A}(\mathbf{D}, \alpha_{ij}) = \begin{pmatrix} \mathbf{D} + \alpha_{11}\mathbf{D}_{I_2} & \alpha_{12}\mathbf{O} \\ -\alpha_{12}\mathbf{O} & \mathbf{D} + \alpha_{22}\mathbf{D}_{I_2} \end{pmatrix}.$$

Here the elements-blocks of the matrix represent tensors of the fourth rank.

Coefficients of expansion of these tensors over the basis $\mathbf{a}_1, \mathbf{a}_{20}, \mathbf{a}_{30}$ form a (6×6)-matrix of the type:

$$\begin{pmatrix} k + \alpha_{11} & 0 & 0 & 0 & 0 & 0 \\ 0 & \mu^{(1)} - \alpha_{11} & 0 & 0 & 0 & \alpha_{12} \\ 0 & 0 & \mu^{(2)} - \alpha_{11} & 0 & -\alpha_{12} & 0 \\ 0 & 0 & 0 & k + \alpha_{22} & 0 & 0 \\ 0 & 0 & -\alpha_{12} & 0 & \mu^{(1)} - \alpha_{22} & 0 \\ 0 & \alpha_{12} & 0 & 0 & 0 & \mu^{(2)} - \alpha_{22} \end{pmatrix},$$

where $\mu^{(1)} = \mu^{(2)} = \mu_1$ for the matrix $\mathbf{A}(\mathbf{D}_1, \alpha_{ij})$, $\mu^{(1)} = \mu^{(2)} = \mu_2$ for the matrix $\mathbf{A}(\mathbf{D}_2, \alpha_{ij})$, and $\mu^{(1)} = \mu_{20}$, $\mu^{(2)} = \mu_{30}$ for the matrix $\mathbf{A}(\mathbf{D}_0, \alpha_{ij})$. The parameters α_{ij} are determined by the requirement of the nonnegativity of the matrices $\mathbf{A}(\mathbf{D}_1, \alpha_{ij})$, $\mathbf{A}(\mathbf{D}_2, \alpha_{ij})$, which leads to the conditions

$$\alpha_{11} \in [-k, \mu_1], \quad \alpha_{22} \in [-k, \mu_1], \quad \alpha_{12}^2 \le (\mu_1 - \alpha_{11})(\mu_1 - \alpha_{22}).$$

Choosing the parameters α_{ij} in the following way

$$\alpha_{11} = \alpha_{22} = -k, \quad \alpha_{12} = \mu_1 + k$$

and repeating the procedure of the previous section we obtain scalar estimate

$$1 + \frac{\mu_1 + k}{\mu_{20} - \mu_1} + \frac{\mu_1 + k}{\mu_{30} - \mu_1} \leq \frac{1}{m_2} \left(1 + 2 \frac{\mu_1 + k}{\mu_2 - \mu_1} \right). \qquad (5.27)$$

Similarly we obtain another estimate

$$1 + \frac{\mu_2 + k}{\mu_{20} - \mu_2} + \frac{\mu_2 + k}{\mu_{30} - \mu_2} \geq \frac{1}{m_1} \left(1 + 2 \frac{\mu_2 + k}{\mu_1 - \mu_2} \right). \qquad (5.28)$$

The inequalities (5.27) and (5.28) bound the invariants of the G-limit tensor of the mixture.

The sharpness of the obtained estimates is proved by constructing the matrix structure realizing them. In Figure 7 such a microstructure is shown. It has layers of the second rank characterized by the angle of 45° between the direction of the normals to layers of the first and second ranks. The estimate (5.27) corresponds to the case when the material $\mathbf{D}_1$ forms a matrix, the material $\mathbf{D}_2$ forms inclusions, and the estimate (5.28) corresponds to the opposite case.

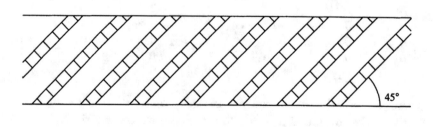

Figure 7. The microstructure realizing the elements of the boundary of the set $G_m U$ with respect to the operator of theory of bending plates; the phases have equal bulk moduli.

Remark 1.

The formulae for the effective elastic moduli can be obtained more easily if the initial phases are distinguished only by bulk moduli and have equal shear moduli μ. Similar to what we said previously, it is easy to conclude that the composite will be an isotropic medium regardless of its microstructure: the shear modulus of this medium is equal to μ and the bulk modulus k_0, as shown in [70], is determined only by the concentrations of phases and is expressed by the formula

$$k_0 = \frac{(k_1 + \mu)(k_2 + \mu)}{m_1(k_2 + \mu) + m_2(k_1 + \mu)} - \mu,$$

where k_1, k_2 are bulk moduli of the phases, and m_1, m_2 their concentrations in the mixture.

Remark 2.

Methods of the theory of the quasiconvexity were used in paper [99] for obtaining the estimates for the effective constants of isotropic elastic composites. These estimates turn out to coincide with known Hashin-Shtrikman inequalities [15] obtained by a different method. In paper [101] the comparative analysis of both methods was performed. As to a question about sharpness of the obtained estimates, it was answered earlier positively for the estimate of the bulk modulus [15]; sharpness of the boundaries in shear modulus was established only recently: in [104] it was done with the help of composite of infinite rank and in [99] — of a composite of finite rank.

5.4. *Two-dimensional polycrystal from the conductive phases [24, 69].*

Let a process be described by equations (3.1) and (3.2) on the plane $(n = 2)$, and U represent the set of tensors $\mathbf{D}$ with fixed eigenvalues d_1, d_2 and all possible orientations of the principal axes:

$$U = \{\mathbf{D} : \mathbf{D} = d_1\mathbf{e}_1\mathbf{e}_1 + d_2\mathbf{e}_2\mathbf{e}_2, \ \forall \mathbf{e}_1, \mathbf{e}_2 : \mathbf{e}_i \cdot \mathbf{e}_k = \delta_{ik}; \ i, k = 1, 2\}. \quad (5.29)$$

It is required to construct the set GU, that is, the set of tensors $\mathbf{D}_0 = \lambda_1\mathbf{e}_{10}\mathbf{e}_{10} + \lambda_2\mathbf{e}_{20}\mathbf{e}_{20}$ of effective heat conductances of all possible polycrystals assembled from the elements of U.

Consider two linearly independent functions $w(1)$ and $w(2)$ of the distribution of the temperature in an inhomogeneous medium assembled from different elements of the set U. Assume that these functions are generated by different sources f or boundary conditions. Let s be an index of subdivision of a domain S into parts occupied by different elements of the set U. As $s \to \infty$, the size of these parts is decreased infinitely so that a polycrystalline composite is formed and the following holds

$$w^s(1) \underset{W_2^1(S)}{\to} w^0(1), \ w^s(2) \underset{W_2^1(S)}{\to} w^0(2).$$

Because of examples 1 and 2 (Section 4.2) we have:

$$\det\left(\nabla w^s(1), \nabla w^s(2)\right) \underset{L_1(S)}{\to} \det\left(\nabla w^0(1), \nabla w^0(2)\right), \tag{5.30}$$

$$\det\left(\mathbf{D}^s \cdot \nabla w^s(1), \mathbf{D}^s \cdot \nabla w^s(2)\right) \underset{L_1(S)}{\to} \det\left(\mathbf{D}_0 \cdot \nabla w^0(1), \mathbf{D}_0 \cdot \nabla w^0(2)\right). \tag{5.31}$$

On the other hand, for (2×2)-matrices the following identity holds

$$\det\left(\mathbf{D} \cdot \nabla w^s(1), \mathbf{D} \cdot \nabla w^s(2)\right) = \det \mathbf{D} \cdot \det\left(\nabla w(1), \nabla w(2)\right). \tag{5.32}$$

But, according to the condition of the problem, the determinant of the matrix $\mathbf{D}^s$ is constant: $\det \mathbf{D}^s = d_1 d_2 = \text{const}(s)$; substitution of (5.32) into (5.31) with taking into account (5.30) shows that

$$\lambda_1 \lambda_2 = \det \mathbf{D}_0 = \det \mathbf{D}^s = d_1 d_2. \tag{5.33}$$

At the same time, the Reuss–Voigt estimates (3.21) imply that

$$\lambda_1 + \lambda_2 \le d_1 + d_2. \tag{5.34}$$

The formulae (5.33) and (5.34) completely characterize the set of invariants of $\mathbf{D}_0$. The first formula shows that in the process of intermixing the value of the second invariant $\lambda_1 \lambda_2$ of the tensor $\mathbf{D}_0$ remains the same regardless of the microstructure, and the second one reflects the fact that this process is irreversible: it leads to media with lower degree of anisotropy then the initial one. In particular, heat conduction of an isotropic polycrystal is determined by a formula [12]:

$$\lambda = \sqrt{d_1 d_2}$$

independently of its microstructure.

Let us note that in contrast with the example in Section 3.3 the set GU is characterized by the segment of hyperbola (5.33), (5.34) and not by the plane configuration as in Figure 2.

Remark 1.

As is obvious from the given proof, the result still holds under more general conditions of the set U: it is sufficient to assume that elements of this set have the same value of $\det \mathbf{D} = d_1 d_2$ and the eigenvalues d_1 and d_2 themselves can be different for different elements of the set U.

Remark 2.

All points of the set GU can be modeled by laminated composites of the first rank for which the normal to the layers coincides with the axes $\mathbf{e}_1$ and $\mathbf{e}_2$ in layers of the first and second type (Figure 8).

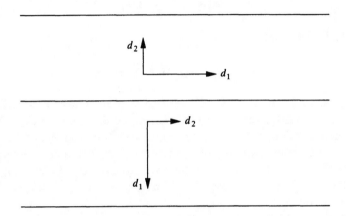

Figure 8.   The microstructure simulating points of the set GU in the problem of a two-dimensional polycrystal.

5.5. *G-closure of the set of anisotropic conductive phases (plane case).*

In this section we give the generalization of the results [33, 93] (see Section 3.4) for the case of a random set of anisotropic phases assembling a composite, that is, constructing the G-closure of such a set with respect to the operator (3.2) on the plane.

First, let us assume that the initial set U consists of two materials:

$$u = \{\mathbf{D}_1, \mathbf{D}_2\}, \ \mathbf{D}_1 = d_1^1 \mathbf{e}_1^1 \mathbf{e}_1^1 + d_2^1 \mathbf{e}_2^1 \mathbf{e}_2^1, \ \mathbf{D}_2 = d_1^2 \mathbf{e}_1^2 \mathbf{e}_1^2 + d_2^2 \mathbf{e}_2^2 \mathbf{e}_2^2,$$

where d_1^i, d_2^i $(d_2^i \geq d_1^i)$ are eigenvalues, and $\mathbf{e}_1^i$, $\mathbf{e}_2^i$ are the corresponding eigenvectors of the tensor $\mathbf{D}_i$ $(i = 1, 2)$. The tensor $\mathbf{D}_0$ of effective constants is characterized, as before, by the eigenvalues λ_1, λ_2 $(\lambda_1 \leq \lambda_2)$. Assume for definiteness that

$$d_1^1 d_2^1 \leq d_1^2 d_2^2.$$

Each of the initial anisotropic materials gives rise to a set of polycrystalline mixtures corresponding to hyperbolic segments on the plane (λ_1, λ_2) (see Section 5.4)

$$\lambda_1\lambda_2 = d_1^1 d_2^1, \ \lambda_1 + \lambda_2 \leq d_1^1 + d_2^1, \tag{5.35}$$

$$\lambda_1\lambda_2 = d_1^2 d_2^2, \ \lambda_1 + \lambda_2 \leq d_1^2 + d_2^2. \tag{5.36}$$

Let us show that the set GU is bounded by the following inequalities

$$d_1^1 d_2^1 \leq \lambda_1\lambda_2 \leq d_1^2 d_2^2. \tag{5.37}$$

Assume first that

$$d_1^1 \leq d_1^2, \ d_2^1 \leq d_2^2. \tag{5.38}$$

Next, consider some composite assembled of the materials $\mathbf{D}_1$, $\mathbf{D}_2$ and possessing the eigenvalues λ_1, λ_2. If we replace the material $\mathbf{D}_1$ with $\mathbf{D}_2$ preserving the orientation of the principal axes then the eigenvalues (and their product) will obviously not decrease. However, such replacement transfers the eigenvalues of the tensor $\mathbf{D}_0$ onto the curve (5.36). The right inequality (5.37) is thus proved, and the left one may be proved in an analogous manner.

Consider now the case

$$d_1^1 \leq d_1^2, \ d_2^1 \geq d_2^2. \tag{5.39}$$

The set GU may only become larger if we include in the set of initially given materials the third material $\mathbf{D}_3$ with the eigenvalues d_1^3, d_2^3; a new set of initial materials is denoted by $\overline{U}$. Let us choose the material $\mathbf{D}_3$ such that (cf. (5.38))

$$d_1^3 d_2^3 = d_1^2 d_2^2, \ d_1^1 \leq d_1^3, \ d_2^1 \leq d_2^3.$$

Then the material $\mathbf{D}_2$ may be considered as polycrystal made of the material $\mathbf{D}_3$, so $GU \subset G\overline{U}$. On the other hand, the set $G\overline{U}$ is bounded by the inequalities (5.37) since for the materials $\mathbf{D}_1$, $\mathbf{D}_3$ the previous reasoning applies.

Thus the inequalities (5.37) describe two components of the boundary of the set GU. Other components of the boundary are given by different analytical expressions for cases (5.38) and (5.39).

Consider two materials whose eigenvalues satisfy the inequalities (5.38) and recall that $d_1^1 \leq d_2^1$, $d_1^2 \leq d_2^2$, $\lambda_1 \leq \lambda_2$. As for the case of two isotropic phases (Section 3.4), the corresponding components of the boundary may be obtained from the consideration of the Reuss–Voigt inequalities, which in this case have the following form (see (5.6))

$$\lambda_2 \leq m_1 d_2^1 + m_2 d_2^2 = m_1 (d_2^1 - d_2^2) + d_2^2, \tag{5.40}$$

$$\lambda_1^{-1} \leq m_1 (d_1^1)^{-1} + m_2 (d_1^2)^{-1} = m_1 [(d_1^1)^{-1} - (d_1^2)^{-1}] + (d_1^2)^{-1}. \tag{5.41}$$

Bearing (5.38) in mind, we eliminate the concentration m from these inequalities $(0 \leq m_1 \leq 1)$; as a result we obtain the following relation

$$d_1^1 \leq \frac{d_1^1 d_1^2 (d_2^2 - d_2^1)}{d_1^2 d_2^2 - d_1^1 d_2^1 - (d_1^2 - d_1^1)\lambda_2} \leq \lambda_1 \leq \lambda_2 \leq d_2^2, \tag{5.42}$$

generalizing the inequalities (3.25) and reducing to them if $d_1^1 = d_2^1 = u_1$, $d_1^2 = d_2^2 = u_2$.

Composites corresponding to the identity

$$\lambda_1 = \frac{d_1^1 d_1^2 (d_2^2 - d_2^1)}{d_1^2 d_2^2 - d_1^1 d_2^1 - (d_1^2 - d_1^1)\lambda_2} \tag{5.43}$$

represent layered microstructures of the first rank; the normal to the layers coincides with the eigenvectors $\mathbf{e}_1^1$, $\mathbf{e}_1^2$ of tensors $\mathbf{D}_1$, $\mathbf{D}_2$, associated with their smallest eigenvalues d_1^1, d_1^2.

We get the set GU now as an intersection of the sets admitted by inequalities (5.37) and (5.42).

Let us turn now to the case when constants of initial materials are connected by inequalities (5.39); for this case elimination of the parameter m_1 from inequalities (5.40) and (5.41) is no longer possible and therefore the construction of the corresponding component of the boundary requires additional considerations. As shown by F. Murat (personal communication, 1985), the indicated component is given by the relationship[7]

$$\lambda_2 = \frac{d_2^1 d_2^2 (d_1^2 - d_1^1)}{d_2^2 d_1^1 - d_2^1 d_1^1 - (d_2^2 - d_2^1)\lambda_1}. \tag{5.44}$$

This equality connects the eigenvalues of a layered microstructure of the first rank with the normal to the layers coinciding with the common eigenvectors $\mathbf{e}_2^1$, $\mathbf{e}_2^2$ of the tensors $\mathbf{D}_1$, $\mathbf{D}_2$ associated in this case with their largest eigenvalues d_2^1, d_2^2.

---

[7]The analysis of the case (5.39) in papers [24, 69, 70] contains an error. The authors are indebted to F. Murat for that observation.

The set GU is now formed as an intersection of the sets admitted by the inequalities (5.37) and by

$$\lambda_2 \leq \frac{d_2^1 d_2^2 (d_1^2 - d_1^1)}{d_2^2 d_1^2 - d_2^1 d_1^1 - (d_2^2 - d_2^1)\lambda_1}. \tag{5.45}$$

The sets GU for the cases (5.38) and (5.39) are shown in Figures 9(a) and (b), respectively, in the form of curvilinear quadrangles $N_1 N_2 Q_2 Q_1$ symmetrically continued across the diagonal by the quadrangles $N_1' N_2' Q_2 Q_1$. Note the difference in shape of the curves $N_1 N_2$: in Figure 9a this curve is upward convex, in Figure 9b, downward. It is connected to the circumstance that in case (a) the curvilinear segment $N_1 N_2$ may be treated as a part of some longer curvilinear segment $P_1 N_1 N_2 P_2$ based on the points P_1, P_2 of a diagonal. These points correspond to two isotropic materials with some eigenvalues α, β that can be calculated using the equalities (5.40) and (5.41). Having those isotropic media at our disposal and assembling of them a layered composite of the first rank, one can obtain the eigenvalues corresponding to any point of the segment $N_1 N_2$ (Figure 9(a)). Similar reasoning is not appropriate for the segment $N_1 N_2$ on the Figure 9(b): there no longer exist isotropic media using which in a layered composite of the first rank it is possible to obtain corresponding pair of eigenvalues.

The obtained results may be illustrated graphically in the coordinates $(\lambda_1, \lambda_2^{-1})$ and $(\lambda_1^{-1}, \lambda_2)$ if we deny the earlier assumption $\lambda_1 \leq \lambda_2$. Any anisotropic material with eigenvalues d_1, d_2 is represented by two points passing to one another under the interchange of symbols d_1 and d_2: in coordinates $(\lambda_1, \lambda_2^{-1})$ these points are (d_1, d_2^{-1}) and (d_2, d_1^{-1}), and in coordinates $(\lambda_1^{-1}, \lambda_2)$, the points (d_1^{-1}, d_2) and (d_2^{-1}, d_1). It is clear that if the point (a, b^{-1}) belongs to the boundary of the set GU in the plane $(\lambda_1, \lambda_2^{-1})$, then the same also holds for the point (b, a^{-1}).

In coordinates $(\lambda_1^{-1}, \lambda_2)$ the material $\mathbf{D}_1$ is represented by two points N_1 and N_1' (Figure 10), the hyperbola (5.35) by a straight segment $N_1 N_1'$ passing through the origin; the material $\mathbf{D}_2$ is represented by the points N_2 and N_2' and the hyperbola (5.36) by a straight segment $N_2 N_2'$.

Now connect the points N_1 and N_2 as well as the points N_1' and N_2' by straight segments. As is obvious from (5.43), the segment $N_1 N_2$ represents layered composites of the first rank with normal to the layers coinciding with the directions of $\mathbf{e}_1^1$, $\mathbf{e}_1^2$ associated with the smallest eigenvalues of phases. The eigenvalue λ_1 corresponds to the principal direction of the tensor $\mathbf{D}_0$ normal to the layers. The segment $N_1' N_2'$ also represents layered composites of the first rank (see (5.44)), whose normal to the layers coincides with the directions of $\mathbf{e}_2^1$, $\mathbf{e}_2^2$ associated with the largest eigenvalues of phases.

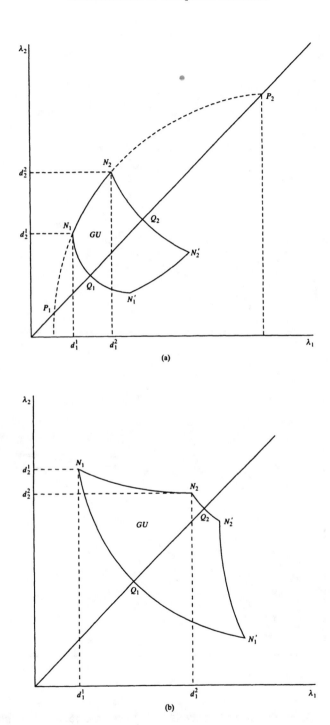

Figure 9. The sets GU generated by two anisotropic phases (plane case).

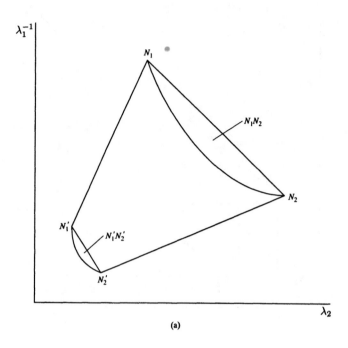

(a)

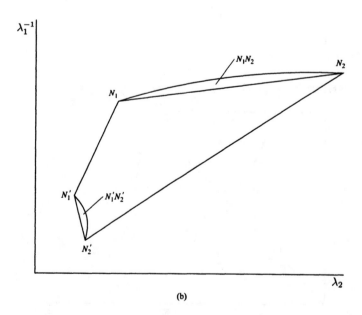

(b)

Figure 10. Construction of the set GU: (a) the case (5.38), (b) the case (5.39).

The normal to the layers then corresponds to the eigenvalue λ_2. The straight segment N_1N_2 represents a component of the boundary of GU in case (5.38), not (5.39), and the segment $N_1'N_2'$ is a component of the boundary of GU in case (5.39), not (5.38). Thus, for each of these two cases three sides of a quadrangle $N_1N_2N_2'N_1'$ form a part of the boundary of GU and the fourth side ($N_1'N_2'$ in case (5.38), N_1N_2 in case (5.39)) belongs to the inner part of GU. The reason is that the straight segments mentioned are not transformed to one another as the coordinates λ_1 and λ_2 interchange.

In order to complement the quadrangle to the set GU, one should build the images of the straight segments N_1N_2 and $N_1'N_2'$ resulting from that interchange operation. The segment N_1N_2 would then be mapped to a curvilinear segment $\widehat{N_1'N_2'}$ (Figure 10), and the segment $N_1'N_2'$ to a curvilinear segment $\widehat{N_1N_2}$. One of the obtained curvilinear segments is inside the quadrangle, the other outside. In case (5.38) segment $\widehat{N_1'N_2'}$ is outside the quadrangle, in case (5.39) the segment $\widehat{N_1N_2}$. The set GU for both cases is represented as a union of quadrangles $N_1N_2N_2'N_1'$ and $\widehat{N_1N_2}N_2'N_1'$.

Using these coordinates it is easy to build the set GU generated by an arbitrary set of initial materials in the plane. Let us prove that the set GU is obtained as a sum of the sets GU borne by any pair of elements of the initial set U considered as initial phases. It suffices to consider the case when the set U consists of three phases represented by the points N_1, N_2, N_3 (Figure 11). The aforementioned sum — the curvilinear polygon $N_1N_2N_3Q_3Q_2Q_1$ – obviously belongs to GU. We show that an inverse inclusion holds.

The polygon $N_1N_2N_3Q_3Q_2Q_1$ is a part of a polygon $N_1N_2N_4Q_3Q_2Q_1$ whose vertex N_4 is determined as a point of intersection of the prolonged segment N_1N_2 and the ray Q_3N_3. This polygon may be treated as the set $G\overline{U}$ borne by two phases represented by the points N_1 and N_4. But it is apparent that $GU \subset G\overline{U}$ since the initial material associated with point N_2 can be obtained as a layered composite assembled of phases N_4 and N_1, and the initial material N_3 as a polycrystal of material N_4. This implies that a curvilinear open polygon $Q_3Q_2Q_1N_1N_2$ is a part of the boundary of $G\overline{U}$; at the same time it represents a part of the boundary of the mentioned sum of sets GU borne by the pairs of phases (N_1, N_2), (N_2, N_3), and (N_3, N_1). Similarly one can prove that the previous reasoning holds also for an open curvilinear polygon $Q_1Q_2Q_3N_3N_2$ from which follows the desired result.

The set GU corresponding to any initial set U of anisotropic phases is now constructed as the least set containing U which is convex in coordinates $(\lambda_1, \lambda_2^{-1})$ and $(\lambda_1^{-1}, \lambda_2)$.

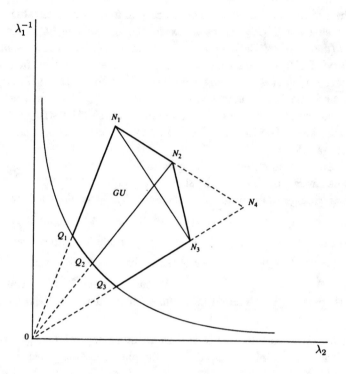

Figure 11. The sets GU generated by three anisotropic phases (plane case).

All points of the set GU may be realized by layered microstructures assembled only of those elements of the set U that belong to the boundary of the indicated convex set. Since the set GU contains composites of arbitrary microstructures, it follows, in particular, that any such material is equivalent to some layered composite in the sense of its effective properties. The boundary points of GU are composites of the first rank and the inner points can be modeled by composites of the second rank.

6. Solutions of some optimal design problems

In this section[8] solutions are given for optimal problems in their regularized statement. In Section 6.1 one solves the problem of an optimal, with respect to rigidity, torsional bar. In Section 6.4 one considers a problem of optimal design of an inhomogeneous heat conducting cylinder, the so-called "thermolens." In addition to the following problems, at the moment there

---

[8]Sections 6.2 and 6.3 are omitted in the translation. Enumeration of the other sections remains original (*editor's comment.*)

exist only a few solved regularized optimal design problems. Let us mention
the papers [61, 62], where the plastic bar design problem was solved, and
also the paper [92], for an interesting comparison of optimal design problems of elastic and plastic bodies. In the paper [34] the regularized control
was constructed for the problems with the operator (3.2). Problems with
similar statements have been solved in [39, 49, 80, 85, 88]; the examples of
regularization of one-dimensional optimal design problems can be found in
[80, 81, 14].

6.1. *Prismatic bar of extremal torsional rigidity.*

The problem (2.1) through (2.3) is considered here in its regularized
statement [23, 16].

Since, according to (2.4), the optimal project is assembled only from
zones where $u = u_1$ or $u = u_2$, the integral restriction can be rewritten in
the following form

$$\int_S u(x)\, dx = u_1 \int_S m_1(x)\, dx + u_2 \int_S m_2(x)\, dx \triangleq u_0 \operatorname{mes} S, \qquad (6.1)$$

where $m_1(x)$, $m_2(x) = 1 - m_1(x)$ are concentrations of materials u_1 and
u_2 at the point x.

As in the problem of Section 2.2, it is required to put these materials throughout the cross-section of the bar in such a way as to maximize
the functional (2.1). Along with this, the problem of minimization of the
functional is also considered.

According to the well-known variational principle [19], the functional
$-I$ equals

$$I = \min_{w \in \mathring{W}_2^1(S)} \int_S (\nabla w \cdot \mathbf{D} \cdot \nabla w - 4w)\, dx. \qquad (6.2)$$

To estimate the first term of the right-hand side of (6.2) one can use the
inequalities (5.4) and (5.5) that estimate the density of the energy $\nabla w \cdot$
$\mathbf{D} \cdot \nabla w$ for any given distribution of the concentrations $m_1(x)$, $m_2(x)$ in
the domain S; after that the only thing left is the problem about optimal
selection of the function $m_1(x)$.

Let us consider a problem of maximal rigidity. We minimize the lower
bound of compliance given by the second inequality (5.6). An augmented
functional, constructed taking into account this inequality, has the form
(recall that u denotes the compliance of the material):

$$J = \int_S [d_\| (\nabla w)^2 + \varkappa (u_2 - u_1) m_1]\, dx, \quad d_\| = [m_1 u_1^{-1} + (1 - m_1) u_2^{-1}]^{-1}, \quad (6.3)$$

where $\varkappa$ is the Lagrange multiple under restriction (6.1).

The conditions of stationarity with respect to the control $m_1(x) \in [0,1]$ have the following type

$$\begin{cases} m_1 = 0, & \text{if } (\nabla w)^2 \leq \varkappa \frac{u_1}{u_2}, \\[2mm] m_1 = 1, & \text{if } (\nabla w)^2 \geq \varkappa \frac{u_2}{u_1}, \\[2mm] d_\parallel^2 (\nabla w)^2 = \varkappa u_1 u_2, & \text{if } \varkappa \frac{u_1}{u_2} \leq (\nabla w)^2 \leq \varkappa \frac{u_2}{u_1}. \end{cases} \tag{6.4}$$

These inequalities show that as for the nonregularized problem (conditions (2.4)), domains of large values $(\nabla w)^2$ are occupied by a rigid material u_1, and domains of small values $(\nabla w)^2$ by a soft material u_2.

In the domain where the value $(\nabla w)^2$ belongs to the interval $(\varkappa u_1/u_2, \varkappa u_2/u_1)$, which is illegal for the nonregularized statement, an intermediate control regime appears, for which $0 < m_1 < 1$. This regime is realized by a layered microstructure; the layers are oriented perpendicular to the vector ∇w. Since the layers infinitely frequently follow each other, their separation line can be interpreted as a generalized curve filling everywhere dense a domain of nonzero measure; this agrees with intuitive ideas expounded in Section 2.2. In Figure 12 a numerically calculated [16] distribution of domains of different regimes for bars of square cross-section is presented for the values of parameters

$$m_0 \triangleq \frac{1}{\text{mes } S} \int_S m_2(x)\, dx = 0.25, \ u_1 = 0.5, \ u_2 = 1.5.$$

It is clear that domains of a soft material are concentrated in the center of the square and close to its angles; domains of a rigid material are situated near the middles of the sides. The layers in a zone of anisotropic regime are oriented so that they form a family of nested ovals surrounding the central domain. The concentration of a soft material in layers increases towards the center. Detailed numerical analysis of this problem was given in [100].

Remark.

In the regularized statement the Weierstrass conditions hold [23]. More precisely, these conditions hold for the value of ∇w averaged over the elementary volume; this value participates in Hooke's law for mixture. On the other hand, fields of stresses and strains that appear on a microscale within phases differ sharply from their averaged values, and are defined by equation (2.2) for the phases. The following question arises: won't the values in phases belong to a prohibited interval in which none of the sta-

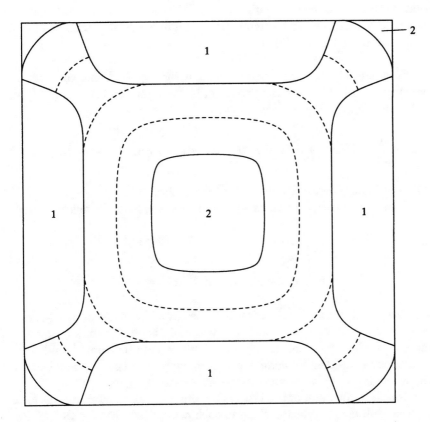

Figure 12. Distribution of materials that maximizes the rigidity of torsion bar of square cross-section.

tionary regimes is optimal? The answer to this question is negative [16]. Indeed, in the optimal layered composite the directions of the vectors ∇w_1 and ∇w_2 are the same in the first and second phases, and the absolute values of these vectors are expressed through averaged value ∇w by the formulae [16]:

$$(\nabla w_1)^2 = \frac{d_\parallel^2}{u_1^2}(\nabla w)^2, \qquad (\nabla w_2)^2 = \frac{d_\parallel^2}{u_2^2}(\nabla w)^2.$$

Substituting the optimal value $(\nabla w)^2$ from the equality (6.4) we obtain that the absolute values of the vectors ∇w_1 (∇w_2) in the intermediate regime have constant values

$$(\nabla w_1)^2 = \varkappa\frac{u_2}{u_1}, \qquad (\nabla w_2)^2 = \varkappa\frac{u_1}{u_2},$$

that coincide with the boundaries of a prohibited interval. The value of the averaged ∇w changes in the intermediate regime only as a consequence of change in the concentration.

Now consider the problem of minimal rigidity I. Using the first inequality (5.6) to estimate the first term in the right-hand side of (5.2), we write an augmented functional in the following form

$$ J = \int_S [d_\perp (\nabla w)^2 + \varkappa (u_2 - u_1)m_1]\, dx, \quad d_\perp = m_1 u_1 + (1 - m_1)u_2, \quad (6.5) $$

where $\varkappa$ is Lagrange multiplier under restriction (5.1).

The stationarity conditions with respect to m_1 have the following form

$$ \begin{cases} m_1 = 0, & \text{if } (\nabla w)^2 \geq \varkappa, \\ m_1 = 1, & \text{if } (\nabla w)^2 \leq \varkappa, \\ 0 < m_1 < 1, & \text{if } (\nabla w)^2 = \varkappa. \end{cases} \qquad (6.6) $$

It follows that in the domain of large values $(\nabla w)^2$ a material with greater compliance must be placed, and in the domain of small values $(\nabla w)^2$ – a material with lower compliance. In the intermediate zone one should place a composite choosing its compliance at every point in such a way so that the value of $(\nabla w)^2$ is constant everywhere in the domain.

For the realization of the last regime one can take a layered composite with moduli $d_\parallel$, $d_\perp$, placing them in such a way that the direction of ∇w coincides with the principal axis along which the compliance is equal to $d_\perp$, that is, orienting the layers parallel to the vector ∇w.

Note that although the layered composite is anisotropic, in reality, the material behaves as if it were isotropic with compliance $d_\perp$. This statement (which holds for the previous problem, too, where, however, the role of $d_\perp$ is played by $d_\parallel$), follows directly from representation (5.5) of the estimate for the energy. In Figure 13 the disposition of zones occupied by different materials and layered composites is represented for a bar of a square cross-section. Directions of layers are shown by dotted lines. This disposition was found numerically [16] for the range of parameters

$$ m_0 = 0.25, \quad u_1 = 0.5, \quad u_2 = 1.5. $$

Let us note that the problem of minimal rigidity of the bar under torsion was solved in [51] under the assumption that the set U of permissible isotropic materials $\mathbf{D} = u\mathbf{E}$ is characterized by the inequalities

$$ u_1 \leq u \leq u_2 \qquad (6.7) $$

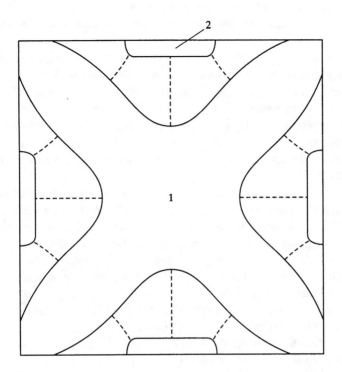

Figure 13. Distribution of materials that minimizes the rigidity of torsion bar of square cross-section.

and

$$\int_S u\,dx = u_0 \operatorname{mes} S, \ u_1 < u_0 < u_2, \tag{6.8}$$

that is, that in addition to materials with compliances u_1 and u_2 a designer also has materials with any intermediate values of compliance. In such a statement the existence of optimal distribution of compliances in the set U was proved in [51]. This example is interesting because of the fact that due to special (variational) character of the functional (2.1) and restrictions (6.7) and (6.8) it is possible to demonstrate the existence of the optimal control in the initial (not G-closed) set U.

If the set U consists of only two materials $u = u_1$ and $u = u_2$, then the intermediate regime of control appearing in the optimal layout (see (6.6)) actually includes the values of compliances u that belong to inner points of the interval (6.7). For comparison let us note that the problem of maximal rigidity with the relationships (6.7), (6.8) does not have a solution (see Section 2.2) when controls are unregularized.

The conclusions that were made in the preceding examples about coincidence of the directions of vector ∇w and one of the eigenvectors of the tensor $\mathbf{D_0}$ were based on the representations (5.4) and (5.5) of estimates of density of energy.

Using a special functional of rigidity I we directly obtained necessary minimal extensions MU of the set U of admissible controls: in the problem of minimal (maximal) rigidity the set MU is described as a set of isotropic media with compliance $u = d_\perp$ $(u = d_\parallel)$ depending on the concentration m_1. Let us emphasize that such materials cannot be realized from initial phases for any microstructure. However, layered composites that are correctly oriented resist the torsion in the same way as the mentioned isotropic materials.

Contrary to this the set $G_m U$, which is also the minimal extension, consists entirely of composites realized by particular microstructures. Hence if the set $G_m U$ is known (for this problem it is described in Section 5.2) then from the beginning it can be considered as the set of admissible controls. Formulae $(3.21)_2$, (5.24), and (5.25) describing $G_m U$ should be considered as additional relations when constructing the necessary conditions of optimality of the functional I with the constraint (6.1). In this case the inequalities $(3.21)_2$, (5.24), (5.25) determining $G_m U$ are used instead of the estimates of the energy (5.4) and (5.5).

In Section 6.4 this approach is applied to the problem of a heat flow through an inhomogeneous medium.

The relative gain, with respect to rigidity, of optimal designs in comparison to an homogeneous bar assembled from the same amounts of materials as the optimal bar, is presented in Table 5. The homogeneous bar was simulated by a double periodic system of inhomogeneous cells of a chess board type.

TABLE 5

#	u_2	u_1	m_0	$\frac{I_{max}-I_0}{I_0}\%$	$\frac{I_0-I_{min}}{I_0}\%$
1	1.5	0.5	0.25	43	16
2	1.5	0.5	0.5	63	17
3	1.5	0.5	0.75	53	10
4	4.5	0.5	0.5	266	–

In addition, the comparison of a bar of maximal rigidity with a bar whose inner part is made of soft material and the periphery from rigid material was undertaken in [16]. If the inner part represents a square then a relative loss with respect to the functional in comparison to the optimal design is 4% for $u_2 = 1.5$, $u_1 = 0.5$, $m_0 = 0.5$.

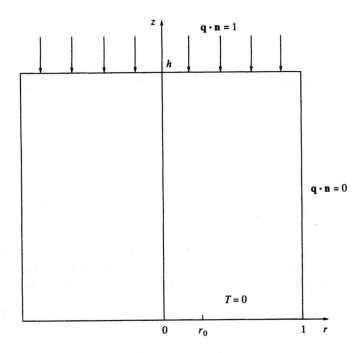

Figure 14. Regarding the statement of the "thermolens" problem.

6.4. The "thermolens" problem.[9].

Consider a circular cylinder $(0 \leq r \leq 1, 0 \leq z \leq h)$ (Figure 14) at the upper face of which there is applied the heat flux $-\mathbf{q} \cdot \mathbf{n}$ of unit intensity. Lateral surface of the cylinder is thermally insulated, and the lower face is kept under zero temperature.

The cylinder is filled by two materials characterized by the values u_1, u_2 of heat conductance. The process of heat transfer is described by a boundary value problem:

$$\mathbf{q} = \mathbf{D}(r, z) \cdot \nabla T, \ \nabla \cdot \mathbf{q} = 0, \ \mathbf{D}(r, z) = u_1 \mathbf{E}\chi_1(r, z) + u_2 \mathbf{E}\chi_2(r, z),$$
$$\mathbf{i}_z \cdot \mathbf{q}|_{z=h} = 1, \ \mathbf{i}_r \cdot \mathbf{q}|_{r=1} = \mathbf{i}_r \cdot \mathbf{q}|_{r=0} = 0, \ T|_{z=0} = 0, \quad (6.21)$$

where $\chi_1(r, z)$, $\chi_2(r, z)$ are characteristic functions of domains occupied by the first and the second material, and $\mathbf{i}_r$, $\mathbf{i}_z$ are the orts of the cylindrical coordinate system.

It is required to place these materials throughout the volume so that

---

[9] Section 6.2 and 6.3 are omitted; enumeration remains original.

a functional

$$I = \int_0^1 \rho(r)\mathbf{i}_z \cdot \mathbf{q}|_{z=0}\, r dr,$$

where $\rho(r) \in L_\infty(0,1)$ is a weight function, has the maximal value. If the weight function $\rho(r)$ has the form

$$\rho(r) = \begin{cases} 1, & 0 \le r \le r_0, \\ 0, & r_0 < r \le 1, \end{cases} \qquad (6.22)$$

then the problem becomes one of maximizing of the heat flux through a circular "window" of radius r_0 on the lower surface of the cylinder.

Assuming that the set of admissible controls contains initial materials as well as mixtures assembled from them, let us extend the set of controls to the set GU whose elements in this case are described by the formulae (3.25). Let us note that, since the overall quantity of materials is not fixed, introduction of sets G_mU is not necessary and it is sufficient to describe only the set GU.

The augmented functional of the problem now has the following form

$$J = \int_0^1 \rho(r)\mathbf{i}_z \cdot \mathbf{q}|_{z=0} r\, dr + 2\pi \int_0^1 \int_0^h \mu \nabla \cdot \mathbf{D}_0 \cdot \nabla T r\, dr dz. \qquad (6.23)$$

Varying (6.23) with respect to the variable T and taking into consideration the boundary conditions (6.22), we get a boundary value problem for the conjugate variable μ:

$$\nabla \cdot \mathbf{D}_0 \cdot \nabla\mu = 0, \ \ \mathbf{D}_0 = \lambda_1 \mathbf{e}_1 \mathbf{e}_1 + \lambda_2 \mathbf{e}_2 \mathbf{e}_2, \ \ \lambda_1, \lambda_2 \in (3.25),$$

$$(\mathbf{D}_0 \cdot \nabla\mu) \cdot \mathbf{n} = 0 \text{ when } z = h \text{ and } r = 1, \qquad (6.24)$$

$$\mu|_{z=0} = \rho(r).$$

Remark.

"Adjoint" boundary value problem (6.24) describes the distribution of "temperature" μ in the initial domain, generated by inhomogeneity on the lower face $z = 0$. We can see that it differs from the initial problem about the distribution of T; it gives an example of a non self-adjoint optimization problem [21]. In this respect this problem is more difficult than those described in the previous sections, where conjugate variables coincided with initial ones due to a specific form of the functional to be minimized. In the problem of Section 6.1 the potential energy was minimized (maximized) which allowed us to directly use the variational principle for the estimate of rigidity without introducing explicitly conjugate variables.

The controls are elements of the tensor $\mathbf{D}_0$, its eigenvalues λ_1, λ_2 and the angle of inclination of the principal axis $\mathbf{e}_1$ to axis Oz.

It is not difficult to establish (cf. [34]) that in the optimal project the eigenvalues λ_1, λ_2 belong to the boundary GU, that is, are connected by the relation (3.24) showing that in addition to initial materials layered composites also appear in the layout.

Thus the problem becomes one of determining two control functions — the angle φ of inclination of layers and the eigenvalue λ_2 of the tensor $\mathbf{D}_0$, this giving the concentration of phases in the composite: $\lambda_2 = m_1 u_1 + m_2 u_2$.

Necessary stationarity conditions of controls have the following form (assume that $u_1 < u_2$)

$$\nabla \mu \cdot \frac{\partial \mathbf{D}_0}{\partial \varphi} \cdot \nabla T = 0,$$

$$\nabla \mu \cdot \frac{d \mathbf{D}_0}{d \lambda_2} \cdot \nabla T = \begin{cases} 0, & \lambda_2 \in (u_1, u_2), \\ \geq 0 & \lambda_2 = u_1, \\ \leq 0, & \lambda_2 = u_2. \end{cases} \tag{6.25}$$

Expand vectors ∇T, $\nabla \mu$ over the eigenvectors $\mathbf{e}_1$, $\mathbf{e}_2$ of the tensor $\mathbf{D}_0$:

$$\begin{aligned} \nabla T &= |\nabla T|(\mathbf{e}_1 \cos \psi + \mathbf{e}_2 \sin \psi), \\ \nabla \mu &= |\nabla \mu|(\mathbf{e}_1 \cos \eta + \mathbf{e}_2 \sin \eta), \end{aligned} \tag{6.26}$$

where ψ and η are angles formed by vectors ∇T, $\nabla \mu$ with axis $\mathbf{e}_1$; recall (see (3.24)) that axis $\mathbf{e}_1$ coincides with a normal to layers forming the composite.

The first equation (6.25) shows that

$$(\lambda_2 - \lambda_1) \sin(\psi + \eta) = 0. \tag{6.27}$$

In the anisotropic regime of control ($\lambda_1 \neq \lambda_2$), as can be seen from (6.27), there exists an equality $\psi = -\eta + k\pi$, $k = 0, 1$, demonstrating that one of the principal axes of the tensor $\mathbf{D}_0$ divides an angle between vectors ∇T, $\nabla \mu$ in half. Selection of these regimes corresponding to maximum or minimum of the functional I has been made with a help of the Weierstrass condition [21], which in this case shows the existence of the equality $\psi = -\eta$. From the second condition (6.25), taking into consideration (3.24), (6.26), (6.27), we obtain

$$\lambda_1 = \lambda_2 = u_1, \qquad \cot^2 \psi \geq \frac{u_2}{u_1},$$

$$\lambda_1 = \lambda_2 = u_2, \qquad \cot^2 \psi \leq \frac{u_1}{u_2},$$

$$\frac{\lambda_1^2}{u_1 u_2} = \tan^2 \psi, \qquad \frac{u_1}{u_2} \leq \cot^2 \psi \leq \frac{u_2}{u_1}.$$

Thus, an optimal distribution of materials is characterized by a a zone of material of low conductivity if the directions of gradients ∇T and $\nabla \mu$ are close to each other, and by zone of material of high conductivity if the angle between directions of gradients ∇T and $\nabla \mu$ is close to π, and by an anisotropic zone if the directions of gradients form the angle close to $\frac{\pi}{2}$. In the last zone the normal to layers divides an angle between gradients in half; the optimal medium tries to turn the direction of the vector of heat flow in the right way.

The distributions of vector lines ∇T (problem (6.21)) and $\nabla \mu$ (problem (6.24)) are shown in Figure 15. It is assumed that the function $\rho(r)$ is defined by the equality (6.22). All vector lines ∇T begin on the lower face and end on the upper face (Figure 15); vector lines $\nabla \mu$ begin on the part of the lower face where $\rho = 0$ and end on the part where $\rho = 1$; exterior vector line $\nabla \mu$ passes a round the lateral surface and the upper face of the cylinder coming back to the lower face along axis Oz (Figure 15) [52].

It is not difficult to see (Figure 16) that the zone A, where the angle ψ is close to zero, adjoins the lateral surface of the cylinder. The zone C is located near the axis of the cylinder; here this angle is close to $\frac{\pi}{2}$. Zones A and C join at the end point of a "window" bb, through which one should pass the maximal quantity of heat; they are divided by the zone B where angle ψ is close to $\frac{\pi}{4}$; this zone adjoins the upper face. In the zone B layered composites are optimal; in Figure 16 the directions of layers are shown by the dotted line. Approaching to the curve bc (resp., ab) from the side of zone B the concentration m_1 of the material u_1 tends to zero (resp., to one).

Physically, the focusing action of the "thermolens" is due to the following effects: (i) from zone A the heat flow is forced out by low conductivity material u_1; (ii) zone B turns the lines ∇T to the necessary direction due to the refractional action of properly oriented layers; (iii) zone C occupied by high conductivity material u_2 concentrates lines ∇T in itself.

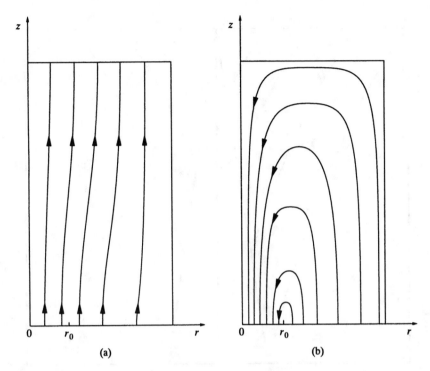

Figure 15. Distributions of vector lines; a) lines ∇T; b) lines $\nabla \mu$.

7. Conclusion

The described method of obtaining the estimates of the sets $G_m U$ can be extended to more complex elliptic operators encountered, for example, in the theory of elasticity. The procedure of constructing the estimates includes the following steps:

1. Determining the maximal number r of independent tests (linearly independent types of loading), defining the tensor $\mathbf{D}_0$ according to the constitutive law. This number is equal to the dimension of a linear space on which the tensor $\mathbf{D}_0$ acts. In the problem of bending of a plate as well as in the plane theory of elasticity this number is $r = 3$, and in the $3d$-problem of the theory of elasticity $r = 6$, because the tensor of elastic moduli acts on a space of symmetric (2×2) or (3×3) tensors.

2. Constructing a function $h(\chi, \mathbf{v})$ that is equal to a sum of energies[10]

---

[10]The energy means the quadratic form of dependent variables, variation of which leads to Euler's equations coinciding with differential equations of the problem.

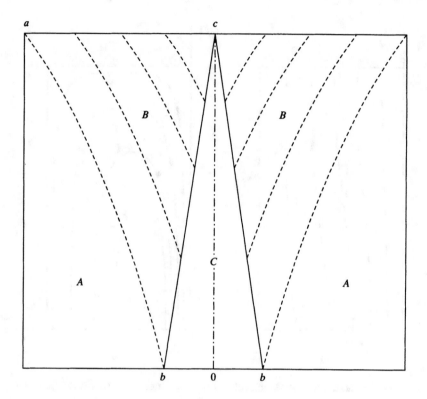

Figure 16. Optimal design of "thermolens."

corresponding to all linear independent tests (cf. (5.18))

$$h(\chi, \mathbf{v}) = \sum_{i=1}^{r} \mathbf{v}(i) \cdot \mathbf{D}(\chi) \cdot \mathbf{v}(i) \triangleq \mathbf{v} \cdot \mathcal{D}(\chi) \cdot \mathbf{v}.$$

Here χ represents a characteristic partition function, $\mathbf{D}(\chi)$ is determined by the relationship (3.24), r^2-vector $\mathbf{v} = (v(1), \ldots, v(r))$ is assembled from vectors $\mathbf{v}(i)$ of dimension r, from which every vector corresponds to an individual test (a stress/ strain state), and $(r^2 \times r^2)$-matrix $\mathcal{D}(\chi)$ is assembled as a direct sum of identical $(r \times r)$-blocks $\mathbf{D}(\chi)$:

$$\mathcal{D}(\chi) \triangleq \sum \otimes \mathbf{D}(\chi).$$

3. Determining quasi-affine and quasiconvex (but not convex) bilinear forms φ_i and ψ_j of an r^2-vector $\mathbf{v}$ (see Examples 1 through 7 in Section 4.2; methods of checking the quasi-convexity are given in [54]). Each of

these forms can be represented as

$$\varphi_i(v) = \mathbf{v} \cdot \boldsymbol{\Phi}_i \cdot \mathbf{v}, \ \psi_j(v) = \mathbf{v} \cdot \boldsymbol{\Psi}_j \cdot \mathbf{v},$$

where $(r^2 \times r^2)$-matrices $\boldsymbol{\Phi}_i$, $\boldsymbol{\Psi}_j$ are not definite.

4. Constructing a "polyconvex hull" of function

$$\lim_{\text{weak}} h(\chi, \mathbf{v})$$

(cf. (4.20)) with the help of the bilinear forms previously obtained. First, let us remark that due to the quasi-affinity of energy (cf. [75, 94])

$$\lim_{\text{weak}} \sum_{i=1}^{r} \mathbf{v}(i) \cdot \mathbf{D}(\chi) \cdot \mathbf{v}(i) = \sum_{i=1}^{r} \mathbf{v}^0(i) \cdot \mathbf{D}_0 \cdot \mathbf{v}^0(i) = \mathbf{v}^0 \cdot \mathcal{D}_0 \cdot \mathbf{v}^0,$$

where

$$\mathbf{v}^0 = (\mathbf{v}^0(1), \ldots, \mathbf{v}^0(r)), \ \mathcal{D}_0 \triangleq \sum \otimes \mathbf{D}_0.$$

Then the function $g_c(m, \mathbf{v}^0, \alpha, \beta)$, contained in (4.20), is built after the function

$$g(\chi, \mathbf{v}, \alpha, \beta) \triangleq \mathbf{v} \cdot [\mathcal{D}(\chi) + \sum_{(i)} \alpha_i \boldsymbol{\Phi}_i + \sum_{(j)} \beta_j \boldsymbol{\Psi}_j] \cdot \mathbf{v} \triangleq \mathbf{v} \cdot A(\mathbf{D}(\chi), \alpha, \beta) \cdot \mathbf{v}.$$

Let us determine the range ε of parameters $\alpha = \{\alpha_i\}$, $\beta = \{\beta_j\}$ by the relation

$$\varepsilon = \{(\alpha, \beta) : (\alpha, \beta) \in (4.19), \ A(\mathbf{D}(\chi), \alpha, \beta) \geq 0$$
$$\forall \mathbf{D}(\chi) = \sum_{i=1}^{N} \mathbf{D}_i \chi_i(x), \ \mathbf{D}_i \in U\}.$$

(Recall that the set U is determined as a set of material properties of phases making up the composite: $U = \{\mathbf{D}_1, \ldots, \mathbf{D}_N\}$).

Then the function $g_c(m, \mathbf{v}^0, \alpha, \beta)$ is determined as (cf. (5.20))

$$g_c(m, \mathbf{v}^0, \alpha, \beta) \triangleq \begin{cases} \mathbf{v}^0 \cdot \left[\sum_{i=1}^{N} m_i A^{-1}(\mathbf{D}_i, \alpha, \beta) \right]^{-1} \cdot \mathbf{v}^0, & \text{if } (\alpha, \beta) \in \varepsilon, \\ -\infty, & \text{if } (\alpha, \beta) \notin \varepsilon. \end{cases}$$

Applying the estimates (4.20) to the function $h(\chi, \mathbf{v})$ and taking into account the preceding remarks, we arrive at the inequality

$$\mathbf{v}^0 \cdot \mathcal{D}_0 \cdot \mathbf{v}^0 \leq \max_{\alpha, \beta \in \varepsilon} \mathbf{v}^0 \cdot \{\left[\sum_{i=1}^{N} m_i \mathbf{A}^{-1}(\mathbf{D}_i, \alpha, \beta)\right]^{-1} - \sum_{(i)} \alpha_i \mathbf{\Phi}_i - \sum_{(j)} \beta_j \mathbf{\Psi}_j\} \cdot \mathbf{v}^0.$$

5. From the obtained estimate for the quadratic forms and using arbitrariness of the vector $\mathbf{v}^0$ we get directly the estimate on the matrix $\mathbf{D}_0$.

6. Similarly, the estimate for tensors $\mathbf{D}_0^{-1}$ is obtained; for this, one should express the energy of each test in terms of a dual, with respect to $\mathbf{v}$, variables $\mathbf{v}^* = \mathbf{D} \cdot \mathbf{v}$. This estimate has the following form

$$\mathbf{D}_0^{-1} - \mathbf{Z}^{-1} \geq 0,$$

$$\forall \mathbf{Z}^{-1} : \mathbf{Z}^{-1} \in \cup_{\alpha^*, \beta^*} \{\left[\sum_{k=1}^{N} m_k (\mathbf{D}_k^{-1} + \sum_{(i)} \alpha_i^* \mathbf{\Phi}_i^* + \sum_{(j)} \beta_j^* \mathbf{\Psi}_j^*)^{-1}\right]^{-1} -$$
$$- \sum_{(i)} \alpha_i^* \mathbf{\Phi}_i^* - \sum_{(j)} \beta_j^* \mathbf{\Psi}_j^*\},$$

where $\mathbf{\Phi}_i^*$, $\mathbf{\Psi}_j^*$ are matrices corresponding to quasi-affine or quasiconvex but not convex quadratic forms of a dual vector $\mathbf{v}^*$; notations are similar to those used earlier.

7. In some cases one should also consider quadratic forms depending both on initial variables $\mathbf{v}$ (in number $r' < r$) and on canonically conjugate variables $\mathbf{v}^*$ (in number $r - r'$). The procedure of obtaining the estimates remains the same.

When $\alpha = \beta = \alpha^* = \beta^* = 0$, the estimates of paragraphs 5 and 6 turn into the inequalities of Reuss and Voigt, respectively; on the other hand, examples of Sections 5.2 and 5.3 represent sharp estimates that can be reached with the help of layered microstructures. One can expect that in other cases G- and G_m-closures would also be exhausted after a finite or infinite number of one-dimensional convexifications or, in other words, after formation of layers of finite or infinite rank. The particular role of layers is connected with the fact that they are the most anisotropic structures and so give an excellent possibility for manoeuvering through their internal parameters. From this point of view the layers represent extremal elements of microstructure and it is not surprising that all elements of G-closures are exhausted with their help.

Upon receiving the description of extended to G_m-closures sets of admissible controls the traditional technique of necessary conditions establishes rules of disposition of optimal media — elements of $G_m U$ — within the volume of construction. Thus the optimal construction has to be assembled not directly from initial materials but from composites previously made from them. An advantage of this approach can be observed in a numerical experiment because the program follows the evolution only of slow variables related to the problem in the large, that is, the varying concentrations of initial materials in the composite and its microstructure. As to the fast variables (i.e., those characterizing the chattering regime) they, similarly to the one-dimensional problem, are averaged in layers, and this averaging is carried out once and for all analytically. Knowing the averaged values of fast variables, it is not difficult to cover the distribution of initial materials on a microscale, that is, within the microstructure of a composite [29, 46].

We thus arrive at the conclusion that the main principle of optimal media design consists of the introduction of a hierarchy of lengthscales, that is, a hierarchy of fast variables. Here this principle is realized with the help of layered microstructures of different ranks, that is, layers assembled from layers. In all examples considered it was sufficient to use layers of some finite rank, that is, a finite number of lengthscales in the hierarchy.

The method of assembling composites described in this paper allows us to reach the main goal of design: material adapts itself at every point to the stress state that must arise in the optimal construction. This is achieved due to the creation of the necessary microstructure at every point that agrees with the initial intuitive ideas about optimal media.

References

[1] J. -L. Armand, *Application of the theory of optimal control of distributed-parameter systems to structural optimization*, Ph.D. Thesis, Department of Aeronautics and Astronautics, Stanford University, 1971

[2] N. V. Banichuk, *Shape optimization of elastic bodies*, (Russian) Nauka Moscow, 1980

[3] N. V. Banichuk, V. M. Kartvelishvili, A. A. Mironov, Numerical solution of two-dimensional optimization problems for elastic plates (Russian) *Izvestiya Akademii Nauk SSSR*, Mekhanika tverdogo tela, 1 1978

[4] N. S. Bakhvalov, Averaging of partial differential equations with rapidly oscillating coefficients, (Russian) *Doklady Akademii Nauk SSSR* **221** 3, 1975, 516–519

[5] V. L. Berdichevskii, *Variational principles of continuum mechanics* (Russian) Nauka, Moscow, 1983

[6] A. S. Bratus', Asymptotic solutions in problems of optimal control of coefficients in elliptic systems, (Russian) *Doklady Akademii Nauk SSSR*

259 5, 1981, 1035–1038

[7] A. S. Bratus', V. M. Kartvelishvili, Approximate analytic solutions in problems of optimizing the stability and vibrational frequencies of thin-walled elastic structures, *Izvestiya Akademii Nauk SSSR*, Mekhanika tverdogo tela, **16** 6, (Russian) 1981, 119–139

[8] J. Warga, *Optimal control of differential and functional equations*, Academic Press, New York, 1972

[9] R. V. Gamkrelidze, Optimal sliding regimes, (Russian) *Doklady Akademii Nauk SSSR* **143** 6, 1962

[10] L. V. Gibianskii, A. V. Cherkaev, *Design of composite plates of extremal rigidity*, (Russian) Preprint 914, A.F.Ioffe Physico-Technical Institute, Academy of Sciences of the USSR, Leningrad 1984

[11] V. B. Grinev, A. P. Filippov, *Optimization of structural elements with respect to mechanical characteristics*, (Russian) Naukova Dumka, Kiev, 1975

[12] A. M. Dykhne, Conductivity of a two-dimensional system, (Russian) *Zhurnal Eksperimental'noi i Teoreticheskoi Fiziki* **59** 7, 1970, 110–116

[13] V. V. Zhikov, S. M. Kozlov, O. A. Oleinik, Kha Thieng Ngoan, Homogenization and G-convergence of differential operators, (Russian) *Uspekhi Matematicheskikh Nauk* **34** 5, (209), 1979

[14] A. A. Zevin, I. M. Volkova, On the theory of optimal structural design with respect to vibrational frequencies, (Russian) in *Fluid mechanics and theory of elasticity*, Dnepropetrovskii Universitet, Dnepropetrovsk, 1982

[15] R. M. Christensen *Mechanics of composite materials*, Wiley, New York, 1979

[16] N. A. Lavrov, K. A. Lurie, A. V. Cherkaev, Non-homogeneous bar of extremal rigidity in torsion, (Russian) *Izvestiya Akademii Nauk SSSR*, Mekhanika Tverdogo Tela, 6, 1980

[17] J. -L. Lions, *Optimal control systems governed by partial differential equations*, Springer-Verlag, Berlin, 1971

[18] V. G. Litvinov, *Optimal control of coefficients in elliptic systems*, (Russian) Preprint 79.4 Institute of Mathematics, Academy of Sciences of the Ukrainian SSR, Kiev, 1979

[19] A. I. Lurie, *Theory of elasticity*, (Russian) Nauka, Moscow, 1970

[20] K. A. Lurie, On the optimal distribution of the resistivity tensor of the working medium in the channel of a MHD generator, (Russian) *Prikladnaya Matematika i Mekhanika* **34** 2, 1970

[21] K. A. Lurie, *Optimal control in problems of mathematical physics*, (Russian) Nauka, Moscow, 1975; transl. English; transl. *Applied optimal control theory of distributed systems*, Plenum Press, New York, 1993

[22] K. A. Lurie, A. V. Cherkaev, Prager theorem application to opti-

mal design of thin plates, (Russian) *Izvestiya Akademii Nauk SSSR,* Mekhanika Tverdogo Tela, 6, 1976

[23] K. A. Lurie, A. V. Cherkaev, Non-homogeneous bar of extremal torsional rigidity, (Russian) in *Nonlinear problems in structural mechanics; Structural optimization,* Kiev, 1978, 64–68

[24] K. A. Lurie, A. V. Cherkaev, *G*-closure of a set of anisotropic conductors in the two-dimensional case, (Russian) *Doklady Akademii Nauk SSSR* **259** 2, 1981

[25] K. A. Lurie, A. V. Cherkaev, Addition of the editors to the book of N.Olhoff, in *Optimal design of constructions (problems of vibration and loss of stability),* (Russian) Mir, Moscow, 1981

[26] K. A. Lurie, A. V. Cherkaev, Exact estimates of conductivity of mixtures composed of two materials taken in prescribed proportion (plane problem), (Russian) *Doklady Akademii Nauk SSSR* **264** 5, 1982, 1128–1130

[27] K. A. Lurie, A. V. Cherkaev, Effective characteristics of composites and problems of optimal design, (Russian) *Uspekhi matematicheskikh nauk* **39** 4 (238), 1984, p. 122

[28] K. A. Lurie, A. V. Cherkaev, Exact estimates of conductivity of a binary mixture of isotropic compounds, (Russian) Preprint-894 A.F.Ioffe Physico-Technical Institute, Academy of Sciences of the USSR, Leningrad, 1984

[29] K. A. Lurie, A. V. Cherkaev, Optimal design of elastic bodies and the problem of regularization, (Russian) in *Dynamics and strength of mechanisms* , 40/84 Vyscha shkola, Kharkov, 1984, 25–31

[30] E. L. Nikolai, Lagrange's problem about the most advantageous shape of columns, (Russian) *Izvestiya of S.-Petersburg's Polytechnic Institute,* 8, 1907

[31] N. Olhoff, *Optimal design of constructions (problems of vibration and loss of stability),* Mir, Moscow, 1981

[32] B. E. Pobedrya, *Mechanics of composite materials,* (Russian) Moscow State University, Moscow, 1984

[33] U. E. Raitums, Extension of extremal problems connected with linear elliptic equations, (Russian) *Doklady Akademii Nauk SSSR* **243** 2, 1978, 282–283

[34] U. E. Raitums, On optimal control problems for linear elliptic equations, *Doklady Akademii Nauk SSSR* **244** 4, 1979

[35] R. T. Rockafellar, *Convex Analysis,* Princeton University Press, Princeton, 1970

[36] J. Rychlewski, On Hooke's law, (Russian) *Prikladnaya matematika i mekhanika* **48** 3, 1984

[37] E. Sanchez-Palencia, *Non-homogeneous media and vibration theory,* Springer-Verlag, Berlin, New York, 1980

[38] V. A. Troitskii, L. V. Petukhov, *Shape optimization of elastic bodies,* (Russian) Nauka, Moscow, 1982

[39] A. V. Fedorov, A. V. Cherkaev, Search of an optimal orientation of the axes of elastic symmetry in an orthotropic plate, (Russian) *Izvestiya Akademii Nauk SSSR,* Mekhanika tverdogo tela **3**, 1983, 135–142

[40] A. F. Filippov, On certain questions in the theory of optimal control, (Russian) *Vestnik Moskovskogo Universiteta, Matematika i Astronomia* **2**, 1959, 25–32

[41] E. J. Haug, J. S. Arora, *Applied optimal design: mechanical and structural systems,* Wiley , New York, 1979

[42] T. D. Šermergor, *Theory of elasticity of micronon-homogeneous media,* (Russian) Nauka, Moscow, 1977

[43] I. Ekeland, R. Temam, *Convex analysis and variational problems,* North-Holland Pub. Co, Amsterdam, 1976

[44] L. S. Young, *Lectures on the calculus of variations and optimal control theory,* Saunders, Philadelphia, 1969

[45] J. -L. Armand, B. Lodier Optimal design of bending elements, *Int. J. Numer. Meth. Eng.* **13** 2, 1978, 373–384

[46] J. -L. Armand, K. A. Lurie, A. V. Cherkaev, Optimal control theory and structural design, in *Optimum Structure Design* vol. 2, eds., R. H. Gallagher, E. Atrek, K. Ragsdell, O. C. Zienkiewicz, Wiley, 1984

[47] J. M. Ball, Convexity conditions and existence theorems in nonlinear elasticity, *Arch. Ration. Mech. Anal.* **63** 4, 1977, 337–403

[48] J. M. Ball, J. C. Currie, P. J. Olver, Null Lagrangians, weak continuity and variational problems of arbitrary order, *J. Funct. Anal.* **41** 1981, 135–174

[49] M. Bendsoe, *Optimization of plates,* Mathematical Institute, The Technical University of Denmark, 1983

[50] D. A. G. Bruggemann, Berechnung verschiedener physikalischer Konstanten von heterogenen Substanzen, *Ann. d. Physik* **22** 1935, 636–679

[51] J. Cea, K. Malanowski, An example of a max-min problem in partial differential equations, *SIAM J. Control* **8** 2, 1970

[52] A. V. Cherkaev, L. V. Gibianskii, K.A. Lurie, *Optimum focusing of heat flux by means of a non-homogeneous heat-conducting medium,* The Danish Center for Applied Mathematics and Mechanics, Report No 305, 1985

[53] T. Clausen, Über die Formarchitektonischer Säulen *Bull. Phys. Math. Acad. St.-Petersburg* **9** 1851, 279–294

[54] B. Dacorogna, Weak continuity and weak lower semicontinuity of nonlinear functionals, *Lect. Notes Math.* **922**, 1982, Springer-Verlag, NY

[55] R. T. Haftka, B. Prasad, Optimum structural design with plate bending elements – a survey, *AIAA J.* **19** 1981, 517–522

[56] Z. Hashin, S. Shtrikman, A variational approach to the theory of the

effective magnetic permeability of multiphase materials, *J. Appl. Phys.* 33, 1962, 3125–3131

[57] E. J. Haug, J. Cea, eds., Optimization of distributed parameter structures, in *Proc. NATO ASI Meeting* **1, 2** Iowa City, 1980; Noordhoff 1981

[58] R. Hill, A self-consistent mechanics of composite materials, *J. Mech. and Phys. Solids* **13** 4, 213, 1965

[59] B. Klosowicz, K. A. Lurie, On the optimal non-homogeneity of a torsional elastic bar, *Arch. mech. stosow.* **24** 2, 1971, 239–249

[60] R. V. Kohn, G. Strang, Structural design optimization, homogenization and relaxation of variational problems, in *Macroscopic Properties of Disordered Media*, R. Burridge, G. Papanicolaou, S. Childress eds., Lect. Notes Phys., 154, Springer-Verlag, 1982

[61] R. V. Kohn, G. Strang, Optimal design for torsional rigidity, in *Hybrid and mixed finite element methods*, S. N. Alturi, R. H. Gallagher, O. C. Zienkiewicz, eds., Wiley, 1983, 281–288

[62] R. V. Kohn, G. Strang, Explicit relaxation of a variational problems in optimal design, *Bull. Amer. Math. Soc.* **9** 1983, 211–214

[63] R. V. Kohn, G. Strang, Optimal design and relaxation of variational problems, I, II, III, *Comm. Pure Appl. Math.* **39** 1986, 113–137, 139–182, 353–377

[64] R. V. Kohn, M. Vogelius, A new model for thin plates with rapidly varying thickness, *Int. J. Solids and Struct.* **20** 1984, 333–350

[65] J. -L. Lions, A. Bensoussan, G. Papanicolaou, *Asymptotic analysis for periodic structures*, North Holland, 1978

[66] V. G. Litvinov, N. G. Medvedev, Some optimal and inverse problems for orthotropic noncircular cylindrical shells *J. Optimiz. Theory and Appl.* **42** 2, 1984, 229–246

[67] K. A. Lurie, A. V. Cherkaev, A. V. Fedorov, Regularization of optimal design problems for bars and plates, I, II, *J. Optimiz. Theory and Appl.* **37** 1982, 499–543

[68] K. A. Lurie, A. V. Cherkaev, A. V. Fedorov, On the existence of solutions to some problems of optimal design for bars and plates, *J. Optimiz. Theory and Appl.* **42** 1984, 247–281

[69] K. A. Lurie, A. V. Cherkaev, *G*-closure of a set of anisotropically conducting media in the two-dimensional case, *J. Optimiz. Theory and Appl.* **42** 1984, 283–304

[70] K. A. Lurie, A. V. Cherkaev, *G*-closure of some particular sets of admissible material characteristics for the problem of bending of thin plates, *J. Optimiz. Theory and Appl.* **42** 1984, 305–315

[71] A. Marino, S. Spagnolo, Un tipo di approssimazione dell' operatore $\sum_{i,j} D_i(a_{ij}(x)D_j)$ con operatori $\sum_j D_j(b(x)D_j)$, *Annali Scuola Norm.*

Sup. Pisa, Sci. Fis. e Mat. **23** 1969, 657–673

[72] J. C. Maxwell, *A treatise on electricity and magnetism* **1**, Oxford 1904

[73] A. G. Michell, The limits of economy of material in frame structures, *Phil. Mag.* **6** 8, 1904, p. 589

[74] C. B. Morrey, Quasiconvexity and the lower semicontinuity of multiple integrals, *Pacific J. Math.* **2** 1952, 25–53

[75] F. Murat, Compacité par compensation, Partie I. Ann, Scuola Norm. Sup. Pisa, **5** 1978, 489–507; Partie II., in *Proc. Int. meeting on recent methods in non-linear analysis*, De Giorgi, Magenes, Mosco, Pitagora, eds., Bologna 1979, 245–256

[76] F. Murat, Control in coefficients, in *Encyclopedia of systems and control*, Pergamon Press, 1983

[77] F. Murat, L. Tartar, Calcul des variations et homogenéisation, in *Homogenization methods: theory and applications in physics*, (Bréau-sans-Nappe, 1983) Collect. Dir. Éudes Rech. Elec. France, 57, Eyrolles, Paris, 1985

[78] F. I. Niordson, *Optimal design of elastic plates with a constraint on the slope of the thickness function*, The Danish Center for Applied Mathematics and Mechanics, Report 225, 1981

[79] F. I. Niordson, N. Olhoff, *Variational methods in optimization of structures*, The Danish Center for Applied Mathematics and Mechanics, Report 161, 1979

[80] N. Olhoff, K. -T. Cheng, An investigation concerning optimal design of solid elastic plates, *Int. J. Solids and Struct.* **17**, 1981, 305–321

[81] N. Olhoff, K. A. Lurie, A. V. Cherkaev, A. V. Fedorov, Sliding regimes and anisotropy in optimal design of vibrating axisymmetric plates *Int. J. Solids and Struct.* **17**, 1981, 931–948

[82] N. Olhoff, J. E. Taylor, On structural optimization, *J. Appl. Mech* **50** 1983, 1139–1151

[83] P. J. Olver, Hyperjacobians, determinantal ideals and weak solutions to variational problems, *Proc. Roy. Soc. Edinburgh* 95A, 1983, 317–340

[84] W. Prager, Introduction to structural optimization *Int. Centre for Mechanical Sci., Udine* **212** 1974, Springer-Verlag, Vienna

[85] S. H. Rasmussen, *Optimering af fiberforstaerkede konstruktioner*, The Danish Center for Applied Mathematics and Mechanics, Rapport S12, 1979

[86] A. Reuss, Berechnung der Fliebgrenze vin Mischkristallen auf Grund der Plastizitätsbedingung für Einkristalle *ZAMM.* **9** 1, 49, 1929

[87] G. I. N. Rozvany, *Optimal design of flexural systems* Pergamon Press, Oxford, 1976

[88] G. I. N. Rozvany, N. Olhoff, K. -T. Cheng, J. E. Taylor, On the solids plate paradox in structural optimization, *J. Struc. Mech.* **10** 1982, 1–32

[89] A. Sawczuk, Z. Mróz, eds., Optimization in structural design, in *Proc. IUTAM Symp.*, Warsaw 1973; Springer-Verlag, Berlin, 1975

[90] K. Schulgasser, Relationship between single-crystal and polycrystal electrical conductivity, *J. Appl. Phys.* **47** 5, 1976

[91] G. Strang, *The polyconvexification of F(∇u)*, Australian Nat. Univ., Research Report CMA-RO9-83, 1983

[92] G. Strang, R. V. Kohn, Optimal design in elasticity and plasticity, *Int. J. Numer. Meth. Eng.* **22** 1, 1986, 183–188

[93] L. Tartar, Problèmes de contrôle des coefficients dans des équations aux dérivées partielles, *Lect. Notes Econ. and Math. Systems* **107** 1975, Springer-Verlag, 420–426

[94] L. Tartar, Compensated compactness and applications to partial differential equations, in *Non-linear Analysis and Mechanics, Heriot-Watt Symp.* IV, R. G. Knops, ed., Pitman Press, 1979

[95] L. Tartar, Estimations fines de coefficientes homogenéisés in *Ennio De Giorgi Colloquium*, (Paris, 1983), Res. Notes in Math. **125** 1985, Pitman, Boston, 168–187

[96] W. Voigt, *Lehrbuch der Kristallphysik*, Teubner, Berlin, 1928

[97] Z. Wasiutynski, A. Brandt, The present state of knowledge in the field of optimum design of structures, *Appl. Mech. Revs.* **16** 5, 1963, 341–350

[98] J. M. Ball, F. Murat $W^{1,p}$ – quasiconvexity and variational problems for multiple integrals, *Journ. Funct. Anal.* **58** 3, 1984, 225–253

[99] G. A. Francfort, F. Murat, Homogenization and optimal bounds in linear elasticity, *Arch. Rat. Mech. Anal.* **94** 4, 1986, 307–334

[100] J. Goodman, R. V. Kohn, L. Reyna, Numerical study of a relaxed variational problem from optimal design, *Comput. Methods Appl. Mech. Engrg.* **57**, 1, 1986, 107–127

[101] R. V. Kohn, G. Milton, On bounding the effective conductivity of anisotropic composites, in *Homogenization and effective moduli of materials and media*, J. L. Ericksen, D. Kinderlehrer, R. Kohn, J.-L. Lions, eds., Springer-Verlag, NY, 1986, 97–125

[102] K. A. Lurie, A. V. Cherkaev, Exact estimates of conductivity of composites, Proc. Roy. Soc. Edinburgh **A99** 1984, 1–2, 71–87

[103] K. A. Lurie, A. V. Cherkaev, Exact estimates of conductivity for a binary mixture of isotropic compounds, *Proc. Roy. Soc. Edinburgh* **A104** , 1986, 1–2, 21–38

[104] K. A. Lurie, A. V. Cherkaev, The problem of formation of an optimal isotropic multicomponent composites, Ioffe Inst. Report 895, A.F.Ioffe Physico-Technical Institute, Academy of Sciences of the USSR, Leningrad, 1984

[105] G. W. Milton, Modelling the properties of composites by laminates, in *Homogenization moduli of materials and media* (Minneapolis, Minn.,

1984/1985), *IMA Math. Appl.* **1**, 1986, Springer-Verlag, New York, Berlin, 150–174

[106] F. Murat, Contre-exemples pour divers problèmes on le contrôle intervent dans les coefficients, *Annali di Mat. Pura et Appl.* **4**, 1977, 112–113, 49–68

[107] A. N. Norris, A differential scheme for the effective moduli of composites, *Mech. Mat.* 4 1985, 1–16

[108] J. Sokolowski, Optimal control in coefficients for weak variational problems in Hilbert space, *Applied Mathemat. and Optimization* 7, 1981, 283–293

[109] S. Spagnolo, Convergence in energy of elliptic operators, in *In Proc. 3rd Symp. Numer. Solut. Part. Equations*, Hutbard, ed., Academic Press, NY, 1976, 469–498

[110] G. Strang, R. V. Kohn, The optimal design of two-way conductor, in *Topics in nonsmooth mechanics*, Birkháuser, Boston, 1988, 143–155 refs.tex

Submitted September 26, 1985

Appendix

Load characteristics of a MHD channel in the case of optimal distribution of the resistivity of the working medium[1]

K. A. Lurie and T. Y. Simkina

Consider a planar channel (Figure A1) of width 2δ whose walls are dielectric everywhere except at the two sections BB' of equal length 2λ along the walls of the channel opposite one another and made of a perfect conductor. The conducting sections are connected through the load R.

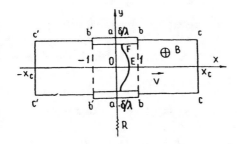

Figure A1

The working medium whose point-to-point variation in resistivity is given by a symmetric tensor function $\rho(x, y)$, is taken to move through the channel with velocity $\mathbf{v}(V, 0)$. If a magnetic field $\mathbf{B}(0, 0, -B(x))$,

---

[1]This Appendix presents an abridged translation of the paper [1] describing an implementation of the idea of artificially assembled anisotropic layouts in the optimal design of electrically conducting medium; in this capacity they were first introduced in [2]. The paper [1] along with the earlier publications [2,3] was aimed at determining the optimal distribution of the specific resistivity for the working medium in the channel of a magnetohydrodynamic power generator. This problem is motivated by the existence of eddy currents occupying the peripheral parts of the channel where the external magnetic field is fading to zero. These currents are confined within the channel and do not contribute to the net electric current through the external load. At the same time, a part of the kinetic energy of the working medium is wasted for their creation, and for this reason they represent losses and should be suppressed. To this end, it has been suggested controlling the distribution of specific resistivity of the working medium in the channel in order to make this distribution *optimal* with respect to the net current produced by the generator.

$B(x) = B(-x) \geq 0$ is applied, an electric current $\mathbf{j}(\zeta^1, \zeta^2)$ is generated in the channel, and a current I equal to

(A.1)
$$I = \int_{-\lambda}^{\lambda} \zeta^2(x, \pm \delta) dx$$

flows through the external resistance R. The distributions of the current $\mathbf{j}$ and the electric potential z^1 in the channel are described by the equations

(A.2)
$$\nabla \cdot \mathbf{j} = 0, \quad \boldsymbol{\rho} \cdot \mathbf{j} = -\nabla z^1 + \frac{1}{c}[\mathbf{v}, \mathbf{B}].$$

By introducing the flow function z^2 with the relation $\mathbf{j} = -\nabla \times \mathbf{i}_3 z^2$, we formulate the boundary conditions

(A.3)
$$\begin{aligned}
z^1(x, \pm \delta) &= z^1_\pm = const, \quad |x| < \lambda \\
z^2(x, \pm \delta) &= z^2_+ = const, \quad x > \lambda \\
z^2(x, \pm \delta) &= z^2_- = const, \quad x < -\lambda \\
z^1_+ - z^1_- &= (z^2_+ - z^2_-)R = IR.
\end{aligned}$$

In the longitudinal direction, the channel will be assumed bounded by a pair of vertical insulating walls CC and $C'C'$, at the distance $x = \pm x_c, x_c >$ λ from the coordinate origin, through which the fluid can penetrate; on these walls,

(A.4)
$$\zeta^1|_{|x|=x_c} = 0.$$

The eigenvalues ρ_1, ρ_2 of the tensor $\boldsymbol{\rho}$ of resistivity are defined by the formulas

$$\rho_1 = \infty, \quad \rho_2 = \rho_{\min}.$$

Under these circumstances, we look for the orientation of the main axes of $\boldsymbol{\rho}$ making the net current I maximum.[1]

---

[1]At the time of its original publication, the paper [1] appeared to be a part of a series of articles aimed at determining the optimal distribution of resistivity in the mhd channel flow. By that time it has already been realized (see [2,3]) that this distribution, (apart from being non-uniform) cannot generally be isotropic, and the appearance of substantially anisotropic regions was believed to be imminent. The required anisotropy could in principle be achieved on a local basis through a variety of means, among them, seeding into the stream such additives as sulfur hexafluoride or water vapor [4]. Another way to produce substantial anisotropy is the insertion of insulating baffles [5] into the flow. An assemblage of locally parallel baffles creates a system of infinitely thin canals for the electric current flow since there can be no such flow in the direction normal to the baffles. This construction was chosen because it permits immediate control over the direction of the current lines which is essential for the maximization of the net current I. Two isotropic materials participate in this type of layout: the original working fluid with the specific resistance $\rho_{\min}$, and the insulating material of baffles with the specific resistance $\rho_{\max} \to \infty$. If we take them, respectively, with the concentration rates $1 - m$ and m and allow for $m \to 0$, $\rho_{\max} \to \infty$, $m\rho_{\max} \to \infty$, then the entire construction will have effective resistances $\rho_1 = \infty$, $\rho_2 = \rho_{\min}$ across and along the baffles. These particular values were chosen as the principal resistances of the anisotropic compound originally introduced in [1-3].

2. As shown in [3], in the optimal range the current lines of **j** coincide with the level curves $L(f)$ of the function f defined as solution to the boundary value problem (Figure A1)

$$\Delta f = (1/2)\bar{\bar{B}}_{\bar{x}} \left[\int_{L(f)} \bar{\bar{B}} d\bar{y} \right]^{-1/2} = (1/2)\bar{\bar{B}}_{\bar{x}} \left[2\int_E^F \bar{\bar{B}} d\bar{y} \right]^{-1/2};$$

(A.5)
$$\int_0^{\delta/\lambda} f_{\bar{x}}|_{\bar{x}=0} d\bar{y} = \sqrt{\frac{1}{2}\frac{\delta}{\lambda}\bar{\bar{B}}(0)}; \quad f|_{\bar{x}=0} = 0;$$

(A.6) $\quad f|_{BC} = f|_{CC} = const = f_+; \quad f_{\bar{y}}|_{\bar{y}=\delta/\lambda, 0<\bar{x}<1} = 0; \quad f_{\bar{y}}|_{\bar{y}=0, 0<\bar{x}<\bar{x}_c} = 0;$

(A.7)
$$\bar{\bar{B}}(\bar{x}) = \bar{B}(\bar{x}) - \frac{\bar{I}\bar{R}\lambda}{2\delta}.$$

Here, we introduced the dimensionless variables and parameters $\bar{x}, \bar{y}, \bar{I}, \bar{B}, \bar{\bar{B}}, \bar{R}$ defined by the formulas (see Figure A2)

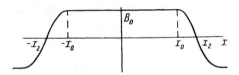

Figure A2

$$x = \bar{x}\lambda, \quad y = \bar{y}\lambda, \quad I = \bar{I}\frac{VB_0\lambda}{c\rho_{\min}}, \quad B = \bar{B}B_0, \quad \bar{\bar{B}} = \bar{\bar{B}}B_0, \quad R = \bar{R}\rho_{\min}.$$

The net current $\bar{I}$ is linked with f by the formula

(A.8)
$$\bar{I} = 2\sqrt{2}\int_0^{f_+} df \left[\int_E^F \bar{\bar{B}} d\bar{y} \right]^{1/2}$$

involving, as the rhs of (A.5), integration along the level curve EF of the function f.

The necessary condition for I to be maximum requires that

(A.9)
$$\bar{\bar{B}}(\bar{x}) \geq 0$$

everywhere in the channel.

3. We now describe the qualitative properties of the optimal load characteristics, i.e. the curves $\bar{I}(\bar{R})$ built for the optimal distribution of the tensor ρ of resistivity in a channel of the length $2x_c$ (Figure A1). In all cases we assume that the function $B(x) = B(-x)$ is constant for $x\epsilon[0, x_0]$, and for $x > x_0$ it decreases monotonically to the value zero taken at the point $x = x_2$, and for $x > x_2$ where x_2 may be greater than, equal to, or less than x_c it also decreases monotonically.

Equations (A.5)-(A.9) describe the distribution of current lines in a channel exposed to a magnetic field $B(x)$ and loaded by the resistance R. As shown in [3], these lines will be preserved in the channel as we assume it to be short-circuited and exposed to a modified external field $\tilde{B}(x)$. We formulate this as the *invariance property* with respect to the transition $(\bar{B}(x), \bar{R}) \to (\tilde{\bar{B}}(x), 0)$. Both the current lines and optimal current $\bar{I}$ will be preserved under this transformation, provided, of course, that Ineq. (A.9) holds.

These properties allow as to give a general description of the optimal load curve $\bar{I}(\bar{R})$ plotted for a fixed field function $\bar{B}(\bar{x})$. We shall assume that Jacobi's condition is satisfied everywhere in the channel [3].

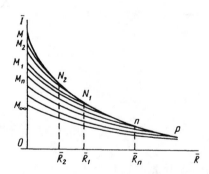

Figure A3

We take a channel of length $2\bar{x}_{c_1}$, with $\bar{x}_{c_1} \leq \bar{x}_2$; the curve $\bar{I} = \bar{I}_1(\bar{R})$ begins at the point M_1 (Figure A3). With increasing $\bar{R}$, the function $\bar{I} = \bar{I}_1(\bar{R})$ decreases monotonically; for relatively small $\bar{R}$, the value of $\bar{I}(\bar{R})$ can be calculated using the invariance of $\bar{I}$ under the replacement of the pair $(\bar{B}, \bar{R})$ by the pair $(\tilde{\bar{B}}, 0)$. At the same time, it is convenient to specify, not $\bar{R}$, but an effective field $\bar{B}(\bar{x})$, which differs from the given field $\bar{B}(\bar{x})$ by a constant: $\bar{B}(\bar{x}) = \tilde{\bar{B}}(\bar{x}) + \bar{b}$; if $\bar{I}$ is the corresponding current, then the resistance $\bar{R}$ is found from (see (A.7))

(A.10)
$$\bar{R} = \frac{2\bar{b}\delta}{\bar{I}\lambda}.$$

The critical value $\bar{R} = \bar{R}_1$ $(\bar{b} = \bar{b}_1)$ after which the above invariance cannot be used corresponds to vanishing of the effective field $\bar{\bar{B}}(\bar{x})$ at the point $\bar{x}_{c_1} : \bar{\bar{B}}(\bar{x}_{c_1}) = 0$ (the point N_1 in Figure A3). For $\bar{R} > \bar{R}_1$ $(\bar{b} > \bar{b}_1)$, optimization requires a reduction in the channel length, i.e., the filling of definite sections adjoining the ends $\mid \bar{x} \mid = \mid \bar{x}_{c_1} \mid$ of the channel with insulating matter (which is equivalent to moving the ends toward the center of the channel). On each occasion, with the specification of $\bar{b} \geq \bar{b}_1$, we must choose the channel length $2\bar{x}_c$ (we shall call this the *critical length*) in such a manner that the equation $\bar{\bar{B}}(\bar{x}_c) = \bar{B}(\bar{x}_c) - \bar{b} = 0$ holds for the given $\bar{b}$; the corresponding values of $\bar{R}$ are found in accordance with (A.10).[2] The points for which $\bar{R} > \bar{R}_1$ fill the part N_1P of the curve $\bar{I}_1(\bar{R})$ (Figure A3) to the right of N_1.

We now repeat the foregoing discussion, increasing the original length of the channel to a value $2\bar{x}_{c_2} > 2\bar{x}_{c_1}$, and leaving the original field $\bar{B}(\bar{x})$ as it was (as before, we assume $\bar{x}_{c_2} < \bar{x}_2$). The curve $\bar{I} = \bar{I}_2(\bar{R})$ begins at the point M_2 (Figure A3); obviously, $\bar{I}_2(0) > \bar{I}_1(0)$. Furthermore, the smallest critical value $\bar{b} = \bar{b}_2$ is less than $\bar{b}_1$; the corresponding value of $\bar{I}_2$ is not less than $\bar{I}_1(\bar{R}_1)$; (A.10) shows that $\bar{R}_2 < \bar{R}_1$. The point N_2 on the curve $\bar{I}_2(\bar{R})$ corresponds to the vanishing of the effective field $\bar{\bar{B}}(\bar{x})$ at $\bar{x} = \bar{x}_{c_2}$. For $\bar{b} > \bar{b}_2$ $(\bar{R} > \bar{R}_2)$, the length of the optimal channel (the critical length) is reduced, and $\bar{I}_2(\bar{R})$ decreases monotonically; for $\bar{R} > \bar{R}_1$ we have the same conditions as in the preceding case, when the original length of the channel was $2\bar{x}_{c_1}$. Therefore, for $\bar{R} > \bar{R}_1$, the curves $\bar{I}_2(\bar{R})$ and $\bar{I}_1(\bar{R})$ coincide.

Thus, the common part N_1P of the curves $\bar{I}_1(\bar{R})$ and $\bar{I}_2(\bar{R})$ is universal in the sense that it belongs to all curves corresponding to channels of length greater than $2\bar{x}_{c_1}$ for one and the same original field $\bar{B}(\bar{x})$.

Continuing this argument, we find that the section N_2P of the curve $\bar{I}_2(\bar{R})$ is universal, belonging to all curves constructed for channels of length greater than $2\bar{x}_{c_2}$, etc. As a result, we obtain the limiting universal curve MN_2N_1P (Figure A3); the point M obviously corresponds to the current $\bar{I}$ taken from the electrodes for $\bar{R} = 0$ from a generator of length $2\bar{x}_2$. The coordinates $(\bar{I}_n, \bar{R}_n)$ of every point n of this curve are such that the current $\bar{I}_n$ is taken from a short-circuited channel of corresponding critical length $2\bar{x}_{c_n}$ for which

$$\bar{\bar{B}}(\bar{x}_{c_n}) = \bar{B}(\bar{x}_{c_n}) - \frac{\bar{I}_n \bar{R}_n \lambda}{2\delta} = 0.$$

---

[2]Obviously, $\bar{x}_{c_1} > \bar{x}_c$ and $\bar{R} > \bar{R}_1$.

As the point n moves along the universal curve from the position M to infinity (the position P), the corresponding critical length $2\bar{x}_{c_n}$ decreases monotonically from $2\bar{x}_2$ to $2\bar{x}_0$. Considering a channel of length $2\bar{x}_{c_n}$, we can construct the curve $\bar{I}(\bar{R})$ for $\bar{R} < \bar{R}_n$. This part of the curve moves away from the point n to the left (Figure A3) and ends at the point M_n corresponding to the short-circuited channel of length $2\bar{x}_{c_n}$. The points lying on $M_n n$ correspond to optimal values of the current taken from a channel of length $2\bar{x}_{c_n}$ for $\bar{R}$ values in the interval $[0, \bar{R}_n]$. Thus, the universal curve MnP generates a one-parameter family of curves of the type $M_n nP$, which consists of a section $M_n n$ below the universal curve, and a section nP of the universal curve. No two curves $M_n n$ of this family have common points: assuming the contrary, we should have channels of different lengths loaded with the same resistances and generating the same optimal current, which is impossible if the function $B(x)$ is given by a graph of the type shown in Figure A2.[3]

As $\bar{R} \to \infty$ the universal curve tends asymptotically to the axis $\bar{R}$; if the point n goes to infinity ($\bar{R}_n \to \infty$), moving along the universal curve, then

$$\frac{\lambda}{2\delta} \bar{I}_n \bar{R}_n \to \bar{B}(x_0) = 1$$

and the curve $M_n n$ tends to the limiting position $M_\infty P$ corresponding to a channel of length $2\bar{x}_0$.

Having at our disposal the family of optimal curves constructed for given field $\bar{B}(\bar{x})$, we can find the optimal current taken from a channel of fixed original length $2\bar{x}_c < 2\bar{x}_2$ for different values of the load $\bar{R}$. Above all, it is necessary to find the point of the universal curve corresponding to the case when the length $2\bar{x}_c$ is critical. This point lies on the intersection of the universal curve and the hyperbola $\bar{I}\bar{R} = 2(\delta/\lambda)\bar{B}(\bar{x}_c)$. Suppose n is such a point and $\bar{I}_n$ and $\bar{R}_n$ are the corresponding critical parameters (see Figure A3). The critical current curve will be the curve corresponding to the family passing through the point n, i.e., the curve $M_n nP$; the channel length is $2\bar{x}_c$ for points of the branch $M_n n$ ($\bar{R} < \bar{R}_n$) and is less than $2\bar{x}_c$ for points of the branch nP; as $\bar{R} \to \infty$, the optimal length of the channel tends to $2\bar{x}_0$.[4] It is easy to see that there can exist only one point of intersection of the hyperbola $\bar{I}\bar{R} = 2(\delta/\lambda)\bar{B}(\bar{x}_c)$ and the universal curve. For sufficiently large $\bar{R}$, the hyperbola does not pass above the universal

---

[3] Touching of the curves of the families can also be ruled out since it contradicts Jacobi's condition.

[4] It follows from these arguments that if the functions $\bar{B}_1(x)$ and $\bar{B}_2(\bar{x})$ both are of the type shown in Figure A2 and differ only for $\bar{x} > \bar{x}_*$, then the corresponding universal curves and families will differ only for $\bar{R} < \bar{R}_*$, where the load $\bar{R}_*$ corresponds to the case when the length $2\bar{x}_*$ is critical.

curve since in the limit $\bar{R} \to \infty$ the limiting relation

$$\bar{I}\bar{R} \to \frac{2\delta}{\lambda}\bar{B}(\bar{x}_0) \geq \frac{2\delta}{\lambda}\bar{B}(\bar{x}_c)$$

is satisfied for the points of this curve. For sufficiently small $\bar{R}$, the hyperbola lies below the universal curve. This proves the possible existence of at least one point of intersection; if $\bar{x}_0 < \bar{x}_2 < \bar{x}_c$, the point is unique since otherwise, proceeding along the section $M_1 1$ corresponding to the channel of length $2\bar{x}_c$ (Figure A4, where continuous line represents the universal curve, and dotted line represents the hyperbola), and then along section 12 of the universal curve, we should arrive at the point 2, corresponding to a channel length $2\bar{x}_{c_2} < 2\bar{x}_{c_1}$, which contradicts the hypothesis.

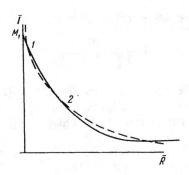

Figure A4

The restriction $2\bar{x}_c < 2\bar{x}_2$ adopted in the foregoing arguments cannot be dropped without violating the conditions of optimality; therefore, a change of the parameter $\bar{x}_c$ from $\bar{x}_0$ to $\bar{x}_2$ corresponds to the curves of the family generated by the universal curve.

It is interesting to elucidate the dependence of the shape of the universal curve and the curves of the family on the variation of parameters $\bar{x}_0$ and $\bar{x}_2$ of the function $\bar{B}(\bar{x})$ (see Figure A2). Suppose $\bar{x}_0$ is fixed and $\bar{x}_2 \to \bar{x}_0$; then the universal curve together with the curves of the family move downward (Figure A3), eventually coalescing with the limiting curve $M_\infty P$, which remains unchanged in this limit. If $\bar{x}_0 \geq 1$, then the equation of the limiting curve is given by ([6])

(A.11)
$$\bar{I} = 2\frac{\delta}{\lambda}\frac{\alpha^*}{1 + R\alpha^*}.$$

Parameter α^* is defined by

$$\alpha^* = \frac{K(\kappa')}{K(\kappa)}, \quad \kappa = \sqrt{1 - \kappa'^2} = dn\left(\frac{\lambda}{\delta} K(k), k'\right),$$

$$k^2 + k'^2 = 1, \quad \frac{K(k')}{K(k)} = \frac{\lambda}{\delta}\bar{x}_0,$$

where $K(k)$ is the complete elliptic integral of the first kind.

It is easy to show that as $\bar{x}_0$ increases from 1 to infinity, the parameter α^* increases monotonically from λ/δ to $\alpha^*_\infty = K(\kappa'_\infty)/K(\kappa_\infty)$, $\kappa_\infty^{-1} = \cosh(\pi\lambda/2\delta)$. In particular, for $\bar{x}_0 = \infty$ $[\bar{B}(\bar{x}) = const = 1]$, the limiting curve (A.11) takes the highest possible position on the plane $(\bar{I}, \bar{R})$ which is determined by the equation

(A.12)
$$\bar{I} = 2\frac{\delta}{\lambda}\frac{\alpha^*_\infty}{1 + \bar{R}\alpha^*_\infty},$$

while for $\bar{x}_0 = 1$ it takes the lowest possible position

(A.13)
$$\bar{I} = \frac{2}{1 + \bar{R}(\lambda/\delta)}$$

in the class of curves corresponding to functions $\bar{B}(\bar{x})$ with a region of constancy that encompasses the electrode region $|\bar{x}| < 1$. If $\bar{x}_0 < 1$, then the equation of the limiting curve $M_\infty P$ is given by ([7])

(A.14)
$$\bar{I} = 2\frac{\bar{x}_0}{1 + \bar{R}(\lambda/\delta)\bar{x}_0}.$$

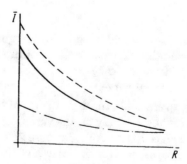

Figure A5

These arguments show that the universal curves $\bar{I}(\bar{R})$ together with the families they determine are located in the part of the $(\bar{I}, \bar{R})$ - plane bounded by the upper limiting universal curve (A.12) and the lower limiting universal curve (A.13) or (A.14) (Figure A5, where dotted line represents the upper limiting universal curve, continuous line represents the universal curve, and puncture line represents the lower limiting universal curve). The limiting universal curves correspond to functions $\bar{B}(\bar{x})$ that have no decreasing sections; therefore, the curves of the corresponding families coincide with the limiting curves themselves. This is confirmed by the fact that the hyperbola $\bar{I}\bar{R} = 2(\delta/\lambda)\bar{B}(\bar{x}_c)$, $\bar{x}_c < \bar{x}_2$, in the limiting cases $\bar{x}_2 = \bar{x}_0$ and $\bar{x}_0 = \infty$, has the equation $\bar{I}\bar{R} = 2(\delta/\lambda)$; this last equation determines a hyperbola that for any finite $\bar{R}$ lies above the upper limiting universal curve (A.12) and coincides with it only asymptotically in the limit $\bar{R} \to \infty$.

What we have presented here has been illustrated by a number of quantitative results for channels of different lengths and different loads. The function $\bar{B}(\bar{x})$ was given by

$$(A.15) \qquad \bar{B}(\bar{x}) = \begin{cases} 1, & |\bar{x}| \le 1 \\ e^{-(|\bar{x}|-1)/\alpha}, & |\bar{x}| \ge 1. \end{cases}$$

The parameter α was taken equal to 1.5; the values of parameters λ/δ and $2\bar{x}_c$ were varied. Numerical solutions of (A.5)-(A.7) were obtained with allowance for the invariance property described above; the functional $\bar{I}$ was calculated in accordance with (A.8).

Figure A6a shows the optimal curves of the current $\bar{I}(\bar{R})$ corresponding to the values $\lambda/\delta = 1$, $\bar{x}_c = 2.5$; the results of calculations are given in Table 1. Curve P is a section of the universal curve of the family; $M_\infty P$ is the limiting curve of the family corresponding to $\bar{x}_c = 1$ or, equivalently, the lower limiting curve corresponding to the value $\alpha = \infty$; this curve is defined by (A.13).

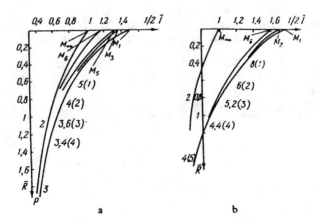

Figure A6

The critical values $2\bar{x}_c$ (the parameters of the family) are given next to the points of the universal curve from which the sections $M_1 1, \ldots, M_5 5$ of the curves of the family initiate.

Figure A6b and Table 2 represent analogous curves and calculated data for $\lambda/\delta = 0.5$, $\bar{x}_c = 4$.

It is of interest to compare the constructed optimal curves with the curves $\bar{I}(\bar{R})$ corresponding to nonoptimal control regimes. In Figure A8, we have plotted the dependences $\bar{I}(\bar{R})$ obtained from the numerical solution of the problem (A.1)-(A.4) for a scalar (isotropic) resistivity $\rho = \rho_{min}$ and parameters $\lambda/\delta = 1$, $\alpha = 1.5$; the curves are parametrized in terms of the channel length $2\bar{x}_c$. Each of the constructed curves lies to the left of the corresponding curve of the family shown in Figure A6a; in particular, the curve $M_1 P$ in this figure characterizes the gain in the current compared with the curve of Figure A7a, which corresponds to $2\bar{x}_c = 5$; the gain is maximal (5%) for $\bar{R} = 0$, equal to 3.5% for $\bar{R} = 1$, and tends monotonically to zero as $\bar{R} \to \infty$.[5]

Table 1

$\bar{x}_c$	1.2	1.2	1.2	1.2	1.2	1.2	1.2	1.4	1.5
b	0	0.1	0.15	0.2	0.25	0.776	0.872	0.766	0
$\bar{R}$	0	0.093	0.149	0.210	0.281	2.958	5.910	2.634	0
$1/2\bar{I}$	1.189	1.070	1.010	0.951	0.891	0.262	0.148	0.290	1.353
$\bar{x}$	1.50	1.5	1.5	1.5	1.7	1.8	1.8	1.8	1.8
b	0.2	0.4	0.617	0.715	0.627	0	0.2	0.35	0.485
$\bar{R}$	0.186	0.502	1.243	1.983	1.282	0	0.179	0.390	0.695
$1/2\bar{I}$	1.075	0.797	0.496	0.361	0.489	1.416	1.119	0.897	0.698
$\bar{x}$	1.8	2.0	2.0	2.0	2.50	2.5	2.5	2.5	
b	0.584	0	0.25	0.512	0	0.1	0.2	0.368	
$\bar{R}$	1/059	0	0.237	0.777	0	0.078	0.177	0.410	
$1/2\bar{I}$	0.552	1.431	1.053	0.659	1.490	1.286	1.132	0.902	

---

[5] The relative gain in the power is twice the relative gain in the current.

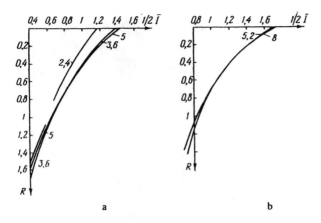

Figure A7

Table 2

$\bar{x}_c$	2.0	2.0	2.0	2.0	2.0	2.0*	2.2
$\bar{b}$	0	0.1	0.2	0.3	0.4	0.512*	0
$\bar{R}$	0	0.136	0.310	0.539	0.857	1.417*	0
$1/2\bar{I}$	1.650	1.471	1.292	1.113	0.934	0.723*	1.699
$\bar{x}_c$	2.2	2.2	2.2	2.2	2.2*	2.6	2.6*
$\bar{b}$	0.1	0.2	0.3	0.4	0.449*	0	0.344*
$\bar{R}$	0.132	0.302	0.531	0.848	1.052*	0	0.647*
$1/2\bar{I}$	1.511	1.323	1.130	0.943	0.852*	1.749	1.063*
$\bar{x}_c$	3.0	3.0*	4.0	4.0	4.0*	3.0	
$\bar{b}$	0	0.263*	0	0.06	0.135*	0.13	
$\bar{R}$	0	0.429	0	0.072	0.180*	0.173	
$1/2\bar{I}$	1.776	1.225	1.785	1.660	1.492*	1.506	

In Figure A7b, we have plotted the nonoptimal dependences of $\bar{I}(\bar{R})$ for $\delta = \delta_{\min} = const$, $\alpha = 1.5$, $\lambda/\delta = 1/2$; the parameter of the curves is again the length $2\bar{x}_c$ of the channel. Comparison of the curves of Figures A6b and A7b corresponding to $2\bar{x}_c = 8$ shows that the gain in the current is maximal (5.5%) for $\bar{R} = 0$, equal to 4.7% for $\bar{R} = 1$, and tends monotonically to zero as $\bar{R} \to \infty$; recalling the foregoing, we find that the gain increases as λ/δ decreases.

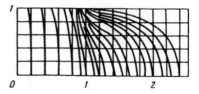

Figure A8

To conclude this section, we give graphs that represent the current lines in the channel for optimal anisotropic control range (Figure A8) and for nonoptimal isotropic control range (Figure A9)

$$\rho = const = \rho_{min}.$$

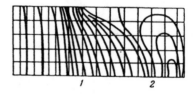

Figure A9

In both cases, we have chosen a channel of length $2\bar{x}_c = 5$ with load resistance $\bar{R} = 0.237$ and the field defined by (A.15) with $\alpha = 1.5$ (only a quarter of the channel is shown in either figure).

It is evident from the figures that the nonoptimal range exhibits well-developed current vortices (eddy currents), whereas there are no such vortices at all in the optimal range.

An exact optimal solution specifies the theoretically achievable maximum value of the cost functional I. Its practical implementation may be accomplished through the introduction of a system of horizontal baffles imitating, with certain degree of accuracy, the optimal pattern of the current lines.

REFERENCES

1. K.A. Lurie, T.Y. Simkina, *Load characteristics of a mhd channel in the case of optimal distribution of the resistivity of the working medium*, Magnitnaya Gidrodinamika (1972), No. 2, 82-90. (Russian).

2. K.A. Lurie, *On the optimal distribution of the resistivity tensor of the working medium in the channel of a mhd generator*, Prikladnaya Matematika i Mekhanika, (1970), **34**, No. 2. (Russian).

3. K.A. Lurie, *Optimal distributions of the resistivity tensor of the working medium in a mhd channel (the case of nonlocal variations)*, Prikladnaya Matematika i Mekhanika, (1971), **35**, No. 2. (Russian).

4. G.G. Cloutier, A.I. Carswell, *Plasma quenching by electro-negative gas seeding*, Physical Review Letters (1963), **10**, No. 8, 327-329.

5. A.B. Vatazhin, N.G. Nemkova, *Some two-dimensional problems of distribution of the electric current within a mhd generator channel with insulating baffles*, Prikladnaya Mekhanika i Tekhnicheskaya Fizika (1964), No. 2. (Russian).

6. A.B. Vatazhin, *Magnetohydrodynamic flow in a planar channel with finite electrodes*, Izvestiya Akademii Nauk SSSR, Otdelenie Tekhnicheskikh Nauk (1962), No. 1. (Russian).

7. K.A. Lurie, *Optimal distribution of the magnetic field along the channel of a mhd generator with electrodes of finite length*, Zhurnal Tekhnicheskoi Fiziki, (1970), **40**, No. 12. (Russian).

Microstructures of Composites of Extremal Rigidity and Exact Bounds on the Associated Energy Density[*]

L.V. Gibiansky and A.V. Cherkaev

Abstract. We study the elastic composites of extremal stiffness, i.e., periodic structures that provide minimal energy density in a given stress field (we call them composites of the minimal compliance or of the maximal stiffness). The composites are assembled from two isotropic materials taken in a prescribed proportion, and may be of an arbitrary microstructure.

Two-sided bounds for the energy density are obtained for composites of maximal stiffness and composites of minimal stiffness. One bound is an inequality valid for all structures independently of their geometry, and the second one corresponds to the energy stored in a composite with a specially chosen microstructure.

Porous media are studied in detail: microstructures of porous composites of maximal stiffness are explicitly determined for all possible external stress fields. Also, the structures of composites that possess minimal stiffness in any given external strains are explicitly found in the case when one of the constituents is absolutely rigid.

We discuss the link of obtained optimal microstructures with the composites that are optimal for the plane problem of elasticity; they have been obtained by the authors earlier.[1] The application of obtained results to the optimal design problems is discussed.

To solve the problem we develop and use the technique of construction of the quasiconvex envelopes for nonconvex integrands of the corresponding multidimensional variational problems.

Introduction.

Finding the extremal effective properties of composite materials is a problem that has recently attracted the attention of researchers in applied mathematics and mechanics. Knowing the quantities and properties of the materials (or phases) of which the composite is built, one can apply the methods of quasiconvexity theory to find bounds for the extremal effective characteristics of the composites [12, 13, 23–25]. The precision of the

---

[*]The present article is a translation of an article originally written in Russian and published as the report of *Ioffe Physico-Technical Institute, Academy of Sciences of USSR, Publication 1115, Leningrad, 1987.*

[1]This material is contained in Chapter 5 of this book.

bounds is then demonstrated by considering special composites for which the effective properties can be expressed analytically. The special structures considered in [14, 15] are families of confocal ellipsoids, whereas the authors of [1, 2, 5, 6, 10, 11, 16, 20, 22, 25] used the rank K laminates, that is, materials with layers that are themselves layered composites.

To date certain problems of this kind have already been solved using these methods. In particular, the set of effective characteristics of all possible microstructures of electricity (heat) conducting composite materials has been described in [5, 6, 25]. For the problems of elasticity theory such descriptions are not yet available due to the more complicated structure of rank four tensors that appear there. Nevertheless, the methods of quasiconvexity have allowed the authors of [22] to determine bounds for the effective moduli of isotropic elastic mixtures and to establish that the bounds for the bulk moduli coincide with the Hashin-Shtrikman bounds, which were obtained by a different method [14].

Along with the previously described problems, some optimal design problems and problems of determination of effective properties and structures of composites of extremal rigidity, or extremal elastic energy, were solved (see [2, 11, 16, 17], and also [1]. In [2], for example, the authors constructed exact bounds for the rigidity of a plate being bent or streched in its plane and presented the optimal structures of the corresponding composites, the choice of which depends on the stress state of the plate.

In this article we consider the problem of three-dimensional elastic composites of extremal rigidity, of which the problem of [2] is a special case. We apply methods of the theory of quasiconvexity to construct a lower bound E_p of the energy E of a composite that is valid for composites of any structure ($E_p \leq E$). At the same time we are interested in the minimal energy E_l of rank K laminate composites. The optimal composites are characterized by the energy E_q that satisfies $E_p \leq E_q \leq E_l$. For the asymptotic case of porous composites we show that the bounds E_p and E_l coincide and determine the analytical expressions for E_q. We also found explicitly the values of parameters of the structure of composites that are optimal for any stress state.

After these results were obtained, the authors learned of some interesting research conducted at the Courant Institute of Mathematical Sciences, USA. Using the ideas of the method of Hashin and Shtrikman [[]14], R. Kohn and R. Lipton [17] determined the bounds of E_p type for the energy of a composite constructed from two elastic incompressible phases and showed that the bounds are attainable by rank K laminate composites. In [11] M. Avellaneda generalized that result for the case of a composite built from arbitrary isotropic materials by considering a problem of minimizing the sum of energies under the action of different kinds of loading.

In §1 of this article we present the main equations; in §2 we formulate the problem of constructing the bounds for the maximal rigidity of a composite which is determined as a minimum of the energy under the action of a given mean stress field. In §3 we construct the lower bound for this quantity by using a method that generalizes the polyconvexity procedure, and in §4 we describe the properties of a set of special mixtures that we call Matrix Laminate Composites (see [10] and [22]). Analogous constructions are presented in §5 for the problem of finding a composite of minimal rigidity. Using the results of previous sections, in §6 we find the bounds for effective characteristics for arbitrary mixtures of two materials taken in given proportions. In §7 we solve the problem of finding a mixture of maximal rigidity consisting of two materials, one of which has zero elastic moduli (this is a problem of finding a porous body of maximal rigidity, i.e., the asymptotic case of a problem of previous sections). For this case we determine the structures of composites of maximal rigidity and corresponding values of the elastic energy density. Some of the cumbersome calculations are given in the Appendix. In §8 we consider the problem of composites of minimal rigidity for a case of one phase of a composite being infinitely rigid. In §9 we discuss the connection between the results of this article and the problem of a composite plate of maximal rigidity solved in [2]. In §10 we discuss the optimal design problem of a body of maximal rigidity. This is the case when one part of space occupied by a body is filled with the elastic material, and the other part is empty, or taken by a "zero moduli material."

Notation.

E is the density of elastic energy. The indices q, p, l refer to the minimal energy density and its lower and upper bounds, correspondingly.

V is the periodic cell of a spatially periodic composite ($meas(V) = 1$).

V_1, V_2 ($V_1 \cup V_2 = V$) are the parts of V filled with the first and second material.

m_1, m_2 are volume fractions of the phases V_1, V_2, $m_1 + m_2 = 1$.

$\mathbf{C}_1(\mathbf{D}_1 = \mathbf{C}_1^{-1})$, $\mathbf{C}_2(\mathbf{D}_2 = \mathbf{C}_2^{-1})$ are compliance (rigidity) tensors of the first and second materials.

κ, μ are the bulk and shear moduli of the material.

u is the vector of displacement.

ε, σ are the tensors of strain and stress.

$$\mathbf{C} \cdot \cdot \sigma = \mathbf{C}_{ijkl}\sigma_{lk},$$
$$\omega_\varepsilon = \frac{\varepsilon_1}{\varepsilon_3}, \eta_\varepsilon = \frac{\varepsilon_2}{\varepsilon_3},$$
$$\omega = \frac{\sigma_1}{\sigma_3}, \eta = \frac{\sigma_2}{\sigma_3},$$

$\varepsilon_1, \varepsilon_2, \varepsilon_3$ $(\sigma_1, \sigma_2, \sigma_3)$ are the eigenvalues of the strain (stress) tensor.

Part I. Two-sided bounds for effective rigidity.

§1 Consider a spatially periodic composite with a unit cubic cell V. Suppose that the cell is divided into two parts V_1 and V_2 $(V_1 \bigcup V_2 = V)$ filled with isotropic elastic materials with compliance tensors $\mathbf{C}_1(\kappa_1, \mu_1)$ and $\mathbf{C}_2(\kappa_2, \mu_2)$, respectively, where κ_1, κ_2 are bulk moduli and μ_1, μ_2 are shear moduli of the first and second materials. Let the volume fractions of V_1, V_2 be $m_1, m_2, (m_1 + m_2 = 1)$. Then the stressed state of each material (or phase) is described by:

$$\nabla \cdot \sigma = 0, \tag{1.1}$$

$$\varepsilon = \frac{1}{2}\left(\nabla u + (\nabla u)^T\right), \tag{1.2}$$

$$\varepsilon = \mathbf{C}_i(\kappa_i, \mu_i) \cdot\cdot\, \sigma. \tag{1.3}$$

where

σ is the 3×3 symmetric stress tensor;

ε is the 3×3 symmetric strain tensor (we assume that the elements of σ, ε are in $L_2(V)$);

u is the three dimensional vector of displacement;

i is the number of a part of a cell (or phase index);

$(\cdot\cdot)$ denotes contraction with respect to two indices;

$(\cdot)$ denotes contraction with respect to one index.

The moduli κ_i, μ_i are given in terms of Young's moduli E_i and Poisson's ratio ν_i by:

$$\kappa_i = \frac{E_i}{3(1 - 2\nu_i)}, \quad \mu_i = \frac{E_i}{2(1 + \nu_i)}, \tag{1.4}$$

assuming that $0 \le \nu_i \le 1/2$, we find:

$$3\kappa_i \ge 2\mu_i, \quad \kappa_i \ge 0, \quad \mu_i \ge 0, \quad i = 1, 2. \tag{1.5}$$

If i, j, k are the unit basis vectors in $\mathbb{R}^3$, we introduce the associated basis for symmetric tensors:

$$e_1 = \mathbf{i} \otimes \mathbf{i}, \qquad\qquad e_2 = \mathbf{j} \otimes \mathbf{j},$$

$$e_3 = \mathbf{k} \otimes \mathbf{k}, \qquad\qquad e_4 = \frac{1}{\sqrt{2}}\left(\mathbf{j} \otimes \mathbf{k} + \mathbf{k} \otimes \mathbf{j}\right), \tag{1.6}$$

$$e_5 = \frac{1}{\sqrt{2}}\left(\mathbf{i} \otimes \mathbf{k} + \mathbf{k} \otimes \mathbf{i}\right), \qquad e_6 = \frac{1}{\sqrt{2}}\left(\mathbf{j} \otimes \mathbf{i} + \mathbf{i} \otimes \mathbf{j}\right).$$

This basis is orthonormal in the sense of contraction $(\cdot\cdot)$, and the tensors σ, ε, $\mathbf{C}_i$ can be written in this basis as [4]:

$$\sigma = \sum_{s=1}^{6} \sigma^s e_s, \quad \varepsilon = \sum_{s=1}^{6} \varepsilon^s e_s, \quad \mathbf{C}_i = \sum_{s,t=1}^{6} C_i^{st} e_s \otimes e_t$$

where

$$\sigma^s = \sigma \cdot\cdot e_s, \quad \varepsilon^s = \varepsilon \cdot\cdot e_s, \quad C_i^{st} = e_s \cdot\cdot \mathbf{C}_i \cdot\cdot e_t.$$

For isotropic materials, the 6×6 matrix $\mathbf{C}_i(k_i, \mu_i)$ is:

$$\mathbf{C}_i(\kappa_i, \mu_i) = \begin{pmatrix} \frac{3\kappa_i + \mu_i}{9\kappa_i \mu_i} & \frac{2\mu_i - 3k_i}{18\kappa_i \mu_i} & \frac{2\mu_i - 3\kappa_i}{18\kappa_i \mu_i} & 0 & 0 & 0 \\ \frac{2\mu_i - 3\kappa_i}{18\kappa_i \mu_i} & \frac{3\kappa_i + \mu_i}{9\kappa_i \mu_i} & \frac{2\mu_i - 3\kappa_i}{18\kappa_i \mu_i} & 0 & 0 & 0 \\ \frac{2\mu_i - 3\kappa_i}{18k_i \mu_i} & \frac{2\mu_i - 3\kappa_i}{18\kappa_i \mu_i} & \frac{3\kappa_i + \mu_i}{9\kappa_i \mu_i} & 0 & 0 & 0 \\ 0 & 0 & 0 & \frac{1}{2\mu_i} & 0 & 0 \\ 0 & 0 & 0 & 0 & \frac{1}{2\mu_i} & 0 \\ 0 & 0 & 0 & 0 & 0 & \frac{1}{2\mu_i} \end{pmatrix}. \qquad (1.7)$$

We assume that the stressed state is described by functions periodic along every axis of the cell with period equal to one. Denote the mean strain and stress in a cell by:

$$\varepsilon_0 = \int_V \varepsilon dV, \quad \sigma_0 = \int_V \sigma dV.$$

The tensor $\mathbf{C}_0$ that relates σ_0 to ε_0,

$$\varepsilon_0 = \mathbf{C}_0 \cdot\cdot \sigma_0, \qquad (1.8)$$

is called the effective tensor of the composite [3,9]. It depends on the tensors $\mathbf{C}_1$, $\mathbf{C}_2$, volume fractions m_1, m_2, as well as on the geometry of V_1, V_2 but it does not depend on the stress state. Consider composites that can be made of materials with compliance tensors $\mathbf{C}_1$, $\mathbf{C}_2$ taken in volume fractions m_1, m_2. The set of tensors $\mathbf{C}_0$ of all such composites is called the G_m-closure [5] of the set $U = \{\mathbf{C}_1, \mathbf{C}_2\}$ of given compliance tensors and is denoted by $G_m U$.

§2 The goal of this paper is to determine particular configurations of phases V_1 and V_2 which would yield a composite of maximal or (minimal) rigidity given that the mean stress (or strain) field in the composite is known. We will call a composite of maximal rigidity a material whose microstructure is such that the density of elastic energy E due to a given mean stress field σ_0 is minimal:

$$E_q = \min_{C_0 \in G_m U} E(\mathbf{C}_0), \tag{1.9}$$

$$E(\mathbf{C}_0) = \sigma_0 \cdot\cdot \mathbf{C}_0 \cdot \sigma_0 = \varepsilon_0 \cdot \cdot \sigma_0.$$

Since the elastic energy equals the work necessary to deform the material, the functional E_q characterizes maximal rigidity of a cell because the minimal strain ε_0 is associated with a given force σ_0. The tensor $\mathbf{C}_0 = C_q$ that minimizes (1.9) depends on the external field σ_0, that is, $C_q = C_q(\sigma_0)$. We construct a set MU of tensors C_q corresponding to all possible fields σ_0. We also describe the corresponding microstructures. To solve the problems (1.9) we will use the principle of the minimum of the complementary energy [4], which states that given the fixed partition of a cell into phases V_1, V_2 and a mean stress field σ_0, the stress state of a cell V is found by solving a variational problem:

$$E(\chi_i) = \inf_{\sigma_0 \in \Sigma} \int_V E(\chi_i, \sigma) dV. \tag{1.10}$$

Here

$$E(\chi_i, \sigma) = \sigma \cdot \cdot (\chi_1 C_1 + \chi_2 C_2) \cdot \cdot \sigma; \tag{1.11}$$

χ_i are the characteristic functions of the regions V_i:

$$\chi_i(x) = \begin{cases} 1, & \text{if } x \in V_i, \\ 0, & \text{if } x \notin V_i, \end{cases}$$

which satisfy

$$\chi_1(x) + \chi_2(x) = 1, \ \forall x \in V \tag{1.12}$$

$$\int_V \chi_i dV = m_i;$$

and the set Σ in (1.10) is:

$$\Sigma = \left\{ \sigma : \ \sigma = \sigma^T, \ \nabla \cdot \sigma = 0, \right.$$

$$\left. \int_V \sigma dV = \sigma_0, \ \sigma(x) \text{ is spatially periodic with period } 1 \right\}. \tag{1.13}$$

Thus the functional E_q is:

$$E_q = \inf_{\chi_i} E(\chi_i) = \inf_{\chi_i} \inf_{\sigma \in \Sigma} \int_V E(\chi_i, \sigma) dV. \qquad (1.14)$$

Since χ_i are either 0 or 1, the functional E_q is not weakly lower semicontinuous and to solve (1.14) it is necessary to construct a relaxed problem by changing the integrand $E(\chi_i, \sigma)$ to its quasiconvex envelope $E_q(m_i, \sigma_0)$ [13,16]. In other words, it is necessary to construct such a function E_q, that:

(1) $\int_V E_q(m_i, \sigma) dV$ is lower semi-continuous;
(2) $Eq(m_i, \sigma)$ is equal to the infimum in (1.14): minimum:

$$\inf_{\chi_i} \inf_{\sigma \in \Sigma} \int_V E_q(\chi_i, \sigma) dV = \min_{\sigma \in \Sigma} \int_V E_q(m_i, \sigma) dV = E_q(m_i, \sigma).$$

To construct E_q, we will establish the two-sided bounds:

$$E_p \leq E_q \leq E_l, \qquad (1.15)$$

with E_p, E_l being constructed in a systematic way. When the lower bound and the upper bound coincide in (1.15), the energy E_q is determined by

$$E_p = E_q = E_l.$$

The lower bound E_p is independent of the cell geometry, whereas E_l is a minimum of energy density over a set of specially chosen composites.

§3 As a lower bound of the quasiconvex envelope $E_q(m_i, \sigma)$ we can use the convex (in both arguments χ_i, σ) envelope E_c of the function $E_q(\chi_i, \sigma)$. For given $\mathbf{C}_1 \geq 0, \mathbf{C}_2 \geq 0$ this envelope is [8]:

$$E_c = \sigma_0 \cdot \cdot (m_1 \mathbf{C}_1^{-1} + m_2 \mathbf{C}_2^{-1}) \cdot \cdot \sigma_0 \leq E_q \qquad (1.16)$$

This bound can be improved if we take into account the differential restrictions (1.1) which ensure that the integrals of certain nonconvex functions $\phi(\sigma)$ are lower semicontinuous.

Lemma 1. *Consider the quadratic form $\phi(\sigma) = \sigma \cdot \cdot \Phi \cdot \cdot \sigma$, where the* *tensor Φ in the basis* (1.6) *is defined by*

$$\Phi = \Phi(a_1, a_2, a_3)$$

$$= \begin{pmatrix} a_1^2 & -a_1 a_2 & -a_1 a_3 & 0 & 0 & 0 \\ -a_1 a_2 & a_2^2 & -a_2 a_3 & 0 & 0 & 0 \\ -a_1 a_3 & -a_2 a_3 & a_3^2 & 0 & 0 & 0 \\ 0 & 0 & 0 & a_2^2 + a_3^2 & 0 & 0 \\ 0 & 0 & 0 & 0 & a_1^2 + a_3^2 & 0 \\ 0 & 0 & 0 & 0 & 0 & a_1^2 + a_2^2 \end{pmatrix}.$$

This $\phi(\sigma)$ is A-quasiconvex [13] *if the symmetric 3×3 tensor σ satisfies the differential restrictions* (1.1).

Proof: In [25] it was shown that the function ϕ' of an $n \times n$ matrix U defined by

$$\phi'(\sigma) = -(\mathbf{Tr}U)^2 + (n-1)\mathbf{Tr}(U \cdot U^T) \tag{1.17}$$

is A-quasiconvex, when the elements u_{ij} satisfy the following differential restrictions

$$\sum_{i=1}^{n} \frac{\partial U_{ij}}{\partial x_i} \text{ is bounded in } L_2, j = 1, \dots, n.$$

Putting $n = 3$, $U = T \cdot \sigma$, with T being an arbitrary diagonal matrix:

$$T = \begin{pmatrix} a_1 & 0 & 0 \\ 0 & a_2 & 0 \\ 0 & 0 & a_3 \end{pmatrix}$$

and computing the coefficients of the quadratic form $\phi'(T \cdot \sigma)$ in the basis (1.6) we get the statement of the lemma.[2] $\square$

We now improve the bound (1.16) in a way analogous to [2,6,16,18]. It is based on a generalization of the procedure of polyconvexification. The bound E_p [16,25] is:

$$E_p = \max_{a_i \in \mathcal{A}} \sigma_0 \cdot \cdot F(\mathbf{C}_1, \mathbf{C}_2, \Phi(a_i)) \cdot \cdot \sigma_0, \tag{1.18}$$

---

[2]One can show that the quadratic form $\hat{\phi}'(u) = -(Tr\ u^S)^2 + (n-1)Tr\ (u^S \cdot u^S)$, where $u^S = (u + u^T)/2$ is the symmetric part of the matrix u, is also A-quasiconvex. By using this form instead of (1.17) one can show the quasiconvexity of the quadratic form of stresses $\hat{\phi}(\sigma) = \sigma \cdot \cdot \hat{\Phi} \cdot \cdot \sigma$ where the matrix $\hat{\Phi}$ coincides with Φ except of the elements $\hat{\Phi}_{44} = (a_2 + a_3)^2/2$, $\hat{\Phi}_{55} = (a_1 + a_3)^2/2$, $\hat{\Phi}_{66} = (a_1 + a_2)^2/2$. By using the quasiconvex quadratic form $\hat{\phi}(\sigma)$ instead of $\phi(\sigma)$ one can ease the restrictions (1.22) and generalize the Theorem 2 for the porous composite made of a matrix with negative Poisson's ratio. - *Authors' comment to the translation*

where $F(\cdot, \cdot, \cdot)$ is defined by:

$$F(\mathbf{C}_1, \mathbf{C}_2, \Phi) = \left[m_1 (\mathbf{C}_1 - \Phi)^{-1} + m_2 (\mathbf{C}_2 - \Phi)^{-1} \right]^{-1} + \Phi, \qquad (1.19)$$

and

$$\mathcal{A} = \{ (a_1, a_2, a_3) : \mathbf{C}_i - \Phi(a_i) \geq 0, \ i = 1, 2 \}. \qquad (1.20)$$

To prove the validity of the bound (1.18) we write $E(\chi_i, \sigma)$ as (cf. (1.11)):

$$E(\chi_i, \sigma) = \chi_1 \sigma \cdot \cdot (\mathbf{C}_1 - \Phi) \cdot \cdot \sigma + \chi_2 \sigma \cdot \cdot (\mathbf{C}_2 - \Phi) \cdot \cdot \sigma + \sigma \cdot \cdot \Phi \cdot \cdot \sigma.$$

Then the bound for the first two terms is given by (1.16) which holds for $a_i \in \mathcal{A}$, whereas the bound for the third term is based on its quasiconvexity. The bound thus derived depends on the parameters a_i; choosing the a_i optimally, we get (1.18).

The inequalities for $a_i \in \mathcal{A}$ can be derived by writing the matrices $\mathbf{C}_i$ and Φ in the basis (1.6). Then the condition of positive definiteness of $\mathbf{C}_i - \Phi$ is equivalent to the inequalities

$$\begin{pmatrix} \frac{3\kappa_i + \mu_i}{9\kappa_i \mu_i} - a_1^2 & \frac{2\mu_i - 3\kappa_i}{18\kappa_i \mu_i} + a_1 a_2 & \frac{2\mu_i - 3\kappa_i}{18\kappa_i \mu_i} + a_1 a_3 \\ \frac{2\mu_i - 3\kappa_i}{18\kappa_i \mu_i} + a_1 a_2 & \frac{3\kappa_i + \mu_i}{9\kappa_i \mu_i} - a_2^2 & \frac{2\mu_i - 3\kappa_i}{18\kappa_i \mu_i} + a_2 a_3 \\ \frac{2\mu_i - 3\kappa_i}{18\kappa_i \mu_i} + a_1 a_3 & \frac{2\mu_i - 3\kappa_i}{18\kappa_i \mu_i} + a_2 a_3 & \frac{3\kappa_i + \mu_i}{9\kappa_i \mu_i} - a_3^2 \end{pmatrix} \geq 0, \qquad (1.21)$$

$$\frac{1}{2\mu_i} - (a_2^2 + a_3^2) \geq 0, \quad \frac{1}{2\mu_i} - (a_1^2 + a_3^2) \geq 0, \quad \frac{1}{2\mu_i} - (a_1^2 + a_2^2) \geq 0. \qquad (1.22)$$

Thus the procedure of constructing the lower bound E_p of the functional E_q is reduced to the operation of finding the maximum with respect to a_i in (1.18).

§4 To get the upper bound E_l of E_q, we consider, as in [2], the energy of a composite of a special microstructure which we call a Matrix Laminate Composite (MLC). Such composites (Figure 1) are built in a multistep process, such that the layer constructed at the ith step is the combination of the initial matrix material and the composite obtained at the $(i-1)$th step. The normals n_i to the layers at every step are different. It is assumed that the widths of layers at the different steps are on different length scales. The number of steps taken to build a particular MLC is called the rank of an MLC. The effective compliance tensor of an MLC is [10]:

$$\mathbf{C}^M = \mathbf{C}_1 + \mathbf{Q}^{-1},$$

$$\mathbf{Q} = \frac{1}{m_2}(\mathbf{C}_2 - \mathbf{C}_1)^{-1} + \frac{m_1}{m_2}\sum_{i=1}^{n}\alpha_i\mathcal{N}_i, \qquad (1.23)$$

$$\mathcal{N}_i = \mathcal{N}(n_i) = \mathbf{P}_i(\mathbf{P}_i^T\mathbf{C}_1\mathbf{P}_i)^{-1}\mathbf{P}_i^T,$$

$$\sum_{i=1}^{n}\alpha_i = 1, \ \alpha_i \geq 0,$$

where $\mathbf{C}_1$ is the compliance tensor of the matrix material, $\mathbf{P}_i = \mathbf{P}(n_i)$ is the projector of the tensor σ on the subspace of components that are discontinuous along the normal n_i (if the properties of the material are discontinuous in the plane with normal n, the components of σ will be $(\sigma_{tt}, \sigma_{bb}, \sigma_{tb})$, where t, b are orthonormal vectors in the plane), and α_i are parameters determining the volume fraction of the material at the ith step of the process. For a rank three MLC, the volume fractions β_i of the matrix material with compliance tensor $\mathbf{C}_1$ at the ith step (Figure 1) are expressed in terms of α_i as [10]:

$$\beta_1 = m_1\alpha_1, \ \beta_2 = \frac{m_1\alpha_2}{1 - m_1\alpha_1}, \ \beta_3 = \frac{m_1\alpha_3}{m_2 + m_1\alpha_3}.$$

The proposed bound E_l is the minimum over the parameters α_i, n_i of the energy density stored in MLC:

$$E_l = \min_{\alpha_i, n_i} \sigma_0 \cdot \cdot \mathbf{C}_M \cdot \cdot \sigma_0 = \sigma_0 \cdot \cdot \mathbf{C}_1 \cdot \cdot \sigma_0 + \min_{\alpha_i, n_i} \sigma_0 \cdot \cdot \mathbf{Q}^{-1} \cdot \cdot \sigma_0 \qquad (1.24)$$

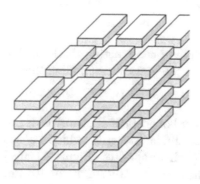

Figure 1.

To obtain the final result explicitly, we need to find the minimum in the second term of the right-hand side of (1.24). Let us limit ourselves to a rank three MLC with mutually orthogonal layers. Assuming n_i to be oriented along the axis x_i ($n_1 = \mathbf{i}$, $n_2 = \mathbf{j}$, $n_3 = \mathbf{k}$) we define, in the basis (1.6):

$$\mathbf{P}^T(n_1) = \mathbf{P}_1^T = \begin{pmatrix} 0 & 1 & 0 & 0 & 0 & 0 \\ 0 & 0 & 1 & 0 & 0 & 0 \\ 0 & 0 & 0 & 1 & 0 & 0 \end{pmatrix},$$

$$\mathbf{P}_1^T \mathbf{C}_1 \mathbf{P}_1 = \begin{pmatrix} \frac{3\kappa_1+\mu_1}{9\kappa_1\mu_1} & \frac{2\mu_1-3\kappa_1}{18\kappa_1\mu_1} & 0 \\ \frac{2\mu_1-3\kappa_1}{18\kappa_1\mu_1} & \frac{3\kappa_1+\mu_1}{9\kappa_1\mu_1} & 0 \\ 0 & 0 & \frac{1}{2\mu_1} \end{pmatrix}$$

The expressions for $\mathbf{P}_2, \mathbf{P}_3$ are analogous. The 6×6 matrix $\sum_{i=1}^{3} \alpha_i \mathcal{N}_i$ is:

$$\sum_{i=1}^{3} \alpha_i \mathcal{N}_i = \frac{2\mu_1}{3\kappa_1 + 4\mu_1} \begin{pmatrix} \mathbf{B}_1 & 0 \\ 0 & \mathbf{B}_2 \end{pmatrix}, \tag{1.25}$$

$$\mathbf{B}_1 = \begin{pmatrix} 2(\alpha_2+\alpha_3)(3\kappa_1+\mu_1) & \alpha_3(3\kappa_1-2\mu_1) & \alpha_2(3\kappa_1-2\mu_1) \\ \alpha_3(3\kappa_1-2\mu_1) & 2(\alpha_1+\alpha_3)(3\kappa_1+\mu_1) & \alpha_1(3\kappa_1-2\mu_1) \\ \alpha_2(3\kappa_1-2\mu_1) & \alpha_1(3\kappa_1-2\mu_1) & 2(\alpha_1+\alpha_2)(3\kappa_1+\mu_1) \end{pmatrix}$$

$$\mathbf{B}_2 = \begin{pmatrix} \alpha_1(3\kappa_1+4\mu_1) & 0 & 0 \\ 0 & \alpha_2(3\kappa_1+4\mu_1) & 0 \\ 0 & 0 & \alpha_3(3\kappa_1+4\mu_1) \end{pmatrix}.$$

Finally, assuming that the axes n_1, n_2, n_3 are oriented along the principal directions of σ_0 so that $\sigma_0 = (\sigma_1, \sigma_2, \sigma_3, 0, 0, 0)$, we compute the energy density E_l as a function of α_i only. In this case the upper left 3×3 block is an eigenblock of matrices $\mathbf{C}_1$, $\mathbf{C}_2$, $\mathcal{N}_i$ and thus

$$E_l = \widehat{\sigma_0} \cdot \widehat{\mathbf{C}_1} \cdot \widehat{\sigma_0} + \max_{\alpha_i} \widehat{\sigma_0} \cdot \widehat{\mathbf{Q}}^{-1} \cdot \widehat{\sigma_0}, \tag{1.26}$$

where

$$\widehat{\sigma_0} = (\sigma_1, \sigma_2, \sigma_3)$$

is the three-dimensional vector of eigenvalues of the tensor σ_0, and $(\widehat{X})$ denotes the operation of obtaining the upper-left 3×3 block of a 6×6 matrix X. Thus we have proved the following.

Proposition 1. *The elastic energy density of a composite of maximal rigidity is no less than E_p of (1.18) and no greater than E_l (1.26).*

§5 We call a composite of minimal rigidity a composite whose microstructure possesses minimal elastic energy density E_q^ε due to the given mean strain field ε_0:

$$E_q^\varepsilon = \min_{\mathbf{D}_0 \in G_m U} \varepsilon_0 \cdot \cdot \mathbf{D}_0 \cdot \cdot \varepsilon_0. \qquad (1.27)$$

Here $\mathbf{D}_0 = \mathbf{C}_0^{-1}$ is the rigidity tensor of a composite. This functional characterizes the minimal rigidity of a cell, because the minimal stress $\sigma_0 = \mathbf{D}_0 \cdot \cdot \varepsilon_0$ is associated with a given strain ε_0. Let us construct the energy bounds independent of a particular cell geometry.

Up to a change in notation $\sigma \to \varepsilon$ this problem is exactly the problem of structures of maximal rigidity and it will be solved by applying the same procedure. However, instead of the principle of minimum complementary energy, we will use the principle of minimum of potential energy [4]. The formal distinction between the two cases is due to the fact that the differential properties of the tensor ε are not identical to those of σ, so the expression for an A-quasiconvex quadratic form $\phi_\varepsilon = \varepsilon \cdot \cdot \Phi_\varepsilon \cdot \cdot \varepsilon$ is different from that for $\phi = \sigma \cdot \cdot \Phi \cdot \cdot \sigma$.

Lemma 2. *Consider the quadratic form $\phi_\varepsilon = \varepsilon \cdot \cdot \Phi_\varepsilon \cdot \cdot \varepsilon$, where the tensor Φ_ε in the basis (1.6) is:*

$$\Phi_\varepsilon = \Phi_\varepsilon(a_1, a_2, a_3) = \begin{pmatrix} 0 & -a_1^2 & -a_2^2 & 0 & 0 & 0 \\ -a_1^2 & 0 & -a_3^2 & 0 & 0 & 0 \\ -a_2^2 & -a_3^2 & 0 & 0 & 0 & 0 \\ 0 & 0 & 0 & a_3^2 & 0 & 0 \\ 0 & 0 & 0 & 0 & a_2^2 & 0 \\ 0 & 0 & 0 & 0 & 0 & a_1^2 \end{pmatrix}.$$

This form ϕ_ε is A-quasiconvex, if the symmetric 3×3 tensor ε satisfies the differential restrictions (1.2).

Proof: The function ϕ_ε is given by

$$\phi_\varepsilon = -2a_1^2 \det \begin{pmatrix} \varepsilon_{x_1 x_1} & \varepsilon_{x_1 x_2} \\ \varepsilon_{x_1 x_2} & \varepsilon_{x_2 x_2} \end{pmatrix}$$

$$- 2a_2^2 \det \begin{pmatrix} \varepsilon_{x_1 x_1} & \varepsilon_{x_1 x_3} \\ \varepsilon_{x_1 x_3} & \varepsilon_{x_3 x_3} \end{pmatrix} - 2a_3^2 \det \begin{pmatrix} \varepsilon_{x_2 x_2} & \varepsilon_{x_2 x_3} \\ \varepsilon_{x_2 x_3} & \varepsilon_{x_3 x_3} \end{pmatrix}.$$

It is a sum of three terms of similiar structure. Let us show that any term, for example, the first, is quasiconvex [2]. We have:

$$-\det\begin{pmatrix} \varepsilon_{x_1 x_1} & \varepsilon_{x_1 x_2} \\ \varepsilon_{x_1 x_2} & \varepsilon_{x_2 x_2} \end{pmatrix} = -\left(\frac{\partial u_1}{\partial x_1}\right)\left(\frac{\partial u_2}{\partial x_2}\right) + \frac{1}{4}\left(\frac{\partial u_1}{\partial x_2} + \frac{\partial u_2}{\partial x_1}\right)^2 \quad (1.28)$$

$$= -\det\begin{pmatrix} \dfrac{\partial u_1}{\partial x_1} & \dfrac{\partial u_2}{\partial x_1} \\ \dfrac{\partial u_1}{\partial x_2} & \dfrac{\partial u_2}{\partial x_2} \end{pmatrix} + \frac{1}{4}\left(\frac{\partial u_1}{\partial x_2} - \frac{\partial u_2}{\partial x_1}\right)^2.$$

The first term is the minor of the matrix ∇u and is quasiconvex [7] whereas the second term is convex in the usual sense. $\qquad\square$

Using the function Φ_ε instead of Φ to obtain the lower bound E_q^ε for E_p^ε, we find that:

$$E_p^\varepsilon = \max_{a_i \in \mathcal{A}_\varepsilon} \varepsilon_0 \cdot \cdot \mathbf{F}(\mathbf{D}_1, \mathbf{D}_2, \Phi_\varepsilon(a_1, a_2, a_3)) \cdot \cdot \varepsilon_0, \qquad (1.29)$$

where

$$\mathcal{A}_\varepsilon = \{(a_1, a_2, a_3) : \mathbf{D}_i - \Phi_\varepsilon(a_1, a_2, a_3) \geq 0, i = 1, 2\} \qquad (1.30)$$

and the function $\mathbf{F}$ is as in (1.19). The construction of the other bound E_l^ε is done with the help of MLC. The matrices $\mathbf{D}^M$ and $\mathbf{C}^M$ are not of the same structure due to the difference between the projectors $\mathbf{P}_i$ of (1.23). $\mathbf{P}^\varepsilon(n)$ is the projector on the subspace of the discontinuous components of ε which in our case are $(\varepsilon_{nn}, \varepsilon_{nt}, \varepsilon_{nb})$. Thus for the normal $n_1 = \mathbf{i}$ in the basis (1.6) we have:

$$(\mathbf{P}^\varepsilon(n_1))^T = (\mathbf{P}_1^\varepsilon)^T = \begin{pmatrix} 1 & 0 & 0 & 0 & 0 & 0 \\ 0 & 0 & 0 & 0 & 1 & 0 \\ 0 & 0 & 0 & 0 & 0 & 1 \end{pmatrix} \qquad (1.31)$$

$$(\mathbf{P}_1^\varepsilon)^T \mathbf{D}_2 \mathbf{P}_1^\varepsilon = \begin{pmatrix} \frac{3\kappa_2 + 4\mu_2}{3} & 0 & 0 \\ 0 & 2\mu_2 & 0 \\ 0 & 0 & 2\mu_2 \end{pmatrix}.$$

The form of the effective elasticity tensor $\mathbf{D}^M$ is analogous to $\mathbf{C}^M$. If $\mathbf{D}_2$ is the rigidity tensor of the matrix material, then

$$\mathbf{D}^M = \mathbf{D}_2 + \left\{ \frac{1}{m_1}(\mathbf{D}_1 - \mathbf{D}_2)^{-1} + \frac{m_2}{m_1}\sum_{i=1}^{3} \alpha_i \mathcal{N}_i^\varepsilon \right\}^{-1}, \qquad (1.32)$$

and the bound E_l is:

$$E_l^\varepsilon = \min_{\alpha_i, n_i} \varepsilon_0 \cdot \cdot \mathbf{D}^M \cdot \cdot \varepsilon_0. \tag{1.33}$$

Thus we obtain the following.

Proposition 2. *The density of the elastic energy of a cell of a composite of minimal rigidity is no less than* E_p^ε *(1.29) and no more than* E_l^ε *(1.33).*

§6 The geometrically independent bounds E_p, E_p^ε (1.18), (1.29) of the elastic energy density, valid without regard to the microstructure of the composite, allow us to determine bounds for the set $G_m U$ of effective tensors. Since for any sets of parameters these bounds are quadratic forms generated by matrices of the form (1.19) and are valid independently of σ, ε we have:

$$\mathbf{F}(\mathbf{C}_1, \mathbf{C}_2, \Phi(a_1, a_2, a_3)) \leq \mathbf{C}_0, \ \forall a_i \in \mathcal{A}, \ \forall \mathbf{C}_0 \in G_m U, \tag{1.34}$$

$$\mathbf{F}(\mathbf{C}_1^{-1}, \mathbf{C}_2^{-2}, \Phi_\varepsilon(a_1, a_2, a_3)) \leq \mathbf{C}_0^{-1}, \ \forall a_i \in \mathcal{A}_\varepsilon, \ \forall \mathbf{C}_0 \in G_m U. \tag{1.35}$$

If we put $a_i = 0$ on the left-hand side of (1.34) and (1.35) we will get rougher bounds for $G_m U$, which are called the Hill inequalities [15] (they are analogous to Reuss-Voigt bounds). The inequalities (1.34) and (1.35) improve the Hill bounds and, unlike the classical Hashin-Shtrikman bounds, they allow us to estimate linear combinations of coefficients of arbitrary anisotropic tensors $\mathbf{C}_0 \in G_m U$. Let us estimate, for example, the bulk part of $\mathbf{C}_0$, that is, the quantity

$$\kappa_0 = \frac{1}{3}(e_s \cdot \cdot \mathbf{C}_0 \cdot \cdot e_s),$$

$$e_s = \frac{1}{\sqrt{3}}(\mathbf{i} \otimes \mathbf{i} + \mathbf{j} \otimes \mathbf{j} + \mathbf{k} \otimes \mathbf{k})$$

If we put $a_1 = a_2 = a_3 = \frac{1}{2\sqrt{\mu}}$ in (1.34) and take a contraction with e_s on both sides, we obtain:

$$\frac{\kappa_0 - \kappa_2}{\kappa_1 - \kappa_2} \leq \frac{m_1}{1 + m_2(\kappa_1 - \kappa_2)/(\kappa_2 + \frac{4}{3}\mu_1)}, \tag{1.36}$$

which coincides with one of the bounds of Hashin-Shtrikman. The second bound for κ_0 can be obtained by putting $a_i = 2\sqrt{\mu}$ and taking a contraction with e_s on both sides in (1.35). The bounds E_l correspond to the set of effective constants of special microstructures, that is, a set which is contained in $G_m U$:

$$\bigcup_{\alpha_i, n_i} \mathbf{C}^M(\alpha_i, n_i) \subset G_m U;$$

they can be used to determine the sharpness of bounds (1.34) and (1.35). In particular, the bulk modulus κ_0 for a rank three MLC with matrix compliance $\mathbf{C}_2$ and layer normals oriented along the principal axes of ε_0, satisfies the inequality (1.36) as equality if we put $\alpha_1 = \alpha_2 = \alpha_3 = 1/3$.

Remark: The bounds of $G_m U$ given by (1.34), (1.35) enclose $G_m U$ better than previously known bounds [3]. Nevertheless, they are not exhaustive. Indeed, to obtain the precise description of $G_m U$ it would be necessary, as in [5,18], to consider the behavior of the composite under several tests, that is, to estimate the minimum of the sum of quadratic forms of the type:

$$\sum_{i=1}^{6} \sigma^i \cdot \cdot \mathbf{C}_0 \cdot \cdot \sigma^i,$$

where $\sigma^i, i = 1, \dots, 6$ are linearly independent tensors.

Part II. The optimal porous body and other asymptotics.

§**7** Consider a mixture of two materials with very different elastic moduli. Namely, we assume that the moduli of the material with compliance tensor $\mathbf{C}_2$ are

$$\mu_2 = 0, \quad \mu_2 = 0. \tag{2.1}$$

Finding the composites of maximal rigidity built from this material and a material with compliance tensor $\mathbf{C}_1$ (κ_1, μ_1 are finite) is equivalent to considering a porous body of maximal rigidity, that is, an asymptotic case of the problem discussed in Part I. The assumption (2.1) makes the expressions for E_q, E_p much easier to handle. It will allow us to present a simple proof of exactness of the bounds, and to describe the optimal cell geometries.

We have for E_p, E_l (cf. (1.18), (1.26)):

$$E_p = \sigma_0 \cdot \cdot \mathbf{C}_1 \cdot \cdot \sigma_0 + \frac{m_2}{m_1} \max_{a_i \in \mathcal{A}} \sigma_0 \cdot \cdot \mathbf{G} \cdot \cdot \sigma_0, \tag{2.2}$$

$$\mathbf{G} = \mathbf{C}_1 - \Phi(a_1, a_2, a_3), \tag{2.3}$$

and

$$E_l = \sigma_0 \cdot\cdot \mathbf{C}_1 \cdot\cdot \sigma_0 + \frac{m_2}{m_1} \min_{\alpha_i \in \mathcal{A}} \sigma_0 \cdot\cdot \mathbf{Q}^{-1} \cdot\cdot \sigma_0, \qquad (2.4)$$

$$\mathbf{Q} = \sum_{i=1}^{3} \alpha_i \mathcal{N}_i.$$

Assume also that from now on:

(1) the basis $e_1, \ldots, e_6$ (cf. (1.6)) of rank four tensors $\mathbf{C}_1$, $\mathbf{\Phi}$, $\mathbf{Q}$ is chosen in such a way that vectors $\mathbf{i}, \mathbf{j}, \mathbf{k}$ point in the principal directions of the tensor σ_0;

(2) the eigenvalues $\sigma_1, \sigma_2, \sigma_3$ of σ_0 are such that

$$\sigma_3 \geq |\sigma_1|, \quad \sigma_3 \geq |\sigma_2| \qquad (2.5)$$

(the fact that $\sigma_3 \geq 0$, which follows from (2.5), causes no loss of generality, because the quadratic forms $E_p(\sigma_0)$, $E_l(\sigma_0)$ are invariant under change of sign of σ_0).

The optimal values of a_i, α_i depend on variables ω, η:

$$\omega = \frac{\sigma_1}{\sigma_3}, \quad \eta = \frac{\sigma_2}{\sigma_3}$$

that take values (cf. (2.5)) in a square region

$$\Omega = \{(\omega, \eta) : \ |\omega| \leq 1, \ |\eta| \leq 1\}.$$

We thus have the following.

Theorem I. *For the porous body of maximal rigidity the following holds: (1) inequalities (1.15) reduce to equalities*

$$E_p(\sigma_0) = E_q(\sigma_0) = E_l(\sigma_0),$$

that is, for any $\sigma_0 \in \Sigma$:

$$E_l(\sigma_0) - E_p(\sigma_0) = \frac{m_2}{m_1} \max_{a_i \in \mathcal{A}} \min_{\alpha_i} \sigma_0 \cdot\cdot \left(\mathbf{Q}^{-1}(\alpha_i) - \mathbf{G}^{-1}(a_i)\right) \cdot\cdot \sigma_0 = 0; \quad (2.6)$$

(2) the minimal elastic energy density E_q is:

$$E_q = \hat{\sigma}_0 \cdot \widehat{\mathbf{C}}_1 \cdot \hat{\sigma}_0 + \frac{m_2}{m_1} \hat{\sigma}_0 \cdot \widehat{\mathbf{G}} \cdot \hat{\sigma}_0, \qquad (2.7)$$

where $\widehat{\mathbf{G}}$ is determined as follows (cf. Fig. 2):

$$\widehat{\mathbf{G}} = \widehat{\mathbf{G}}_1 = \frac{3\kappa + 4\mu}{36\kappa\mu} \begin{pmatrix} 1 & 1 & 1 \\ 1 & 1 & 1 \\ 1 & 1 & 1 \end{pmatrix}, \quad \textit{if } \sigma_0 \in \Omega_1,$$
$$\Omega_1 = \{\sigma_0 : \omega \leq 1, \ \eta \leq 1, \ \omega + \eta \geq 1\} \tag{2.7a}$$

$$\widehat{\mathbf{G}} = \widehat{\mathbf{G}}_2 = \frac{3\kappa + 4\mu}{36\kappa\mu} \begin{pmatrix} 1 & 1 & -(1+2\nu) \\ 1 & 1 & -(1+2\nu) \\ -(1+2\nu) & -(1+2\nu) & (1+2\nu)^2 \end{pmatrix}, \quad \textit{if } \sigma_0 \in \Omega_2$$
$$\Omega_2 = \{\sigma_0 : \omega \geq -1, \ \eta \geq -1, \ \omega - \eta + (1 - 2\nu) \geq 0, \\ \omega - \eta - (1 - 2\nu) \leq 0, \ \omega + \eta + (1 - 2\nu) \leq 0\}, \tag{2.7b}$$

$$\widehat{\mathbf{G}} = \widehat{\mathbf{G}}_3 = \frac{3\kappa + 4\mu}{36\kappa\mu} \begin{pmatrix} 1 & -(1+2\nu) & 1 \\ -(1+2\nu) & (1+2\nu)^2 & -(1+2\nu) \\ 1 & -(1+2\nu) & 1 \end{pmatrix}, \quad \textit{if } \sigma_0 \in \Omega_3$$
$$\Omega_3 = \{\sigma_0 : \omega \leq 1, \ \eta \geq -1, \ \omega - (1 - 2\nu)\eta - 1 \geq 0\}, \tag{2.7c}$$

$$\widehat{\mathbf{G}} = \widehat{\mathbf{G}}_4 = \frac{3\kappa + 4\mu}{36\kappa\mu} \begin{pmatrix} (1+2\nu)^2 & -(1+2\nu) & -(1+2\nu) \\ -(1+2\nu) & 1 & 1 \\ -(1+2\nu) & 1 & 1 \end{pmatrix}, \quad \textit{if } \sigma_0 \in \Omega_4,$$
$$\Omega_4 = \{\sigma_0 : \omega \geq -1, \ \eta \leq 1, \ (1 - 2\nu)\omega - \eta + 1 \leq 0\}, \tag{2.7d}$$

$$\widehat{\mathbf{G}} = \widehat{\mathbf{G}}_5 = \frac{3\kappa + \mu}{9\kappa\mu} \begin{pmatrix} 1 & 1 & -\nu \\ 1 & 1 & -\nu \\ -\nu & -\nu & 1 \end{pmatrix}, \quad \textit{if } \sigma_0 \in \Omega_5,$$
$$\Omega_5 = \{\sigma_0 : \omega \geq 0, \ \eta \geq 0, \ \omega + \eta \leq 1\} \cup \\ \{\sigma_0 : \omega \leq 0, \ \eta \leq 0, \ \omega + \eta + (1 - 2\nu) \geq 0\}, \tag{2.7e}$$

$$\widehat{\mathbf{G}} = \widehat{\mathbf{G}}_6 = \frac{3\kappa + \mu}{9\kappa\mu} \begin{pmatrix} 1 & 2\nu^2 - 1 & -\nu \\ 2\nu^2 - 1 & 1 & -\nu \\ -\nu & -\nu & 1 \end{pmatrix}, \quad \textit{if } \sigma_0 \in \Omega_6,$$
$$\Omega_6 = \{\sigma_0 : \omega \geq 0, \ -1 \leq \eta \leq 0, \ \omega - (1 - 2\nu)\eta - 1 \leq 0\}, \tag{2.7f}$$

$$\widehat{\mathbf{G}} = \widehat{\mathbf{G}}_7 = \widehat{\mathbf{G}}_6, \quad \textit{if } \sigma_0 \in \Omega_7$$
$$\Omega_7 = \{\sigma_0 : -1 \leq \omega \leq 0, \ \eta \geq 0, \ (1 - 2\nu)\omega - \eta + 1 \geq 0\}, \tag{2.7g}$$

$$\widehat{\mathbf{G}} = \widehat{\mathbf{G}}_8 = \frac{3\kappa + \mu}{9\kappa\mu} \begin{pmatrix} 1 & -\nu & 2\nu^2 - 1 \\ -\nu & 1 & -\nu \\ 2\nu^2 - 1 & -\nu & 1 \end{pmatrix}, \quad \textit{if } \sigma_0 \in \Omega_8$$
$$\Omega_8 = \{\sigma_0 : \omega \leq 0, \ \eta \geq -1, \ \omega - \eta - (1 - 2\nu) \geq 0\}, \tag{2.7h}$$

$$\widehat{G} = \widehat{G}_9 = \frac{3\kappa + \mu}{9\kappa\mu} \begin{pmatrix} 1 & -\nu & -\nu \\ -\nu & 1 & 2\nu^2-1 \\ -\nu & 2\nu^2-1 & 1 \end{pmatrix}, \ if \ \sigma_0 \in \Omega_9,$$

$$\Omega_9 = \{\sigma_0 : \omega \geq -1, \ \eta \leq 0, \ \omega - \eta + (1 - 2\nu) \leq 0\}. \tag{2.7i}$$

Here $\kappa = \kappa_1$, $\mu = \mu_1$, and $\nu = \nu_1 = (3\kappa 2\mu)/(6\kappa + 2\mu)$, are the bulk and shear moduli of the material, and its Poisson ratio.

(3) The optimal values of a_i are given by:

$$a_1 = a_2 = a_3 = \frac{1}{2\sqrt{\mu}}, \ if \ \sigma_0 \in \Omega_1, \tag{2.8a}$$

$$a_1 = a_2 = \frac{1}{2\sqrt{\mu}}, \ a_3 = \frac{2\nu - 1}{2\sqrt{\mu}}, \ if \ \sigma_0 \in \Omega_2, \tag{2.8b}$$

$$a_1 = a_3 = \frac{1}{2\sqrt{\mu}}, \ a_2 = \frac{2\nu - 1}{2\sqrt{\mu}}, \ if \ \sigma_0 \in \Omega_3, \tag{2.8c}$$

$$a_1 = \frac{2\nu - 1}{2\sqrt{\mu}}, \ a_2 = a_3 = \frac{1}{2\sqrt{\mu}}, \ if \ \sigma_0 \in \Omega_4, \tag{2.8d}$$

$$a_1 = a_2 = \frac{1}{2\sqrt{\mu}}, \ a_3 = \frac{1}{2\sqrt{\mu}}(\omega + \eta), \ if \ \sigma_0 \in \Omega_5, \tag{2.8e}$$

$$a_1 = \frac{1}{2\sqrt{\mu}}, \ a_2 = \frac{2\nu - 1}{2\sqrt{\mu}}, \ a_3 = \frac{\omega - (1 - 2\nu)\eta}{2\sqrt{\mu}}, \ if \ \sigma_0 \in \Omega_6, \tag{2.8f}$$

$$a_1 = \frac{2\nu - 1}{2\sqrt{\mu}}, \ a_2 = \frac{1}{2\sqrt{\mu}}, \ a_3 = \frac{\eta - (1 - 2\nu)\omega}{2\sqrt{\mu}}, \ if \ \sigma_0 \in \Omega_7, \tag{2.8g}$$

$$a_1 = \frac{1}{2\sqrt{\mu}}, \ a_2 = \frac{\omega - (1 - 2\nu)}{2\sqrt{\mu}\eta}, \ a_3 = \frac{2\nu - 1}{2\sqrt{\mu}}, \ if \ \sigma_0 \in \Omega_8, \tag{2.8h}$$

$$a_1 = \frac{\eta - (1 - 2\nu)}{2\sqrt{\mu}\omega}, \ a_2 = \frac{1}{2\sqrt{\mu}}, \ a_3 = \frac{2\nu - 1}{2\sqrt{\mu}}, \ if \ \sigma_0 \in \Omega_9. \tag{2.8i}$$

(4) The optimal structures for regimes (a) through (d) are rank three MLCs with layer normals oriented along the principal directions of σ_0, and such that:

$$\alpha_1 = \frac{1 - \omega + \eta}{1 + \omega + \eta}, \quad \alpha_2 = \frac{1 + \omega - \eta}{1 + \omega + \eta}, \quad \alpha_3 = \frac{\omega + \eta - 1}{1 + \omega + \eta}, \tag{2.9a}$$

if $\sigma_0 \in \Omega_1$;

$$\alpha_1 = \frac{-\omega + \eta - (1 - 2\nu)}{(1 - 2\nu)(\omega + \eta - (1 + 2\nu))}, \quad \alpha_2 = \frac{\omega - \eta - (1 - 2\nu)}{(1 - 2\nu)(\omega + \eta - (1 + 2\nu))},$$
$$\alpha_3 = \frac{\omega + \eta + (1 - 2\nu)}{\omega + \eta - (1 + 2\nu)}, \tag{2.9b}$$

if $\sigma_0 \in \Omega_2$;

$$\alpha_1 = \frac{1 - \omega - (1 - 2\nu)\eta}{(1 - 2\nu)(1 + \omega - (1 + 2\nu)\eta)}, \quad \alpha_2 = \frac{1 + \omega + (1 - 2\nu)\eta}{1 + \omega - (1 + 2\nu)\eta}, \tag{2.9c}$$
$$\alpha_3 = \frac{\omega - (1 - 2\nu)\eta - 1}{(1 - 2\nu)(1 + \omega - (1 + 2\nu)\eta)},$$

if $\sigma_0 \in \Omega_3$;

$$\alpha_1 = \frac{\eta - 1 - \omega(1 - 2\nu)}{(1 - 2\nu)(1 - (1 + 2\nu)\omega - \eta)}, \quad \alpha_2 = \frac{1 - \omega(1 - 2\nu) - \eta}{(1 - 2\nu)(1 - (1 + 2\nu)\omega - \eta)},$$
$$\alpha_3 = \frac{1 + \omega(1 - 2\nu) + \eta}{1 - (1 + 2\nu)\omega - \eta}, \tag{2.9d}$$

if $\sigma_0 \in \Omega_4$.

For regimes (e) through (i) the optimal structures are rank two MLCs with layer normals oriented along the principal axes of σ_0 and:

$$\alpha_1 = \frac{\eta}{\omega + \eta}, \quad \alpha_2 = \frac{\omega}{\omega + \eta}, \quad \alpha_3 = 0, \; \text{if } \sigma_0 \in \Omega_5; \tag{2.9e}$$

$$\alpha_1 = -\frac{\eta}{\omega - \eta}, \quad \alpha_2 = \frac{\omega}{\omega - \eta}, \quad \alpha_3 = 0, \; \text{if } \sigma_0 \in \Omega_6; \tag{2.9f}$$

$$\alpha_1 = -\frac{\eta}{\omega - \eta}, \quad \alpha_2 = \frac{\omega}{\omega - \eta}, \quad \alpha_3 = 0, \; \text{if } \sigma_0 \in \Omega_7; \tag{2.9g}$$

$$\alpha_1 = \frac{1}{1 - \omega}, \quad \alpha_2 = 0, \quad \alpha_3 = \frac{\omega}{\omega - 1}, \; \text{if } \sigma_0 \in \Omega_8; \tag{2.9h}$$

$$\alpha_1 = 0, \quad \alpha_2 = \frac{1}{1-\eta}, \quad \alpha_3 = -\frac{\eta}{1-\eta}, \quad \text{if } \sigma_0 \in \Omega_9. \tag{2.9i}$$

The proof of the theorem is given in the appendix. The subregions Ω_1 through Ω_9 are shown in Figure 2. We note that:

(1) although the optimal values of a_i depend on ω and η, the expressions for E_p are still quadratic forms of the eigenvalues of the tensor σ_0;

(2) for the subregions Ω_1–Ω_4 that contain the vertices of the square region Ω, that is, the points where $|\sigma_1| = |\sigma_2| = |\sigma_3|$, the optimal structures are rank three MLCs. For subregions Ω_5–Ω_9 where one of the eigenvalues of σ_0 is considerably less than the remaining two, the optimal structures are rank two MLCs. On the lines $\sigma_1 = 0$, $\sigma_2 = 0$ the optimal structures are simple layered composites, and in addition, $E_p = E_l = E_c$ (cf. (1.16)). The lines on which the parameters α_i of the composites vanish are shown in Figure 2;

(3) the relation between the parameters a_i of the bound and the parameters α_i of the optimal structure of the composite is as follows. Of the six parameters, three have values corresponding to a boundary of their admissible intervals, and the remaining three lie inside them. As the optimal values of a_i approach the boundaries of their admissible intervals, the optimal values of α_i move inside their admissible intervals. The rank of a composite that realizes the best bound E_p equals the number of parameters a_i whose values lie on the boundaries of the corresponding intervals (three in the case of Ω_1–Ω_4, and two in the case of Ω_5–Ω_9);

(4) since $\Phi(a_1, a_2, a_3) = \Phi(-a_1, -a_2, -a_3)$, the optimal values of the parameters listed in (3) of the theorem do not change if we simultaneously change the sign of parameters a_i;

(5) it follows from the theorem that any tensor $\mathbf{C}_0 \in G_m U$ satisfies

$$\mathbf{C}_0 \geq \mathbf{C}_1 + \frac{m_2}{m_1} \mathbf{G}_i, \quad i = 1, \ldots, 9,$$

and hence the invariants of $\mathbf{C}_0$ are bounded below by the invariants of any of the tensors $\mathbf{C}_1 + \frac{m_2}{m_1} \mathbf{G}_i$, $i = 1, \ldots, 9$;

(6) the results are especially simple for the case when Poisson's ratio ν is equal to zero, that is, when $3\kappa = 2\mu$ (cf. Figure 3). In this case E_q is the following function of $|\sigma_i|$, $i = 1, 2, 3$:

$$E_q = \begin{cases} \sigma_0 \cdot\cdot \mathbf{C}_1 \cdot\cdot \sigma_0 + \frac{m_2}{2m_1} \left(|\sigma_1| + |\sigma_2| + |\sigma_3| \right)^2, & \text{if } |\omega| + |\eta| \geq 1, \\ \sigma_0 \cdot\cdot \mathbf{C}_1 \cdot\cdot \sigma_0 + \frac{m_2}{m_1} \left(|\sigma_1| + |\sigma_2| \right)^2 + |\sigma_3|^2, & \text{if } |\omega| + |\eta| \leq 1, \end{cases} \tag{2.10}$$

(we assign here $3\kappa = 2\mu = 1$). The subregions Ω_8, Ω_9 reduce to points and the optimal values of parameters a_i are:

$$a_1 = \begin{cases} \frac{1}{\sqrt{2}}, & \text{if } \sigma_1\sigma_2 \geq 0, \\ -\frac{1}{\sqrt{2}}, & \text{if } \sigma_1\sigma_2 \leq 0, \end{cases}$$

$$a_2 = \begin{cases} \frac{1}{\sqrt{2}}, & \text{if } \sigma_2\sigma_3 \geq 0, \\ -\frac{1}{\sqrt{2}}, & \text{if } \sigma_2\sigma_3 \leq 0, \end{cases} \tag{2.11}$$

and

$$a_3 = \begin{cases} \frac{1}{\sqrt{2}}, & \text{if } |\omega| + |\eta| \geq 1, \\ \frac{|\omega|+|\eta|}{\sqrt{2}}, & \text{if } |\omega| + |\eta| \leq 1, \end{cases} \tag{2.12}$$

The optimal structures are either rank three MLCs (in the region $|\omega| + |\eta| \geq 1$) with parameters α_i given by

$$\alpha_1 = \frac{1 - |\omega| + |\eta|}{1 + |\omega| + |\eta|},$$
$$\alpha_2 = \frac{1 + |\omega| - |\eta|}{1 + |\omega| + |\eta|}, \tag{2.13}$$
$$\alpha_3 = \frac{-1 + |\omega| + |\eta|}{1 + |\omega| + |\eta|},$$

or rank two MLCs (in the region $|\omega| + |\eta| \leq 1$) oriented along the directions of the maximal in magnitude eigenvalue σ_3 of the tensor σ_0 and such that

$$\alpha_1 = \frac{|\eta|}{|\omega| + |\eta|}, \quad \alpha_2 = \frac{|\omega|}{|\omega| + |\eta|}, \quad \alpha_3 = 0. \tag{2.14}$$

The formulae (2.10) through (2.14) follow immediately from (2.6) through (2.9), and they also may be easily derived directly.

§8 We now focus on the problem of a body of minimal rigidity. We consider a composite with one phase having a much greater rigidity than the other, that is, we let $\kappa_1 = \mu_1 \to \infty$. The expression for the bound E_p^ε then becomes:

$$E_p^\varepsilon = \varepsilon_0 \cdot\cdot \mathbf{D_2} \cdot\cdot \varepsilon_0 + \frac{m_1}{m_2} \max_{a_i \in \mathcal{A}_\varepsilon} \varepsilon_0 \cdot\cdot (\mathbf{D_2} - \Phi_\varepsilon(a_i)) \cdot\cdot \varepsilon_0, \tag{2.15}$$

where

$$\mathcal{A}_\varepsilon = \{(a_1, a_2, a_3) : \mathbf{D_2} - \Phi_\varepsilon \geq 0\},$$

the tensor $\mathbf{D}_2$ in the basis (1.6) is

$$\mathbf{D}_2 = \begin{pmatrix} \frac{3\kappa+4\mu}{3} & \frac{3\kappa-2\mu}{3} & \frac{3\kappa-2\mu}{3} & 0 & 0 & 0 \\ \frac{3\kappa-2\mu}{3} & \frac{3\kappa+4\mu}{3} & \frac{3\kappa-2\mu}{3} & 0 & 0 & 0 \\ \frac{3\kappa-2\mu}{3} & \frac{3\kappa-2\mu}{3} & \frac{3\kappa+4\mu}{3} & 0 & 0 & 0 \\ 0 & 0 & 0 & 2\mu & 0 & 0 \\ 0 & 0 & 0 & 0 & 2\mu & 0 \\ 0 & 0 & 0 & 0 & 0 & 2\mu \end{pmatrix},$$

and the set $\mathcal{A}_\varepsilon$ is described by the inequalities:

$$\begin{pmatrix} \frac{3\kappa+4\mu}{3} & \frac{3\kappa-2\mu}{3}+a_1^2 & \frac{3\kappa-2\mu}{3}+a_2^2 \\ \frac{3\kappa-2\mu}{3}+a_1^2 & \frac{3\kappa+4\mu}{3} & \frac{3\kappa-2\mu}{3}+a_3^2 \\ \frac{3\kappa-2\mu}{3}+a_2^2 & \frac{3\kappa-2\mu}{3}+a_3^2 & \frac{3\kappa+4\mu}{3} \end{pmatrix} \geq 0,$$

$$2\mu - a_1^2 \geq 0, \ 2\mu - a_2^2 \geq 0, \ 2\mu - a_3^2 \geq 0.$$

(We denote here as $\mu = \mu_2$, $\kappa = \kappa_2$ and $\nu = (3\kappa - 2\mu)/(6\kappa + 2\mu)$ the shear and bulk moduli and Poisson's ratio of the nonrigid material).

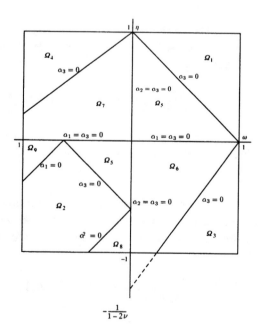

Figure 2.

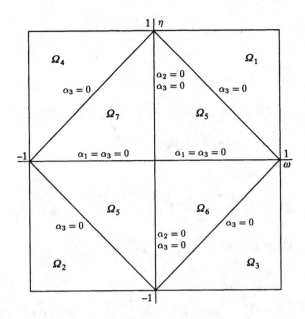

Figure 3.

Let $\widehat{\varepsilon}_0 = (\varepsilon_1, \varepsilon_2, \varepsilon_3)$ be the vector of eigenvalues of the tensor ε_0. We assume that

$$\varepsilon_3 \geq |\varepsilon_1|, \quad \varepsilon_3 \geq |\varepsilon_2|, \quad \omega_\varepsilon = \frac{\varepsilon_1}{\varepsilon_3}, \quad \eta_\varepsilon = \frac{\varepsilon_2}{\varepsilon_3}.$$

Writing the right-hand side of (2.15) in the eigenbasis of the tensor ε_0, we obtain:

$$E_p^\varepsilon = \frac{1}{m_2} \widehat{\varepsilon}_0 \cdot \widehat{\mathbf{D}}_2 \cdot \widehat{\varepsilon}_0 + \frac{2m_1}{m_2} \max_{a_i \in \mathcal{A}_\varepsilon} \left(a_1^2 \varepsilon_1 \varepsilon_2 + a_2^2 \varepsilon_1 \varepsilon_3 + a_3^2 \varepsilon_2 \varepsilon_3 \right). (2.16)$$

(recall that $\widehat{\mathbf{D}}_2$ is the upper-left 3×3 block of a 6×6 matrix $\mathbf{D}_2$). The optimal values of the parameters a_i are then:

$$a_1^2 = \begin{cases} 2\mu, & \text{if } \varepsilon_1 \varepsilon_2 \geq 0, \\ 0, & \text{if } \varepsilon_1 \varepsilon_2 < 0 \end{cases},$$

$$a_2^2 = \begin{cases} 2\mu, & \text{if } \varepsilon_1 \varepsilon_3 \geq 0, \\ 0, & \text{if } \varepsilon_1 \varepsilon_3 < 0 \end{cases}, \qquad (2.17)$$

$$a_3^2 = \begin{cases} 2\mu, & \text{if } \varepsilon_2 \varepsilon_3 \geq 0, \\ 0, & \text{if } \varepsilon_2 \varepsilon_3 < 0 \end{cases},$$

and the bound E_p^ε is

$$E_p^\varepsilon = \widehat{\varepsilon}_0 \cdot \widehat{\mathbf{D}}_2 \cdot \widehat{\varepsilon}_0 + \frac{m_1}{m_2} \widehat{\varepsilon}_0 \cdot \widehat{\mathbf{G}}^\varepsilon \cdot \widehat{\varepsilon}_0, \qquad (2.18)$$

where (cf. Figure 4)

$$\widehat{\mathbf{G}}^\varepsilon = \widehat{\mathbf{G}}_1^\varepsilon = \frac{3\kappa + 4\mu}{3} \begin{pmatrix} 1 & 1 & 1 \\ 1 & 1 & 1 \\ 1 & 1 & 1 \end{pmatrix}, \text{ if } \varepsilon_0 \in \Omega_1^\varepsilon, \qquad (2.18a)$$
$$\Omega_1^\varepsilon = \{\varepsilon_0 : \omega_\varepsilon \geq 0, \eta_\varepsilon \geq 0\},$$

$$\widehat{\mathbf{G}}^\varepsilon = \widehat{\mathbf{G}}_2^\varepsilon = \frac{3\kappa + 4\mu}{3} \begin{pmatrix} 1 & \frac{\nu}{1-\nu} & \frac{\nu}{1-\nu} \\ \frac{\nu}{1-\nu} & 1 & 1 \\ \frac{\nu}{1-\nu} & 1 & 1 \end{pmatrix}, \text{ if } \varepsilon_0 \in \Omega_2^\varepsilon,$$
$$\Omega_2^\varepsilon = \{\varepsilon_0 : \omega_\varepsilon \leq 0, \eta_\varepsilon \geq 0\}, \qquad (2.18b)$$

$$\widehat{\mathbf{G}}^\varepsilon = \widehat{\mathbf{G}}_3^\varepsilon = \frac{3\kappa + 4\mu}{3} \begin{pmatrix} 1 & 1 & \frac{\nu}{1-\nu} \\ 1 & 1 & \frac{\nu}{1-\nu} \\ \frac{\nu}{1-\nu} & \frac{\nu}{1-\nu} & 1 \end{pmatrix}, \text{ if } \varepsilon_0 \in \Omega_3^\varepsilon,$$
$$\Omega_3^\varepsilon = \{\varepsilon_0 : \omega_\varepsilon \leq 0, \eta_\varepsilon \leq 0\}, \qquad (2.18c)$$

$$\widehat{\mathbf{G}}^\varepsilon = \widehat{\mathbf{G}}_4^\varepsilon = \frac{3\kappa + 4\mu}{3} \begin{pmatrix} 1 & \frac{\nu}{1-\nu} & 1 \\ \frac{\nu}{1-\nu} & 1 & \frac{\nu}{1-\nu} \\ 1 & \frac{\nu}{1-\nu} & 1 \end{pmatrix}, \text{ if } \varepsilon \in \Omega_4^\varepsilon,$$
$$\Omega_4^\varepsilon = \{\varepsilon_0 : \omega_\varepsilon \geq 0, \eta_\varepsilon \leq 0\}. \qquad (2.18d)$$

When the eigenvalues of ε_0 are of the same sign (i.e., case (2.18a)), $E_p^\varepsilon = E_q^\varepsilon = E_l^\varepsilon$ for rank three MLCs with mutually orthogonal layers (rank three orthogonal MLC). Assuming that the normals to layers point in the principal directions of the tensor ε_0, we obtain, in the notation of §**5**:

$$E_l^\varepsilon = \min_{\alpha_i} \widehat{\varepsilon}_0 \cdot \widehat{\mathbf{D}}^M \cdot \widehat{\varepsilon}_0 = \widehat{\varepsilon}_0 \cdot \widehat{\mathbf{D}}_2 \cdot \widehat{\varepsilon}_0 + \frac{m_1}{m_2} \min_{\alpha_i} \widehat{\varepsilon}_0 \cdot \widehat{\mathbf{Q}}_\varepsilon^{-1} \cdot \widehat{\varepsilon}_0, \qquad (2.19)$$

where

$$\mathbf{Q}^\varepsilon = \sum_{i=1}^{3} \alpha_i \mathcal{N}_i^\varepsilon = \frac{3}{3\kappa + 4\mu} \begin{pmatrix} \alpha_1 & & \\ & \alpha_2 & \\ & & \alpha_3 \end{pmatrix}$$

Choosing α_i as

$$\alpha_1 = \frac{\omega_\varepsilon}{1 + \omega_\varepsilon + \eta_\varepsilon},$$

$$\alpha_2 = \frac{\eta_\varepsilon}{1 + \omega_\varepsilon + \eta_\varepsilon},$$

$$\alpha_3 = \frac{1}{1 + \omega_\varepsilon + \eta_\varepsilon},$$

we see that

$$\widehat{\varepsilon}_0 \cdot \widehat{\mathbf{Q}}_\varepsilon^{-1} \cdot \widehat{\varepsilon}_0 = \widehat{\varepsilon}_0 \cdot \widehat{\mathbf{G}}_1^\varepsilon \cdot \widehat{\varepsilon}_0, \ \text{if} \ \omega_\varepsilon \geq 0, \ \eta_\varepsilon \geq 0.$$

When the eigenvalues of the tensor ε_0 are of different signs, that is, when $\varepsilon_0 \in \Omega_2^\varepsilon \cup \Omega_3^\varepsilon \cup \Omega_4^\varepsilon$, the bounds E_p and E_l for the orthogonal MLCs with layer normals directed along the principal axes of ε_0 do not coincide. In this case one has to look for the optimal structures among a wider class of composites with layers that are not necessarily orthogonal which is a much more difficult task. Nevertheless, it is possible to "guess" some of the parameters of the optimal structure and prove that for the structure in question the bounds E_p and E_l coincide.

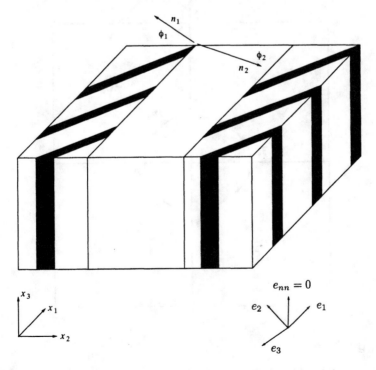

Figure 4.

Let us prove, for example, that $E_p = E_l$ in the subregion Ω_3^ε, where $\varepsilon_1 \leq 0$, $\varepsilon_2 \leq 0$, $\varepsilon_3 \geq 0$. In this case the bound E_p is:

$$E_p^\varepsilon = \widehat{\varepsilon}_0 \cdot \widehat{\mathbf{D}}_\varepsilon \cdot \widehat{\varepsilon}_0 + 4\mu \frac{m_1}{m_2} \varepsilon_1 \varepsilon_2. \tag{2.20}$$

To construct the bound E_l^ε we consider a rank two MLC with layers that are not orthogonal. The rigid phase of this MLC forms cylindrical inclusions, and the rigidity of an MLC is infinite in the direction $\mathbf{n}$ of the axis of the cylinder. Therefore this axis should be chosen in such a way that $\varepsilon_{nn} = 0$, which is always possible for $\varepsilon_2 \varepsilon_3 \leq 0$. Let us compute the matrix $\mathbf{D}^M$ of this composite with respect to some basis associated with a set of axes X_1, X_2, X_3, such that the direction of X_3 coincides with that of $\mathbf{n}$:

$$\mathbf{D}^M = \mathbf{D}_2 + \frac{m_1}{m_2} \left(\alpha_1 \mathcal{N}_1^\varepsilon(\varphi_1) + \alpha_2 \mathcal{N}_2^\varepsilon(\varphi_2) \right)^{-1}, \quad \alpha_1 + \alpha_2 = 1, \tag{2.21}$$

where $\varphi_1(\varphi_2)$ is the angle in the plane $X_1 O X_2$ between the normal to the first (second) layer and the axis X_1 (cf. Figure 5).

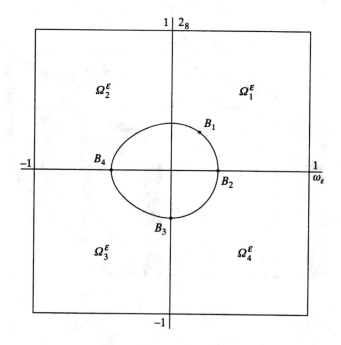

Figure 5.

In the basis (1.6) generated by the unit normals of X_1, X_2, X_3, the nonzero coefficients of the matrix $\mathcal{N}(\varphi_i)$ are:

$$\mathcal{N}_{11} = \frac{3}{3\kappa + 4\mu}c_i^4 + \frac{1}{\mu}s_i^2 c_i^2,$$

$$\mathcal{N}_{22} = \frac{3}{3\kappa + 4\mu}s_i^4 + \frac{1}{\mu}s_i^2 c_i^2,$$

$$\mathcal{N}_{12} = \mathcal{N}_{21} = \left(\frac{3}{3\kappa + 4\mu} - \frac{1}{\mu}\right)s_i^2 c_i^2,$$

$$\mathcal{N}_{44} = \frac{s_i^2}{2\mu},$$

$$\mathcal{N}_{45} = \mathcal{N}_{54} = -\frac{c_i s_i}{2\mu}, \tag{2.22}$$

$$\mathcal{N}_{55} = \frac{c_i^2}{2\mu},$$

$$\mathcal{N}_{61} = \mathcal{N}_{16} = -\frac{3\sqrt{2}}{3\kappa + 4\mu}s_i c_i^3 + \frac{1}{\mu}(c_i^2 - s_i^2)c_i s_i,$$

$$\mathcal{N}_{62} = \mathcal{N}_{26} = -\frac{3\sqrt{2}}{3\kappa + 4\mu}s_i c_i^3 + \frac{1}{\mu}(s_i^2 - c_i^2)c_i s_i,$$

$$\mathcal{N}_{66} = \frac{6}{3\kappa + 4\mu}s_i^2 c_i^2 + \frac{1}{2\mu}(c_i^2 - s_i^2)^2,$$

where we denote $c_i = \cos\varphi_i$, $s_i = \sin\varphi_i$, $i = 1, 2$. If we put $\varphi_1 = -\varphi_2 = \varphi$ and $\alpha_1 = \alpha_2 = \frac{1}{2}$, then

$$\mathbf{D}^M = \mathbf{D}_2 + \frac{m_1}{m_2}\mathbf{Q}_\varepsilon^{-1},$$

and the matrix $\mathbf{Q}_\varepsilon^{-1}$ is:

$$\mathbf{Q}_\varepsilon^{-1} = \begin{pmatrix} \frac{\mu}{c^2} + \frac{3\kappa + \mu}{3} & \frac{3\kappa + \mu}{3} & 0 & 0 & 0 & 0 \\ \frac{3\kappa + \mu}{3} & \frac{\mu}{s^2} + \frac{3\kappa + \mu}{3} & 0 & 0 & 0 & 0 \\ 0 & 0 & \infty & 0 & 0 & 0 \\ 0 & 0 & 0 & \frac{2\mu}{s^2} & 0 & 0 \\ 0 & 0 & 0 & 0 & \frac{2\mu}{c^2} & 0 \\ 0 & 0 & 0 & 0 & 0 & \frac{1}{\mathcal{N}_{66}} \end{pmatrix}. \tag{2.23}$$

Now we choose the axes X_1, X_2, X_3 in such a way that the tensor ε_0 is presented in a basis associated with these axes as

$$\varepsilon_0 = \begin{pmatrix} \varepsilon_1 & 0 & 0 \\ 0 & \varepsilon_1 + \varepsilon_2 & \sqrt{-\varepsilon_2 \varepsilon_3} \\ 0 & \sqrt{-\varepsilon_2 \varepsilon_3} & 0 \end{pmatrix}, \tag{2.24}$$

so that the elements of ε_0 in the basis $e_1, \ldots, e_6$ are

$$\varepsilon^1 = \varepsilon_1, \quad \varepsilon^2 = \varepsilon_2 + \varepsilon_3, \quad \varepsilon^3 = 0, \quad \varepsilon^4 = \sqrt{-2\varepsilon_2\varepsilon_3}, \quad \varepsilon^5 = 0, \quad \varepsilon^6 = 0.$$

(This is possible since $\varepsilon_2\varepsilon_3 \leq 0$.) We obtain for the energy density of this composite

$$\begin{aligned}
E_M^\varepsilon &= \varepsilon_0 \cdot \cdot \mathbf{D}_2 \cdot \cdot \varepsilon_0 \\
&\quad + \frac{m_1}{m_2}\left[\frac{3\kappa + \mu}{3}(\varepsilon_1 + \varepsilon_2 + \varepsilon_3)^2 + \mu\left(\frac{\varepsilon_1^2}{c^2} + \frac{(\varepsilon_2 - \varepsilon_3)^2}{s^2}\right)\right] \\
&= \frac{1}{m_2}\varepsilon_0 \cdot \cdot \mathbf{D}_2 \cdot \cdot \varepsilon_0 \\
&\quad + \mu\frac{m_1}{m_2}\left[\frac{\varepsilon_1^2}{c^2} + \frac{(\varepsilon_2 - \varepsilon_3)^2}{(1 - c^2)}\right. \\
&\quad \left. + 4(\varepsilon_1\varepsilon_2 + \varepsilon_1\varepsilon_3 + \varepsilon_2\varepsilon_3) - (\varepsilon_1 + \varepsilon_2 + \varepsilon_3)^2\right].
\end{aligned} \tag{2.25}$$

The minimum of (2.25) with respect to φ is reached for the angle φ_0 which satisfies

$$\cos^2\varphi_0 = \frac{\varepsilon_1}{\varepsilon_1 + \varepsilon_2 - \varepsilon_3}. \tag{2.26}$$

When $\varphi = \varphi_0$, the energy $\varepsilon_0 \cdot \cdot \mathbf{D}^M(\varphi_0) \cdot \cdot \varepsilon_0$ equals E_p^ε (cf. (2.20)). Changing the indices of ε_i, we find in an analogous way the function E_l^ε for the subregions Ω_2^ε and Ω_4^ε, which coincides with E_p^ε and therefore with E_q^ε as well.

We have thus proved the following.

Theorem II. *For a composite of minimal rigidity with one of the phases having infinite rigidity the following assertions hold:*

(1) $E_p^\varepsilon = E_q^\varepsilon = E_l^\varepsilon$;

(2) the minimum of the energy E_q^ε is determined by the formulae (2.18);

(3) the optimal values of the parameters a_i are determined by the formulae (2.17);

(4) the composites of optimal structures are either a rank three MLC with normals to layers having the same direction as the principal axes of the strain tensor ε_0 (in the subregion Ω_1^ε) or rank two MLC with a particular spatial orientation of layers (cf. (2.24)), a variable angle φ_0 between the layer normals (cf. (2.25)), and parameters $\alpha_1 = \alpha_2 = \frac{1}{2}$, $\alpha_3 = 0$.

Let us examine how the optimal structure of a composite changes as we follow the curve $B_1 B_2 B_3 B_4$ in the $(\omega_\varepsilon, \eta_\varepsilon)$ plane (cf. Figure 4). At the point B_1 the optimal composite is a rank three MLC with the normals to layers directed along the principal axes of ε_0. As the point B_2 is reached, the optimal structure degenerates to that of a rank two MLC with orthogonal layers, the layer normals being directed along the principal axes of ε, and the axis n of the cylindrical inclusions being the axis X_1. As we follow the curve further into the interior of the subregion Ω_4^ε, the axes of the cylindrical inclusions start following the direction $\varepsilon_{nn} = 0$. Furthermore, the layers cease to be orthogonal, while the angle $2\varphi_0$ between the normals increases, reaching eventually a value of π at the point B_3 which is the case $\varepsilon_2\varepsilon_3 \leq 0$, $\varepsilon_1 = 0$. The optimal structure at the point B_3 is that of a rank one MLC, that is, a rank one laminates such that in the plane of the layers all elements of the tensor ε_0 are equal to zero. Past the point B_3 the optimal structure is again that of a rank two MLC. Here the direction of the axis of the cylindrical inclusions changes in a jump-like fashion (this direction is not defined at the point B_3 itself). Further along the curve, the described behaviour repeats, up to a change in axes, in reverse order. We remark that the solution for the subregions Ω_2^ε, Ω_3^ε, and Ω_4^ε is not unique. Indeed, a rank two MLC, with the configuration of layers being such that in the basis associated with axes X_1, X_2, X_3 the strain tensor is

$$\varepsilon_0 = \begin{pmatrix} \varepsilon_1 + \varepsilon_2 & 0 & \sqrt{-\varepsilon_1\varepsilon_3} \\ 0 & \varepsilon_2 & 0 \\ \sqrt{-\varepsilon_1\varepsilon_3} & 0 & 0 \end{pmatrix},$$

$$(\varepsilon^1 = \varepsilon_1 + \varepsilon_3,\ \varepsilon^2 = \varepsilon_2,\ \varepsilon^3 = 0,\ \varepsilon^4 = 0,\ \varepsilon^5 = \sqrt{-2\varepsilon_1\varepsilon_3},\ \varepsilon^6 = 0),$$

(2.27)

also yields the optimal value of the energy E_i^ε, provided the angle φ_0' satisfies

$$\cos^2 \varphi_0' = \frac{\varepsilon_2}{\varepsilon_1 + \varepsilon_2 - \varepsilon_3}. \tag{2.28}$$

§9 Let us study the two-dimensional version of the problem of a composite of maximal rigidity. Let a state of a body be described by the equations of the two-dimensional elasticity theory (plane stress state) [4]:

$$\widetilde{\nabla}\cdot\widetilde{\sigma} = 0,$$
$$\widetilde{\varepsilon} = \widetilde{C}\cdot\widetilde{\sigma}, \tag{2.29}$$
$$\widetilde{\varepsilon} = \frac{1}{2}\left(\widetilde{\nabla}\widetilde{u} + \left(\widetilde{\nabla}\widetilde{u}\right)^T\right),$$

where $\widetilde{\nabla}$ is a two-dimensional gradient operator, and

$$\widetilde{\varepsilon} = \{\widetilde{\varepsilon}_{ij}\},\ (\varepsilon_{ij} \in L_2(V))$$
$$\widetilde{\sigma} = \{\widetilde{\sigma}_{ij}\},\ (\sigma_{ij} \in L_2(V))$$

are symmetric strain and stress tensors, $\tilde{u} = (u_1, u_2)$ is a vector of displacement, and $\widetilde{\mathbf{C}}(\tilde{\kappa}, \mu)$ is a compliance tensor of the material. We represent $\tilde{\varepsilon}, \tilde{\sigma}, \widetilde{\mathbf{C}}$ in the basis

$$
\begin{aligned}
\tilde{e}_1 &= \frac{1}{\sqrt{2}} \left(\mathbf{i} \otimes \mathbf{i} + \mathbf{j} \otimes \mathbf{j} \right), \\
\tilde{e}_2 &= \frac{1}{\sqrt{2}} \left(\mathbf{i} \otimes \mathbf{i} - \mathbf{j} \otimes \mathbf{j} \right), \\
\tilde{e}_3 &= \frac{1}{\sqrt{2}} \left(\mathbf{i} \otimes \mathbf{j} + \mathbf{j} \otimes \mathbf{i} \right).
\end{aligned}
\tag{2.30}
$$

The compliance tensor of an isotropic material is given by

$$
\widetilde{\mathbf{C}}(\tilde{\kappa}, \mu) = \frac{1}{2\tilde{\kappa}} \tilde{e}_1 \otimes \tilde{e}_1 + \frac{1}{2\mu} \left(\tilde{e}_2 \otimes \tilde{e}_2 + \tilde{e}_3 \otimes \tilde{e}_3 \right),
\tag{2.31}
$$

where

$$
\tilde{\kappa} = \kappa + \frac{\mu}{3},
$$

is the bulk modulus for planar strain. The periodic cell of the composite in this case is a unit square $\tilde{V}$, divided into parts $\tilde{V}_1$ and $\tilde{V}_2$ filled by materials with compliance tensors $\widetilde{\mathbf{C}}_1(\tilde{\kappa}_1, \mu_1)$ and $\widetilde{\mathbf{C}}_2(\tilde{\kappa}_2, \mu)$. We assume that

$$
\tilde{\kappa}_1 \geq \tilde{\kappa}_2, \ \mu_1 \geq \mu_2.
$$

The bound E_p of the minimal elastic energy density E_q is given by (1.18), where we shall put

$$
\begin{aligned}
\mathbf{C}_i &= \widetilde{\mathbf{C}}_i(\tilde{\kappa}_i, \mu_i) \\
\Phi &= \tilde{\Phi}(a) = \widetilde{\mathbf{C}}(-a, a).
\end{aligned}
$$

Indeed, letting the dimension of the space equal to 2 in Lemma **1**, we establish the quasiconvexity of the quadratic form

$$
\tilde{\sigma} \cdot \cdot \tilde{\Phi}(a) \cdot \cdot \tilde{\sigma} = -2a \left(\tilde{\sigma}_{11} \tilde{\sigma}_{22} - \tilde{\sigma}_{12}^2 \right).
$$

Using this, we construct the bound

$$
\begin{aligned}
E_p &= \max_{a \in \tilde{A}} \tilde{\sigma}_0 \cdot \cdot \mathbf{F} \left(\widetilde{\mathbf{C}}_1, \widetilde{\mathbf{C}}_2, \tilde{\Phi}(a) \right) \cdot \cdot \tilde{\sigma}_0, \\
\tilde{A} &= \left\{ a : \widetilde{\mathbf{C}}_i - \tilde{\Phi}(a) \geq 0, \ i = 1, 2 \right\}.
\end{aligned}
\tag{2.32}
$$

Since the basis (2.30) is an eigenbasis of the tensors $\tilde{\mathbf{C}}_1(\tilde{\kappa}_1, \mu_1)$, $\tilde{\mathbf{C}}_2(\tilde{\kappa}_1, \mu_2)$, and $\tilde{\Phi}$ appearing in (2.32), we have that

$$\mathbf{F}\left(\tilde{\mathbf{C}}_1, \tilde{\mathbf{C}}_2, \tilde{\Phi}(a)\right) = \gamma\left(\tilde{e}_1 \otimes \tilde{e}_1\right) + \rho\left(\tilde{e}_2 \otimes \tilde{e}_2 + \tilde{e}_3 \otimes \tilde{e}_3\right), \qquad (2.33)$$

where

$$\gamma = \left[m_1 \left(\frac{1}{2\tilde{\kappa}_1} + a\right)^{-1} + m_2 \left(\frac{1}{2\tilde{\kappa}_2} + a\right)^{-1}\right]^{-1} - a,$$

$$\rho = \left[m_1 \left(\frac{1}{2\mu_1} - a\right)^{-1} + m_2 \left(\frac{1}{2\mu_2} - a\right)^{-1}\right]^{-1} + a,$$

and the parameter a belongs to the interval

$$-\frac{1}{2\tilde{\kappa}_1} \leq a \leq \frac{1}{2\mu_1}. \qquad (2.34)$$

Taking the maximum in (2.32), we obtain

$$E_p = \begin{cases} \tilde{\sigma}_0 \cdot \cdot \tilde{\mathbf{C}}_I \cdot \cdot \tilde{\sigma}_0, & \text{if } \tilde{\sigma}_0 \in \tilde{\Omega}_1, \\ \tilde{E}_{II}, & \text{if } \tilde{\sigma}_0 \in \tilde{\Omega}_2, \\ \tilde{\sigma}_0 \cdot \cdot \tilde{\mathbf{C}}_{III} \cdot \cdot \tilde{\sigma}_0, & \text{if } \tilde{\sigma}_0 \in \tilde{\Omega}_3, \end{cases} \qquad (2.35)$$

where

$$\tilde{\mathbf{C}}_I = \tilde{\mathbf{C}}\left(\tilde{\kappa}_1, \Theta_I\right), \qquad (2.35a)$$

$$\Theta_I = \left\{\left[m_1 \left(\frac{1}{\mu_1} + \frac{1}{\tilde{\kappa}_1}\right)^{-1} + m_2 \left(\frac{1}{\mu_2} + \frac{1}{\tilde{\kappa}_1}\right)^{-1}\right]^{-1} - \frac{1}{\tilde{\kappa}_1}\right\}^{-1}$$

$$\tilde{E}_{II} = \left(\frac{m_1}{\tilde{\kappa}_1} + \frac{m_2}{\tilde{\kappa}_2}\right)\left(\frac{\tilde{\sigma}_1 + \tilde{\sigma}_2}{2}\right)^2 + \left(\frac{m_1}{\mu_1} + \frac{m_2}{\mu_2}\right)\left(\frac{\tilde{\sigma}_1 - \tilde{\sigma}_2}{2}\right)^2$$

$$- \frac{m_1 m_2}{4}\left(\frac{m_1}{\tilde{\kappa}_2} + \frac{m_2}{\tilde{\kappa}_1} + \frac{m_1}{\mu_2} + \frac{m_2}{\mu_1}\right)^{-1}\left[\left(\frac{1}{\tilde{\kappa}_2} - \frac{1}{\tilde{\kappa}_1}\right)|\tilde{\sigma}_1 + \tilde{\sigma}_2|\right.$$

$$\left. + \left(\frac{1}{\mu_2} - \frac{1}{\mu_1}\right)|\tilde{\sigma}_1 - \tilde{\sigma}_2|\right]^2, \qquad (2.35b)$$

$$\tilde{\mathbf{C}}_{III} = \tilde{\mathbf{C}}\left(\Theta_{III}, \mu_1\right),$$

$$\Theta_{III} = \left\{\left[m_1\left(\frac{1}{\tilde{\kappa}_1} + \frac{1}{\mu_1}\right)^{-1} + m_2\left(\frac{1}{\tilde{\kappa}_2} + \frac{1}{\mu_1}\right)^{-1}\right]^{-1} - \frac{1}{\mu_1}\right\}^{-1},$$

(2.35c)

and

$$\tilde{\Omega}_1 = \left\{\tilde{\sigma}_0 : \left|\frac{\tilde{\sigma}_1 - \tilde{\sigma}_2}{\tilde{\sigma}_1 + \tilde{\sigma}_2}\right| \geq \xi_1\right\}, \ \xi_1 = \frac{m_1/\mu_2 + m_2/\mu_1 + 1/\tilde{\kappa}_1}{m_1\left(1/\mu_2 - 1/\mu_1\right)},$$

$$\tilde{\Omega}_2 = \left\{\tilde{\sigma}_0 : \left|\frac{\tilde{\sigma}_1 - \tilde{\sigma}_2}{\tilde{\sigma}_1 + \tilde{\sigma}_2}\right| \in [\xi_2, \xi_1]\right\}, \ \xi_2 = \frac{(1/\tilde{\kappa}_2 - 1/\tilde{\kappa}_1)\, m_1}{m_1/\tilde{\kappa}_2 + m_2/\tilde{\kappa}_1 + 1/\mu_1}, \quad (2.36)$$

$$\tilde{\Omega}_3 = \left\{\tilde{\sigma}_0 : \left|\frac{\tilde{\sigma}_1 - \tilde{\sigma}_2}{\tilde{\sigma}_1 + \tilde{\sigma}_2}\right| \leq \xi_2\right\},$$

and $\tilde{\sigma}_1, \tilde{\sigma}_2$ are the eigenvalues of the tensor $\tilde{\sigma}_0$. The bound $\tilde{\mathbf{E}}_{II}$ corresponds to the critical value of the parameter a which belongs to the interval (2.34) and is equal to

$$a_0 = \frac{\left(\frac{1}{\tilde{\kappa}_1} - \frac{1}{\tilde{\kappa}_2}\right)\left(\frac{m_1}{\mu_2} + \frac{m_2}{\mu_1}\right)|\tilde{\sigma}_1 + \tilde{\sigma}_2| - \left(\frac{1}{\mu_1} - \frac{1}{\mu_2}\right)\left(\frac{m_1}{\tilde{\kappa}_2} + \frac{m_2}{\tilde{\kappa}_1}\right)|\tilde{\sigma}_1 - \tilde{\sigma}_2|}{2\left[\left(\frac{1}{\tilde{\kappa}_1} - \frac{1}{\tilde{\kappa}_2}\right)|\tilde{\sigma}_1 + \tilde{\sigma}_2| + \left(\frac{1}{\mu_1} - \frac{1}{\mu_2}\right)|\tilde{\sigma}_1 - \tilde{\sigma}_2|\right]},$$

whereas the bounds $\tilde{\sigma}_0 \cdot\cdot \mathbf{C}_I \cdot\cdot \tilde{\sigma}_0$ and $\tilde{\sigma}_0 \cdot\cdot \mathbf{C}_{III} \cdot\cdot \tilde{\sigma}_0$ correspond to the left and right endpoints of (2.34). These bounds are exact. The optimal structures are a rank two MLC with orthogonal layers such that the normals to the layers are directed along the principal axes of the tensor $\tilde{\sigma}_0$.

Indeed, in the basis

$$\mathbf{i} \otimes \mathbf{i} = \frac{1}{\sqrt{2}}\left(\tilde{e}_1 + \tilde{e}_2\right), \ \mathbf{j} \otimes \mathbf{j} = \frac{1}{\sqrt{2}}\left(\tilde{e}_1 - \tilde{e}_2\right), \ \tilde{e}_3$$

the matrices $\mathcal{N}_1$, and $\mathcal{N}_2$ (cf. (1.23)) are

$$\mathcal{N}_1 = \begin{pmatrix} 0 & 0 & 0 \\ 0 & \frac{4\tilde{\kappa}_1\mu_1}{\tilde{\kappa}_1+\mu_1} & 0 \\ 0 & 0 & 0 \end{pmatrix}, \ \mathcal{N}_2 = \begin{pmatrix} \frac{4\tilde{\kappa}_1\mu_1}{\tilde{\kappa}_1+\mu_1} & 0 & 0 \\ 0 & 0 & 0 \\ 0 & 0 & 0 \end{pmatrix}. \quad (2.37)$$

So the matrix $\tilde{\mathbf{C}}^M$ is

$$\tilde{\mathbf{C}}^M = \tilde{\mathbf{C}}_1 + \left[\frac{1}{m_2}\left(\tilde{\mathbf{C}}_2 - \tilde{\mathbf{C}}_1\right)^{-1} + \frac{m_1}{m_2}\left(\alpha_1\mathcal{N}_1 + \alpha_2\mathcal{N}_2\right)\right]^{-1}$$

$$= \tilde{\mathbf{C}}_1 + m_2 \begin{pmatrix} d_{11} & d_{12} & 0 \\ d_{21} & d_{22} & 0 \\ 0 & 0 & \frac{2\mu_1\mu_2}{\mu_1-\mu_2} \end{pmatrix},$$

(2.38)

where

$$d_{11} = \frac{\tilde{\kappa}_1 \tilde{\kappa}_2}{\tilde{\kappa}_1 - \tilde{\kappa}_2} + \frac{\mu_1 \mu_2}{\mu_1 - \mu_2} + \frac{4m_1 \tilde{\kappa}_1 \mu_1}{\tilde{\kappa}_1 + \mu_1} \alpha_2,$$

$$d_{12} = d_{21} = \frac{\tilde{\kappa}_1 \tilde{\kappa}_2}{\tilde{\kappa}_1 - \tilde{\kappa}_2} - \frac{\mu_1 \mu_2}{\mu_1 - \mu_2},$$

$$d_{22} = \frac{\tilde{\kappa}_1 \tilde{\kappa}_2}{\tilde{\kappa}_1 - \tilde{\kappa}_2} + \frac{\mu_1 \mu_2}{\mu_1 - \mu_2} + \frac{4m_1 \tilde{\kappa}_1 \mu_1}{\tilde{\kappa}_1 + \mu_1} \alpha_1.$$

Choosing the parameters

$$\alpha_1 = \frac{1}{2} - \frac{\tilde{\sigma}_1 + \tilde{\sigma}_2}{2m_1 (\tilde{\sigma}_1 - \tilde{\sigma}_2)} \left(\frac{1/\tilde{\kappa}_1 + 1/\mu_1}{1/\mu_2 - 1/\mu_1} + m_1 \right), \quad \text{if } \tilde{\sigma}_0 \in \tilde{\Omega}_1, \qquad (2.39)$$

$$\alpha_2 = 1 - \alpha_1$$

or

$$\begin{cases} \alpha_1 = 1, \ \alpha_2 = 0, & \text{if } |\tilde{\sigma}_1| \le |\tilde{\sigma}_2| \\ \alpha_1 = 0, \ \alpha_2 = 1, & \text{if } |\tilde{\sigma}_1| \ge |\tilde{\sigma}_2| \end{cases}, \quad \text{if } \tilde{\sigma}_0 \in \tilde{\Omega}_2, \qquad (2.40)$$

or

$$\alpha_1 = \frac{1}{2} - \frac{\tilde{\sigma}_1 - \tilde{\sigma}_2}{2m_1 (\tilde{\sigma}_1 + \tilde{\sigma}_2)} \left(\frac{1/\tilde{\kappa}_1 + 1/\mu_1}{1/\tilde{\kappa}_2 - 1/\tilde{\kappa}_1} + m_1 \right), \quad \text{if } \tilde{\sigma}_0 \in \tilde{\Omega}_3, \qquad (2.41)$$

$$\alpha_2 = 1 - \alpha_1$$

we obtain the expressions for the quadratic form E_q for which $E_p = E_q = E_l$:

$$E_q = \frac{1}{\tilde{\kappa}_1} \left(\frac{\tilde{\sigma}_1 + \tilde{\sigma}_2}{2} \right)^2 +$$

$$\left\{ \left[m_1 \left(\frac{1}{\tilde{\kappa}_1} + \frac{1}{\mu_1} \right)^{-1} + m_2 \left(\frac{1}{\tilde{\kappa}_1} + \frac{1}{Gm_2} \right)^{-1} \right]^{-1} - \frac{1}{\tilde{\kappa}_1} \right\} \left(\frac{\tilde{\sigma}_1 - \tilde{\sigma}_2}{2} \right)^2, \ \cdot \text{ if } \tilde{\sigma}_0 \in \tilde{\Omega}_1;$$

$$E_q = \tilde{E}_{II}, \quad \text{if } \tilde{\sigma}_0 \in \tilde{\Omega}_2; \qquad (2.42)$$

$$E_q = \frac{1}{\mu_1} \left(\frac{\tilde{\sigma}_1 - \tilde{\sigma}_2}{2} \right)^2$$

$$+ \left\{ \left[m_1 \left(\frac{1}{\tilde{\kappa}_1} + \frac{1}{\mu_1} \right)^{-1} + m_2 \left(\frac{1}{\tilde{\kappa}_2} + \frac{1}{\mu_1} \right)^1 \right]^{-1} - \frac{1}{\mu_1} \right\} \left(\frac{\tilde{\sigma}_1 + \tilde{\sigma}_2}{2} \right)^2, \quad \text{if } \tilde{\sigma}_0 \in \tilde{\Omega}_3.$$

In the case of a porous material ($\tilde{\kappa}_2 = 0$, $\mu_2 = 0$) we have

$$E_q = \begin{cases} \tilde{\sigma}_0 \cdot\cdot \tilde{C}(\tilde{\kappa}_1, \mu') \cdot\cdot \tilde{\sigma}_0, & \text{if } \left|\frac{\tilde{\sigma}_1 - \tilde{\sigma}_2}{\tilde{\sigma}_1 + \tilde{\sigma}_2}\right| \geq 1, \\ \tilde{\sigma}_0 \cdot\cdot \tilde{C}(\kappa', \mu_1) \cdot\cdot \tilde{\sigma}_0, & \text{if } \left|\frac{\tilde{\sigma}_1 - \tilde{\sigma}_2}{\tilde{\sigma}_1 + \tilde{\sigma}_2}\right| \leq 1, \end{cases}$$

$$\mu' = \frac{m_1 \tilde{\kappa}_1 \mu_1}{\tilde{\kappa}_1 + m_2 \mu_1}, \quad \kappa' = \frac{m_1 \tilde{\kappa}_1 \mu_1}{m_2 \tilde{\kappa}_1 + \mu_1}. \tag{2.43}$$

The solution for this asymptotic case differs from the general case by the fact that the subregion $\tilde{\Omega}_2$, which corresponds to simple laminate composites, reduces to the point $|\tilde{\sigma}_1 - \tilde{\sigma}_2| / |\tilde{\sigma}_1 + \tilde{\sigma}_2| = 1$ where one of the eigenvalues of the tensor $\tilde{\sigma}_0$ is equal to zero.

We can guess that the three dimensional problem of two nonzero phases (cf. §3, §4) differs from its asymptotic case $\kappa_2 = \mu_2 = 0$ (cf. §7) by the appearance in the (ω, η) plane of a new subregion of simple laminate composites in the neighborhood of the axes $\omega = 0$ and $\eta = 0$ (cf. Figure 2).

Remark I. Comparing the formulae (2.9e)–(2.9g) with (2.39), we notice that the MLC that are optimal in the two-dimensional case are optimal in the three-dimensional case as well, if the axis X_3 can be chosen in the direction of the eigenvector associated with σ_3 and

$$|\sigma_3| \gg |\sigma_1|, \quad |\sigma_3| \gg |\sigma_2|.$$

If one of the eigenvalues of σ_0 is zero, for example $\sigma_1 = 0$ (plane stress state), then the optimal structure for the three-dimensional case is a rank one laminate with layers parallel to the plane of the load $\sigma_2\sigma_3$. In this case the bound (2.7) becomes a Reuss type bound (1.16).

Therefore the three-dimensional case with $\sigma_1 = 0$ is not equivalent to a corresponding plane stress problem; for the latter problem only composites that vary the properties in the directions x_2, x_3 are allowed. This requirement prohibits composites with layers in the direction of the x_1 axis, which are optimal in a three-dimensional case.

The problem of a composite of minimal rigidity has its two-dimensional analogue as well. The corresponding bounds are given by (1.29) with ε_0 changed to $\tilde{\varepsilon}_0$, D changed to the tensor $\tilde{D} = \tilde{C}^{-1}$, and Φ_ε to the tensor $\tilde{\Phi}_\varepsilon$, which in the basis (2.30) is

$$\tilde{\Phi}_\varepsilon(a) = \begin{pmatrix} -a^2 & 0 & 0 \\ 0 & a^2 & 0 \\ 0 & 0 & a^2 \end{pmatrix}. \tag{2.44}$$

The condition $\tilde{D}_i - \tilde{\Phi}_\varepsilon \geq 0$ is equivalent to the requirement $a^2 \in$

$[0, 2\mu_2]$, and the bound E_p^ε is

$$E_p^\varepsilon = \left(\frac{m_1}{\widetilde{\kappa}_1} + \frac{m_2}{\widetilde{\kappa}_2}\right)^{-1} (\widetilde{\varepsilon}_1 + \widetilde{\varepsilon}_2)^2 + \left(\frac{m_1}{\mu_1} + \frac{m_2}{\mu_2}\right)^{-1} (\widetilde{\varepsilon}_1 - \widetilde{\varepsilon}_2)^2, \text{ if } \widetilde{\varepsilon}_0 \in \widetilde{\Omega}_1^\varepsilon,$$

$$E_p^\varepsilon = (m_1\widetilde{\kappa}_1 + m_2\widetilde{\kappa}_2)(\widetilde{\varepsilon}_1 + \widetilde{\varepsilon}_2)^2 + (m_1\mu_1 + m_2\mu_2)(\widetilde{\varepsilon}_1 - \widetilde{\varepsilon}_2)^2$$
$$- \frac{m_1m_2((\widetilde{\kappa}_2 - \widetilde{\kappa}_1)|\widetilde{\varepsilon}_1 + \widetilde{\varepsilon}_2| + (\mu_2 - \mu_1)|\widetilde{\varepsilon}_1 - \widetilde{\varepsilon}_2|)^2}{m_1(\widetilde{\kappa}_2 + \mu_2) + m_2(\widetilde{\kappa}_1 + \mu_1)}, \text{ if } \widetilde{\varepsilon}_0 \in \widetilde{\Omega}_2^\varepsilon,$$

$$E_p^\varepsilon = \left[\left[\frac{m_1}{\widetilde{\kappa}_1 + \mu_2} + \frac{m_2}{\widetilde{\kappa}_2 + \mu_2}\right]^{-1} - \mu_2\right](\widetilde{\varepsilon}_1 + \widetilde{\varepsilon}_2)^2 + \mu_2(\widetilde{\varepsilon}_1 - \widetilde{\varepsilon}_2)^2, \text{ if } \widetilde{\varepsilon}_0 \in \widetilde{\Omega}_3^\varepsilon,$$
$$(2.45)$$

where

$$\widetilde{\Omega}_1^\varepsilon = \left\{\widetilde{\varepsilon}_0 : \left|\frac{\widetilde{\varepsilon}_1 - \widetilde{\varepsilon}_2}{\widetilde{\varepsilon}_1 + \widetilde{\varepsilon}_2}\right| \geq \zeta_1\right\}, \ \zeta_1 = \frac{(m_2\mu_1 + m_1\mu_2)(\widetilde{\kappa}_2 - \widetilde{\kappa}_1)}{(m_2\widetilde{\kappa}_1 + m_1\widetilde{\kappa}_2)(\mu_2 - \mu_1)},$$

$$\widetilde{\Omega}_2^\varepsilon = \left\{\widetilde{\varepsilon}_0 : \left|\frac{\widetilde{\varepsilon}_1 - \widetilde{\varepsilon}_2}{\widetilde{\varepsilon}_1 + \widetilde{\varepsilon}_2}\right| \in [\zeta_2, \zeta_1]\right\}, \ \zeta_2 = \frac{m_2(\widetilde{\kappa}_2 - \widetilde{\kappa}_1)}{m_1\widetilde{\kappa}_2 + m_2\widetilde{\kappa}_1 + \mu_2}, \quad (2.46)$$

$$\widetilde{\Omega}_3^\varepsilon = \left\{\widetilde{\varepsilon}_0 : \left|\frac{\widetilde{\varepsilon}_1 - \widetilde{\varepsilon}_2}{\widetilde{\varepsilon}_1 + \widetilde{\varepsilon}_2}\right| \leq \zeta_2\right\}.$$

In the subregion $\widetilde{\Omega}_2^\varepsilon$ the bound corresponds to the critical value of the parameter a^2, which equals

$$a_0^2 = \frac{(\widetilde{\varepsilon}_1 + \widetilde{\varepsilon}_2)(m_1\mu_2 + m_2\mu_1) - (\mu_2 - \mu_1)(m_1\widetilde{\kappa}_2 + m_2\widetilde{\kappa}_1)(\widetilde{\varepsilon}_1 - \widetilde{\varepsilon}_2)}{2[(\widetilde{\kappa}_2 - \widetilde{\kappa}_1)(\widetilde{\varepsilon}_1 + \widetilde{\varepsilon}_2) + (\mu_2 - \mu_1)(\widetilde{\varepsilon}_1 - \widetilde{\varepsilon}_2)]}.$$
$$(2.47)$$

In the subregion $\widetilde{\Omega}_3^\varepsilon$ it corresponds to $a = 0$, and in the subregion $\widetilde{\Omega}_1^\varepsilon$ to $a^2 = 2\mu$. Here also the bounds are exact, and the optimal structures are known. In the subregion $\widetilde{\Omega}_3^\varepsilon$ the optimal structure is that of an MLC for which matrices $\mathcal{N}_1^\varepsilon, \mathcal{N}_2^\varepsilon$ in the basis (2.30) are

$$\mathcal{N}_1^\varepsilon = \begin{pmatrix} \frac{1}{\widetilde{\kappa}_2 + \mu_2} & 0 & 0 \\ 0 & 0 & 0 \\ 0 & 0 & \frac{1}{2\mu_2} \end{pmatrix}, \ \mathcal{N}_2^\varepsilon = \begin{pmatrix} 0 & 0 & 0 \\ 0 & \frac{1}{\widetilde{\kappa}_2 + \mu_2} & 0 \\ 0 & 0 & \frac{1}{2\mu_2} \end{pmatrix}, \quad (2.48)$$

and the optimal values of the parameters α_1, α_2 are

$$\alpha_1 = \frac{1}{2} + \frac{\widetilde{\varepsilon}_1 - \widetilde{\varepsilon}_2}{2(\widetilde{\varepsilon}_1 + \widetilde{\varepsilon}_2)m_2}\left(\frac{\widetilde{\kappa}_2 + \mu_2}{\widetilde{\kappa}_1 - \widetilde{\kappa}_2} + m_2\right), \ \alpha_2 = 1 - \alpha_1.$$

In the subregion $\widetilde{\Omega}_2^\varepsilon$ (which disappears when $\widetilde{\kappa}_1 \to \infty$, $\mu_1 \to \infty$), the optimal structures are rank one laminates with normals to layers pointing along the principal directions of the tensor $\widetilde{\varepsilon}_0$. As for the subregion $\widetilde{\Omega}_1^\varepsilon$

for which the eigenvalues of the tensor $\tilde{\varepsilon}_0$ have different signs, the optimal composites are rank one laminates with orientation of the layers such that $\tilde{\sigma}_{tt} = 0$, where t is the direction tangent to layers. One can check that the angle ψ between the direction of the eigenvector associated with the eigenvalue $\tilde{\varepsilon}_1$ and the axis $\mathbf{n}$ equals

$$\psi = \frac{1}{2} \arccos \left[-\frac{(\tilde{\kappa}_2 - \tilde{\kappa}_1)(m_2\mu_1 + m_1\mu_2)(\tilde{\varepsilon}_1 + \tilde{\varepsilon}_2)}{(\mu_2 - \mu_1)(m_2\tilde{\kappa}_1 + m_1\tilde{\kappa}_2)(\tilde{\varepsilon}_1 - \tilde{\varepsilon}_2)} \right].$$

Since the other elements σ_{nn} and σ_{nt} of the tensor $\tilde{\sigma}_0$ are continuous in the laminate composite, the tensor $\tilde{\sigma}_0$ is also continuous. Thus it is clear that the quadratic form of the energy of such a composite yields a bound of a Reuss-Voigt type (cf. (1.16)).

In the asymptotic case $\tilde{\kappa}_1 \to \infty$, $\mu_1 \to \infty$ the subregion $\tilde{\Omega}_2^\varepsilon$ reduces to the point $|\tilde{\varepsilon}_1 - \tilde{\varepsilon}_2| / |\tilde{\varepsilon}_1 + \tilde{\varepsilon}_2| = 1$, which is the case of uniaxial loading, and the bound becomes

$$E_p^\varepsilon = \begin{cases} \frac{1}{m_2}\kappa_2\left(\tilde{\varepsilon}_1 + \tilde{\varepsilon}_2\right)^2 + \frac{\mu_2}{m_2}\left(\tilde{\varepsilon}_1 - \tilde{\varepsilon}_2\right)^2, & \text{if } \left|\frac{\tilde{\varepsilon}_1 - \tilde{\varepsilon}_2}{\tilde{\varepsilon}_1 + \tilde{\varepsilon}_2}\right| \geq 1, \\ \frac{\kappa_2 + m_1\mu_2}{m_2}\left(\tilde{\varepsilon}_1 + \tilde{\varepsilon}_2\right)^2 + \mu_2\left(\tilde{\varepsilon}_1 - \tilde{\varepsilon}_2\right)^2, & \text{if } \left|\frac{\tilde{\varepsilon}_1 - \tilde{\varepsilon}_2}{\tilde{\varepsilon}_1 + \tilde{\varepsilon}_2}\right| \leq 1 \end{cases} \tag{2.49}$$

with the optimal structures being those for which

$$\begin{cases} \alpha_1 = 1, \ \alpha_2 = 0, & \text{if } |\tilde{\varepsilon}_1| \geq |\tilde{\varepsilon}_2|, \ \left|\frac{\tilde{\varepsilon}_1 - \tilde{\varepsilon}_2}{\tilde{\varepsilon}_1 + \tilde{\varepsilon}_2}\right| \geq 1, \\ \alpha_1 = 0, \ \alpha_2 = 1, & \text{if } |\tilde{\varepsilon}_1| \leq |\tilde{\varepsilon}_2|, \ \left|\frac{\tilde{\varepsilon}_1 - \tilde{\varepsilon}_2}{\tilde{\varepsilon}_1 + \tilde{\varepsilon}_2}\right| \geq 1, \\ \alpha_1 = \frac{\tilde{\varepsilon}_1}{\tilde{\varepsilon}_1 + \tilde{\varepsilon}_2}, \ \alpha_2 = \frac{\tilde{\varepsilon}_2}{\tilde{\varepsilon}_1 + \tilde{\varepsilon}_2}, & \text{if } \left|\frac{\tilde{\varepsilon}_1 - \tilde{\varepsilon}_2}{\tilde{\varepsilon}_1 + \tilde{\varepsilon}_2}\right| < 1. \end{cases}$$

Comparing these exact bounds with the results for the three dimensional case, one sees that the solution for the region Ω_1^ε is the asymptotic case of the corresponding three-dimensional solutions for $\varepsilon_1 \to 0$. The corresponding bound becomes the second bound of (2.49) when $\varepsilon_1 \to 0$, and the angle φ approaches $\frac{\pi}{2}$ and the optimal structure reduces to that of a laminate composite with the angle between the layers and the principal axes of the tensor $\tilde{\varepsilon}_0$ being given by

$$\varphi = \frac{1}{2} \arccos \left[-\frac{\tilde{\varepsilon}_1 + \tilde{\varepsilon}_2}{\tilde{\varepsilon}_1 - \tilde{\varepsilon}_2} \right].$$

As in the problem of a composite of maximal rigidity, one can expect that the general three dimensional case differs from its asymptotic cases by the presence of subregions for which the optimal structures are rank one laminates with layers oriented in the principal directions of the tensor $\tilde{\varepsilon}_0$.

In [2] the authors constructed quasiconvex envelopes for the elastic energy density of the type (1.18) and (1.29) for the problem of the bending of plates with properties changing along its plane. Since the operator of the theory of bending of plates ($\widetilde{\nabla}\widetilde{\nabla} \cdot \cdot \mathbf{D} \cdot \cdot \widetilde{\nabla}\widetilde{\nabla}$, where $\widetilde{\mathbf{D}}$ is a self-adjoint tensor of rigidity of the plate) is also the operator of the plane elasticity theory with respect to the Airy function, the solution of [2] is valid for the problem just discussed and the results of [2] coincide with the results of this section up to a change in notation.

Part III. Applying the results to problems of optimal design.

§10 Consider an elastic body W whose boundary is the surface ∂W. Suppose that the body is fixed at some part ∂W_2 of ∂W, and that the forces acting on the body are given by a system of loads $\mathbf{q}$ being applied to the remaining part ∂W_1 of the boundary ∂W. We are also given two materials in quantities

$$\int_{\partial W} m_1 dW = W_1, \quad \int_{\partial W} m_2 dW = W_2 \tag{3.1}$$

with compliance tensors $\mathbf{C}_1$ and $\mathbf{C}_2$. Our task is to construct W so as to minimize the compliance $\mathbf{I}$, or the work of external forces, given by

$$\mathbf{I} = \int_{\partial W_1} \mathbf{q} \cdot \mathbf{u} \, d(\partial W),$$

In particular, when $\mathbf{C}_2 \to \infty$, the problem becomes that of finding an optimal "shape" of the body, or a problem of how to pack a quantity W_1 of an elastic material with compliance $\mathbf{C}_1$ in a volume $W(W \geq W_1)$ in order to get a body of maximal rigidity.

The variational principle of complementary work [4] allows an equivalent formulation of the problems described. That is, we are looking for functions $\chi_1(x)$, $\chi_2(x) = 1 - \chi_1(x)$ such that the functional

$$\mathbf{I}(\chi_1, \chi_2) = \min_{\sigma \in \Sigma(W)} \int_W [\sigma \cdot \cdot \mathbf{C}(\chi_1, \chi_2) \cdot \cdot \sigma + \lambda \chi_1] \, dW, \tag{3.2}$$

is minimal, where

$$\Sigma(W) = \left\{ \sigma : \nabla \cdot \sigma = 0, \ \sigma = \sigma^T, \ \sigma \cdot \mathbf{n}|_{\partial W_1} = \mathbf{q}, \ \mathbf{t} \cdot \mathbf{C} \cdot \cdot \sigma|_{\partial W_2 = 0} \right\}.$$

Here λ is a Lagrange multiplier of the constraints (3.1), $\mathbf{n}$ is the normal, and $\mathbf{t}$ is the tangent to ∂W.

This problem can be solved in two steps. First we solve the problem for the known but arbitrary distributions $m_1(x)$, $m_2(x)$, of the "local averages" of $\chi_1(x)$, $\chi_2(x)$. Obviously, the body found in this step consists of the composites of maximal rigidity of §7. At the second step we choose the distribution of m_1 and m_2 by solving

$$\inf_{\chi_1,\chi_2} \mathbf{I}(\chi_i) = \min_{m_1,m_2} \widetilde{\mathbf{I}}(m_i), \tag{3.3}$$

where

$$\widetilde{\mathbf{I}}(m_i) = \int_W (\mathbf{E}_q(m_i,\sigma) + \lambda m_1) \ d(\partial W).$$

Changing (3.2) to (3.3) is equivalent to the decomposition of $\chi_i(x)$ into rapidly and slowly changing parts. In (3.3) we are looking for the averages of $\chi_1(x)$, $\chi_2(x)$ over regions of small measure. The functional $\widetilde{\mathbf{I}}(m_i)$ is lower semicontinuous in m_i and σ_0 due to quasiconvexity of the integrand, hence the solution in (3.3) exists *a priori*. The problem of finding optimal structures treated earlier is a problem of averaging, or choosing a suitable rapidly varying part of $\chi_1(x)$.

Solving the variational problem with the integrand $E_q(m_i,\sigma)$, we obtain the quasiconvex envelope for the nonconvex integrand:

$$E(\chi_i,\sigma) = \sigma \cdot \cdot \mathbf{C}(\chi_i) \cdot \cdot \sigma. \tag{3.4}$$

In the case $\mathbf{C}_2 \to \infty$ (which is a shape optimization problem) the envelope is found from the condition (cf. (2.7)):

$$\frac{\partial}{\partial m_1} (\mathbf{E}_q(m_i,\sigma) + \lambda m_1) =$$

$$\frac{\partial}{\partial m_1} \left(\widehat{\sigma} \cdot \left(\widehat{\mathbf{C}}_1 + \frac{1-m_1}{m_1} \widehat{\mathbf{G}} \right) \cdot \widehat{\sigma} + \lambda m_1 \right) = 0, \tag{3.5}$$

whence

$$m_1 = \begin{cases} 1, & \text{if } \sqrt{\widehat{\sigma} \cdot \widehat{\mathbf{G}} \cdot \widehat{\sigma}/\lambda} \geq 1, \\ \sqrt{\widehat{\sigma} \cdot \widehat{\mathbf{G}} \cdot \widehat{\sigma}/\lambda}, & \text{if } \sqrt{\widehat{\sigma} \cdot \widehat{\mathbf{G}} \cdot \widehat{\sigma}/\lambda} < 1, \end{cases} \tag{3.6}$$

and $E_q(m(\sigma),\sigma)$ is

$$E_q(\sigma) = \begin{cases} \widehat{\sigma} \cdot \widehat{\mathbf{C}}_1 \cdot \widehat{\sigma} + \lambda, & \text{if } \sqrt{\widehat{\sigma} \cdot \widehat{\mathbf{G}} \cdot \widehat{\sigma}/\lambda} \geq 1, \\ \widehat{\sigma} \cdot (\widehat{\mathbf{C}}_1 - \widehat{\mathbf{G}}) \cdot \widehat{\sigma} + 2\sqrt{\widehat{\sigma} \cdot \widehat{\mathbf{G}} \cdot \widehat{\sigma}/\lambda}, & \text{if } \sqrt{\widehat{\sigma} \cdot \widehat{\mathbf{G}} \cdot \widehat{\sigma}\lambda} < 1. \end{cases} \tag{3.7}$$

Thus we need to solve a variational problem in $\sigma(x)$ of the following kind

$$\widetilde{I} = \min_{\sigma \in \Sigma(\partial W)} \int_{\partial W} E_q(\sigma)dW. \tag{3.8}$$

The Lagrange multiplier λ must be chosen so as to ensure that the isoperimetric contstraint (3.1) holds. Having found $\sigma(x)$, we can recover both the distribution of the material in the volume W (i.e., the function $m(x)$, cf. (3.6)) and the parameters of the optimal microstructure (cf. (2.9)).

Remark. The procedure just described was applied to the problem of bending of plates [2] and allowed the authors to find the optimal design for soft and rigid plates.

Appendix.

1. To prove the statement of the theorem we need to choose parameters $a_i \in \mathcal{A}$ (cf. (1.20)–(1.22)) such that

$$\hat{\sigma} \cdot \left(\widehat{\mathbf{C}}_1 + \widehat{\mathbf{G}}(a_1, a_2, a_3)\right) \cdot \hat{\sigma} = E_q$$

(cf. (2.2), (2.7))

The constraints $a_i \in \mathcal{A}$ form in the space of all parameters a closed symmetric region $\overline{A}$ with optimal values of a_i lying on its boundary. We make a guess that the optimal values are either the vertices of $\overline{A}$, given by

$$(a_1, a_2, a_3) = \left(\frac{1}{2\sqrt{\mu}}, \frac{1}{2\sqrt{\mu}}, \frac{1}{2\sqrt{\mu}}\right),$$

$$(a_1, a_2, a_3) = \left(\frac{1}{2\sqrt{\mu}}, \frac{1}{2\sqrt{\mu}}, \frac{2\nu - 1}{2\sqrt{\mu}}\right)$$

or they belong to the edges of $\overline{A}$:

$$(a_1, a_2, a_3) = \left(\frac{1}{2\sqrt{\mu}}, \frac{1}{2\sqrt{\mu}}, \frac{C}{2\sqrt{\mu}}\right), C \in [2\nu - 1, 1]$$

$$(a_1, a_2, a_3) = \left(\frac{1}{2\sqrt{\mu}}, \frac{2\nu - 1}{2\sqrt{\mu}}, \frac{C}{2\sqrt{\mu}}\right), C \in [0, 1],$$

or to the points symmetric to the preceding ones. We prove that the bounds E_p constructed with this choice of a_i coincide with the quasiconvex envelope E_q, which will justifies our guess about a_i.

Let us now construct the bounds.

(a) Let $a_1 = a_2 = \frac{1}{2\sqrt{\mu}}$. We find a_3 which satisfies

$$a_3 \in \left[\frac{2\nu - 1}{2\sqrt{\mu}}, \frac{1}{2\sqrt{\mu}}\right],$$

and maximizes the following quadratic form:

$$\hat{\sigma}_0 {\cdot} \mathbf{G}\left(\frac{1}{2\sqrt{\mu}}, \frac{1}{2\sqrt{\mu}}, a_3\right) \cdot \hat{\sigma}_0 = \sigma_0 \cdot\cdot \mathbf{C}_1 \cdot\cdot \sigma_0$$
$$- \sigma_3{}^2 \left(\frac{(\omega - \eta)^2}{4\mu} - \frac{\omega + \eta}{\sqrt{\mu}} a_3 + a_3{}^2\right) \qquad (A.1)$$

From the condition

$$\frac{\partial}{\partial a_3} \hat{\sigma}_0 \cdot \mathbf{G} \cdot \hat{\sigma}_0 = 0$$

we find the critical value $a_3{}^0$ of a_3 is equal to:

$$a_3{}^0 = \frac{1}{2\sqrt{\mu}}(\omega + \eta), \text{ if } a_3{}^0 \in \left[\frac{2\nu - 1}{2\sqrt{\mu}}, \frac{1}{2\sqrt{\mu}}\right]. \qquad (A.2)$$

If (A.2) holds, the bound E_p is given by (2.7e); it is given by (2.7b) if a_3 equals the left endpoint of the segment in (A.2), and it is as in (2.7a) if a_3 coincides with the right endpoint of the same segment.

(b) Now let $a_1 = \frac{1}{2\sqrt{\mu}}$, $a_2 = \frac{2\nu-1}{2\sqrt{\mu}}$. In this case the constraints on a_3 are given by

$$a_3 \in \left[0, \frac{1}{2\sqrt{\mu}}\right] \qquad (A.3)$$

as follows from (1.20) through (1.22). If the critical value of a_3 is inside the interval in (A.3), then the expression for E_p corresponding to the value $a_3{}^0$ of a_3 is given by (2.7d), whereas if a_3 is the endpoint $a_3 = \frac{1}{2\sqrt{\mu}}$ of the interval, the bound E_p is given by (2.7c). All other expressions for E_p are obtained by interchanging $a_i \to a_j$. Thus for the case of the subregion Ω_7 the optimal a_i belong to the interval

$$(a_1, a_2, a_3) = \left(\frac{2\nu - 1}{2\sqrt{\mu}}, \frac{1}{2\sqrt{\mu}}, \frac{C}{2\sqrt{\mu}}\right), \quad C \in [0, 1]. \qquad (A.4)$$

For the case of the subregion Ω_4 parameters a_i are given by (4) with $C = 1$

whereas for the case of subregions Ω_8 and Ω_9, a_i are in the intervals

$$\left(\frac{1}{2\sqrt{\mu}}, \frac{C}{2\sqrt{\mu}}, \frac{2\nu - 1}{2\sqrt{\mu}}\right), \ C \in [0, 1]$$

$$\left(\frac{C}{2\sqrt{\mu}}, \frac{1}{2\sqrt{\mu}}, \frac{2\nu - 1}{2\sqrt{\mu}}\right), \ C \in [0, 1],$$

accordingly.

2. To prove the other statements of the theorem it is necessary for every vector $\hat{\sigma}_0 = (\sigma_1, \sigma_2, \sigma_3)$, that is, for every point (ω, η) of a unit square $\Omega = \{|\omega| \leq 1, |\eta| \leq 1\}$, to find such values of the parameters α_i that

$$\hat{\sigma}_0 \cdot \left(\hat{\mathbf{G}} - \hat{\mathbf{Q}}^{-1}(\alpha_i)\right) \cdot \hat{\sigma}_0 = 0 \tag{A.5}$$

(recall that by $(\hat{X})$ we denote the upper-left 3×3 block of a 6×6 matrix X). The proof that we give goes in the other direction; that is, for fixed α_i we find a $\hat{\sigma}_0$ satisfying (A.5) for every subregion Ω_i, i.e. for every bound E_p.

Let, for example, $\sigma_0 \in \Omega_1$. For the matrix $\hat{\mathbf{G}} - \hat{\mathbf{Q}}^{-1}(\alpha_i)$ we need to find an eigenvector $\hat{\sigma}_0(\alpha_1, \alpha_2, \alpha_3) \in \Omega_1$ with zero eigenvalue and then check that there exists a set of values of α_i such that

$$\sum_{i=1}^{3} \alpha_i = 1, \ \alpha_i \geq 0, \tag{A.6}$$

and that the set of the corresponding vectors $\hat{\sigma}_0(\alpha_1, \alpha_2, \alpha_3)$ is exactly the subregion Ω_1. We fix α_i and find $\hat{\sigma}_0(\alpha_1, \alpha_2, \alpha_3)$:

$$\left[\hat{\mathbf{G}}_1 - \hat{\mathbf{Q}}^{-1}(\alpha_i)\right] \cdot \hat{\sigma} = \hat{\mathbf{Q}}^{-1}\left[\hat{\mathbf{Q}}\hat{\mathbf{G}}_1 - I\right] \cdot \hat{\sigma} = 0 \tag{A.7}$$

where $\mathbf{Q}$ is given by (2.4), (1.25), and $\hat{\mathbf{G}}_1$ is given by (2.7a). We find that one of the eigenvalues of the matrix $\hat{\mathbf{Q}}\hat{\mathbf{G}}_1$ is equal to 1 for all values of α_i ($\sum \alpha_i = 1$), and the corresponding eigenvector $\hat{\sigma}_0$ is given by

$$\hat{\sigma}(\alpha_1, \alpha_2, \alpha_3) = \frac{\hat{\sigma}_1 + \hat{\sigma}_2 + \hat{\sigma}_3}{2}(\alpha_2 + \alpha_3, \alpha_1 + \alpha_3, \alpha_1 + \alpha_2).$$

By solving the preceding equations we get the values of α_i:

$$\alpha_1 = \frac{1 - \omega + \eta}{1 + \omega + \eta}, \ \alpha_2 = \frac{1 + \omega - \eta}{1 + \omega + \eta}, \ \alpha_3 = \frac{\omega + \eta - 1}{1 + \omega + \eta}.$$

These values satisfy the conditions (A.6) for the case of the subregion Ω_1 and the boundary $\omega + \eta = 1$ of this subregion corresponds to the boundary of the set of admissible α_3. Analogously, for the subregion Ω_2, we find

$$\alpha_1 = \frac{-\omega + \eta - (1 - 2\nu)}{(1 - 2\nu)(\omega + \eta - (1 + 2\nu))}, \quad \alpha_2 = \frac{\omega - \eta - (1 - 2\nu)}{(1 - 2\nu)(\omega + \eta - (1 + 2\nu))},$$
$$\alpha_3 = \frac{\omega + \eta + (1 - 2\nu)}{\omega + \eta - (1 + 2\nu)},$$

and the boundaries of Ω_2 correspond to the values of one of the α_i lying on the boundaries of the set of admissible α_i (cf. Figure 2). In Ω_3 we have:

$$\alpha_1 = \frac{1 - \omega - (1 - 2\nu)\eta}{(1 - 2\nu)(1 + \omega - (1 + 2\nu)\eta)}, \quad \alpha_2 = \frac{1 + \omega + (1 - 2\nu)\eta}{1 + \omega - (1 + 2\nu)\eta},$$
$$\alpha_3 = \frac{\omega - 1 - (1 - 2\nu)\eta}{(1 - 2\nu)(1 + \omega - (1 + 2\nu)\eta)},$$

and for the case of Ω_4:

$$\alpha_1 = \frac{\eta - 1 - (1 - 2\nu)\omega}{(1 - 2\nu)(1 - (1 + 2\nu)\omega + \eta)}, \quad \alpha_2 = \frac{1 - (1 - 2\nu)\omega - \eta}{(1 - 2\nu)(1 - (1 + 2\nu)\omega + \eta)},$$
$$\alpha_3 = \frac{1 + (1 - 2\nu)\omega + \eta}{1 - (1 + 2\nu)\omega + \eta}.$$

Similarly we obtain the results for cases (2.7d) through (2.7i). The optimal composites for these cases are rank two MLC's with cylindrical inclusions. The effective properties of such composites depend only on one parameter $\alpha = \alpha_1$, because α_2 is found from $\alpha_1 + \alpha_2 = 1$ and $\alpha_3 = 0$. This makes calculations much easier and allows us to determine the optimal cylindrical MLC directly.

Indeed, let $\alpha_3 = 0$, $\alpha_1 = \alpha$, $\alpha_2 = 1 - \alpha$. Then (cf. (1.25) and (1.4))

$$\hat{Q}(\alpha, 1 - \alpha, 0) = \frac{4\mu(3\kappa + \mu)}{3\kappa + 4\mu} \begin{pmatrix} 1 - \alpha & 0 & \nu(1 - \alpha) \\ 0 & \alpha & \alpha\nu \\ \nu(1 - \alpha) & \alpha\nu & 1 \end{pmatrix}$$

and

$$\hat{Q}^{-1}(\alpha, 1 - \alpha, 0) = \frac{3\kappa + \mu}{9\kappa\mu} \begin{pmatrix} \frac{1 - \nu^2\alpha}{1 - \alpha} & \nu^2 & -\nu \\ \nu^2 & \frac{1 - \nu^2(1 - \alpha)}{\alpha} & -\nu \\ -\nu & -\nu & 1 \end{pmatrix}.$$

Minimizing the energy density of the composite with respect to α, we find, that

$$\alpha = \alpha_1{}^0 = \frac{|\sigma_2|}{|\sigma_1| + |\sigma_2|}, \quad 1 - \alpha = \alpha_2{}^0 = \frac{|\sigma_1|}{|\sigma_1| + |\sigma_2|},$$

and the minimum of the quadratic form $\widehat{\sigma}_0 \cdot \widehat{\mathbf{Q}}^{-1} \cdot \widehat{\sigma}_0$ is

$$\sigma_0 \cdot \widehat{\mathbf{Q}}^{-1} \cdot \sigma_0 =$$

$$\left\{ (|\sigma_1| + |\sigma_2|)^2 + \sigma_3^2 + 2\nu^2 (\sigma_1\sigma_2 - |\sigma_1\sigma_2|) - 2\nu (\sigma_1 + \sigma_2) \sigma_3 \right\} \frac{3\kappa + \mu}{9\kappa\mu}.$$

Let us now check that the expression for the bound (2.7e) (i.e., the subregion Ω_5 of Figure 2) does correspond to the case $\sigma_1\sigma_2 \geq 0$. Indeed, here $|\sigma_1\sigma_2| = \sigma_1\sigma_2$ and

$$\widehat{\mathbf{Q}}^{-1}(\alpha_1{}^0, \alpha_2{}^0, 0) = \frac{3\kappa + \mu}{9\kappa\mu} \begin{pmatrix} 1 & 1 & -\nu \\ 1 & 1 & -\nu \\ -\nu & -\nu & 1 \end{pmatrix}.$$

If, however, $\sigma_1\sigma_2 \leq 0$, then $|\sigma_1\sigma_2| = -\sigma_1\sigma_2$ and

$$\widehat{\mathbf{Q}}^{-1}(\alpha_1{}^0, \alpha_2{}^0, 0) = \frac{3\kappa + \mu}{9\kappa\mu} \begin{pmatrix} 1 & 2\nu^2 - 1 & -\nu \\ 2\nu^2 - 1 & 1 & -\nu \\ -\nu & -\nu & 1 \end{pmatrix}.$$

This is the case of the subregions Ω_6 and Ω_7. For the subregions Ω_8 and Ω_9 the results are easily obtained after an obvious interchange of the indices of α_i.

We shall check, finally, that the parameters α_i change continuously on the boundaries of subregions Ω_1 through Ω_4 of rank three MLC's and subregions Ω_5 through Ω_9 of rank two MLCs. Consider, for example, the boundary of Ω_1 and Ω_5 ($\omega \geq 0$, $\eta \geq 0$). Here $\omega + \eta = 1$, and

$$\alpha_1 = \frac{1 - \omega + \eta}{1 + \omega + \eta} = \frac{\eta}{\omega + \eta} = \alpha_1{}^0,$$

$$\alpha_2 = \frac{1 + \omega - \eta}{1 + \omega + \eta} = \frac{\omega}{\omega + \eta} = \alpha_2{}^0,$$

$$\alpha_3 = \frac{1 - \omega - \eta}{1 + \omega + \eta} = 0 = \alpha_3{}^0$$

where $\alpha_1(\alpha_1{}^0)$, $\alpha_2(\alpha_2{}^0)$, $\alpha_3(\alpha_3{}^0)$ are the limit values of the parameters as we approach the boundary from inside the subregion $\Omega_1(\Omega_5)$. One can easily check the same conditions analogously for the other subregions.

References

[1] Gibiansky L.V., Cherkaev A.V., The problem of design of a viscoelastic rod under torsion loading, notes of the second seminar "Problems of optimization in machine industry", Khar'kov, 1986, p.45 (in Russian).

[2] Gibiansky, L.V., Cherkaev A.V., *Design of composite plates of extremal rigidity*, preprint No. 914, Physico-Technical Institute USSR Academy of Sciences, Leningrad, 1984, (in Russian); see chapter 5 of this volume.

[3] Christensen, R.M., *Mechanics of Composite Materials*, Wiley-Interscience, New-York, 1979.

[4] Lurie A.I., *Theory of elasticity*, Moscow, Nauka, 1970, (in Russian), 939pp

[5] Lurie K.A., Cherkaev A.V., G-closure of a set of anisotropic conducting media for a case of two dimensions, *Dokladii USSR Academy of Sciences*, v. 259, No. 2, 1981 (in Russian).

[6] Lurie K.A. Cherkaev A.V., Effective characteristics of composites and problems of optimal design, *Uspekhi Mat. Nauk*, 1984, **39** 4(238), p.122. (in Russian); see chapter 7 of this volume

[7] Reshetnyak, U.G. General theorems on semi-continuity and quasiconvexity of functionals, *Siberian Mat. Zhurnal*, 1967, **8**, p.801–816 (in Russian).

[8] Rockefeller R., *Convex Analysis*, Princeton, N.J., Princeton University Press, 1970.

[9] Sanchez-Palencia E., Nonhomogeneous Media and Vibration Theory, *Lecture Notes in Physics* **127**, Springer-Verlag, 1980.

[10] Cherkaev A.V., Relaxation of non-convex multidimensional problems, *Uspekhi Mat. Nauk* **41**, 1986, 4(250), p.194 (in Russian).

[11] Avellaneda M. Optimal bounds and microgeometries for elastic two-phase composites, SIAM J. Appl. Math. **47** (1987), 1216–1228.

[12] Ball J.M. Currie J.C., Olver P.J., Null Lagrangians, weak continuity and variational problems of arbitrary order, *J. Funct. Anal* **41** (1981), 135–174.

[13] Dacorogna B., Weak continuity and weak lower semicontinuity fo nonlinerar functionals, *Lecture Notes in Math.* **922**, Springer-Verlag, 1982.

[14] Hashin Z., Shtrikman S., A variational approach to the theory of the elastic behaviour of multiphase materials. *J. Mech. and Phys .Solids*, 1963, VII, pp.127-140.

[15] Hill R., A self-consistent mechanics of composite materials, *J. Mech.Phys. Solids*, **13**, 4, 1965. p.213.

[16] Kohn R.V. and Strang, G., Structural design optimization, homogenization, and relaxation of variational problems, *Comm. Pure and Appl. Math.*, 1986, **39**,No1, 113–137; No 2,139–182; No 3, 353–378.

[17] Kohn, R.V., Lipton R., The effective viscosity of a mixture of two

Stokes fluids, G. Papanicolaou, ed., *SIAM*, 1986.

[18] Lurie K.A. and Cherkaev A.V., Exact estimates of conductivity of composites formed by two isotropically conducting media, taken in prescribed proportion, *Proc. Royal Soc. Edinburgh*, 1984, 99a, 71–87.

[19] Lurie K.A. and Cherkaev A.V., G-closure of some particular sets of admissible material characteristics for the problem of bending of thin plates, *J. Opt. Th. Appl.*, 1984, 305–316.

[20] Milton G.W., Modelling the properties of composites by laminates, in *Homogenization and Effective Moduli of Materials and Media*, Springer-Verlag, 1986.

[21] Morrey C.B., Quasiconvexity and the lower semicontinuity of multiple integrals, *Pacific J. Math.*, 1952, 225–253.

[22] Murat F. and Francfort G.A., Homogenization and optimal bounds in linear elastisity, *Arch. Rat. Mech. Anal.*, 1986, **94** 4, 307–334.

[23] Strang G., The polyconvexification of $F(\nabla u)$, The Austr. Nat. University, Research Rep. NCNA-R09-83, 1983, 21p.

[24] Tartar L., Compensated compactness and applications to partial differential equations, in *Non Linear Mech. Anal., Heriot Watt Symposium*, 1976, vol.39, 136–212.

[25] Tartar L., Estimation fines do coefficients homogénisés, in *Ennio De Giordgi Coll.*, Pitman Press, P.Kree, ed., 1985, v.125, 168–187.

Printed in the United States
By Bookmasters